T0269231

CAMBRIDGE STUDIES IN MAGNETISM:1

EDITED BY
David Edwards
Department of Mathematics, Imperial College of Science and Technology
and
David Melville
Vice Rector, Lancashire Polytechnic

Spin glasses

Spin glasses

K. H. FISCHER

*Professor of Physics, Institut für Festkörperforschung
der Kernforschungsanlage Jülich,
and Technische Hochschule, Aachen*

J. A. HERTZ

Professor, Nordisk Institut for Teoretisk Atomfysik (NORDITA), Copenhagen

CAMBRIDGE
UNIVERSITY PRESS

CAMBRIDGE UNIVERSITY PRESS
Cambridge, New York, Melbourne, Madrid, Cape Town, Singapore,
São Paulo, Delhi, Dubai, Tokyo, Mexico City

Cambridge University Press
The Edinburgh Building, Cambridge CB2 8RU, UK

Published in the United States of America by
Cambridge University Press, New York

www.cambridge.org
Information on this title: www.cambridge.org/9780521447775

First published 1991
First paperback edition 1993

A catalogue record for this publication is available from the British Library

Library of Congress Cataloguing in Publication Data

Spin glasses / K.H. Fisher, J.A. Hertz.   II. title.
      p.   cm.
ISBN 0-521-34296-1
1. Spin glasses.   I. Hertz, J. A. II Title.
QC176.8.S68F57 1991
530.4'1–dc20

ISBN 978-0-521-34296-4 Hardback
ISBN 978-0-521-44777-5 Paperback

Contents

Preface

Spin glasses are a fascinating new topic in condensed matter physics which developed essentially after the middle of the 1970's. The aim of this book is to give an introduction to it which will both attract the newcomer to the field (say, a student with a basic knowledge of solid state physics and statistical mechanics) and give a comprehensive survey to the expert who perhaps has worked on a very specific problem. It is a field which is still open to new ideas and concepts and in which important new experiments can certainly still be done.

Our understanding of spin glasses is based on three approaches: theory, experiment, and computer simulation. We have tried to present the most important developments in all of them. One possibility is to take the theory as a guide and to check it by comparison with experimental data and simulations. This is roughly what we do in the first part of this book (Chapters 3 to 6), after introducing the basic experiments, models and concepts which define what we are talking about. (Spin glasses are disordered systems, so we have to introduce several concepts which are unknown in the 'classical' theory of ideal solids.)

In Chapters 3 to 6 we discuss a mean field theory, which is so far the only well-established spin glass theory. It turns out to be highly nontrivial and has been developed over more than a decade. Its underlying ideas have also proved to be fruitful in optimization problems and the theory of neural networks. This led us to include a brief account of these subjects in Chapter 14 (entitled 'the physics of complexity').

However, the mean field theory gives only a hint about what happens in real spin glasses, and in Chapters 7 to 11 we rely more and more on experiment and computer simulation. Here the concepts of scaling and renormalization permit us considerable insight into the spin glass phase and the transition between it and the paramagnetic phase, and the idea of 'frustration' gives at least a feeling of the fundamental difference between ideal periodic solids and disordered ones.

This book is not a review. In the early and mid-1980's more than 400 papers per year were written on spin glasses (altogether more than 4000), and it would be completely hopeless to discuss or even mention them all. Rather, we have tried to present the most important ideas and developments in the field. Naturally, this is a very personal selection, and we want to apologize to the thousands of authors whose interesting papers we could not mention. Some of these papers have been discussed in the excellent review of Binder and Young (1986), in the somewhat older reviews by one of us (Fischer, 1983c, 1985), in the Heidelberg Colloquia on spin glasses and on glassy dynamics (van Hemmen and Morgenstern, 1983, 1986), and in the books by Chowdhury (1986) and Mézard et al (1987).

Our understanding of spin glasses has grown over many years, and it is a pleasure for us to thank the large number of our colleagues who contributed to it. We especially want to thank Philip Anderson, Alan Bray, Cyrano De Dominicis, Anil Khurana, Wolfgang Kinzel, Richard Klemm, Hans Maletta, Mike Moore, Richard Palmer, Hans-Jürgen Sommers, Peter Young, and Annette Zippelius.

We are also very grateful to Mrs Ch. Hake, who typed a large part of this book in TEX.

1
Introduction

Questo é quel pezzo di calamita:
Pietra mesmerica, ch'ebbe l'origine nell'Alemagna,
Che poi sì celebre là in Francia fu.[1]

Lorenzo da Ponte, *Così fan Tutte*, Act I

One of the dominant themes in the history of physics in this century has
been the effort to understand condensed states of matter. This began with
very simple systems — the Van der Waals description of the liquid–gas
transition and the Weiss mean field theory of ferromagnetism — and has
gradually developed to include more and more complex and subtle states
and phenomena. Spin glasses are the current frontier in this development,
the most complex kind of condensed state encountered so far in solid state
physics.

In trying to understand these systems, experimentalists have used a
wide spectrum of probes in ingenious ways, and theorists have invented an
equally wide variety of models and new theoretical concepts. The resulting
developments have had an impact, not only on other parts of physics, but
also on other fields such as computer science, mathematics, and biology. It
is because of this widespread influence and the interest in spin glasses that
it has aroused that we are writing this book.

We expect that many people who read this book will be condensed mat-

---

[1] This is a magnet: the mesmerizing stone discovered in Germany and then so
famous in France.

ter physicists. However, we also have in mind as a typical reader someone from another area in physics, or perhaps a graduate student looking for a research topic, who wants to find out what all the excitement is about. She need not be a condensed matter physicist, or even a physicist at all, though we do assume a reasonable knowledge of basic statistical mechanics. With her in mind, we begin with some basic questions.

First: What is a spin glass? The simplest answer (which we will naturally have to improve on in the course of succeeding paragraphs and chapters) is that it is a collection of spins (i.e. magnetic moments) whose low-temperature state is a frozen disordered one, rather than the kind of uniform or periodic pattern we are accustomed to finding in conventional magnets. It appears that in order to produce such a state, two ingredients are necessary: There must be competition among the different interactions between the moments, in the sense that no single configuration of the spins is uniquely favoured by all the interactions (this is commonly called 'frustration'), and these interactions must be at least partially random. These facts suggest that the spin glass state is intrinsically different from conventional forms of order and requires new formal concepts to describe it. This challenge has been the fundamental motivation for theorists in this field.

Experimentally, it does not seem to be hard to find spin glasses. Quite the contrary, spin glass behaviour has been seen in virtually every kind of system which people have been able to make that satisfies these requirements.

The first kind of system to be studied widely consisted of dilute solutions of magnetic transition metal impurities in noble metal hosts. The impurity moments produce a magnetic polarization of the host metal conduction electrons around them which is positive at some distances and negative at others. Other impurity moments then feel the local magnetic field produced by the polarized conduction electrons and try to align themselves along it. Because of the random placement of the impurities, some of interactions are positive (i.e. favouring parallel alignment of the moments) and some are negative. Thus we clearly have random, competing interactions.

At one time, many people believed that spin glass behaviour was sensitively dependent on particular features of this special class of systems. But we now know that this is not so. Spin glass states have also been found in magnetic insulators and in amorphous alloys, where the dependence of the interactions on the distance between the moments is entirely different from that in the above crystalline metallic systems. The 'spin' degrees of freedom need not even be magnetic. Properties analogous to those of spin glasses, with the electric dipole moment taking the place of the magnetic one, have been seen in ferroelectric–antiferroelectric mixtures, and a kind of orientational freezing has been observed in disordered molecular crystals in which the electric quadrupole moment plays the role of the spin. This universal nature of the observed phenomena is another reason for thinking this is an important problem to study. The next fundamental question we ask is how we observe such a state.

The description 'frozen disorder' suggests that we are dealing with a state where the local spontaneous magnetization $m_i = \langle S_i \rangle$ at a given site i is nonzero, though the average magnetization $M = N^{-1} \sum_i m_i$, as well as any 'staggered' magnetization $M_K = N^{-1} \sum_i e^{-i\mathbf{K}\cdot\mathbf{r}_i} m_i$, vanishes. That the low-temperature state was not an antiferromagnet was indicated by neutron scattering experiments, which showed no magnetic Bragg peaks which would have indicated long range order. (Here and henceforth $\langle S_i \rangle$ means the conventional thermal average, and the magnetization is in units of $-g\mu_B$.)

The local spontaneous magnetizations make their presence felt in an experiment because they reduce the susceptibility from the value it otherwise would have. This effect is in fact familiar from antiferromagnets, where a sharp reduction in the susceptibility from its extrapolated high-temperature form signals the onset of antiferromagnetic order. The same thing happens in spin glasses, and Fig. 1.1 shows examples of some susceptibility measurements that played a key role in arousing the interest in this field that exploded in the mid-1970's. They exhibit a marked cusp at a temperature which is rather sharply defined, suggesting a second-order phase transition between the disordered paramagnetic state and a spin glass state characterized by nonvanishing local spontaneous magnetizations m_i. The difference between the measured susceptibility and the extrapolation of the high-temperature form should be some measure of the degree of freezing. Immediately, people wanted to know in what ways this transition (if, indeed, it was a sharp transition) was like ordinary second-order phase transitions and in what ways it might be different.

The connection between the susceptibility and the existence of frozen moments can be made more explicit by supposing we have a system of Ising spins $(S_i = \pm 1)$ and considering the single-site susceptibility χ_{ii} defined as the amount of magnetization m_i induced at site i by an external field $B_i = -h_i/g\mu_B$ acting only on this site:

$$\chi_{ii} \equiv \frac{\partial m_i}{\partial h_i} \qquad (1.1)$$

A fundamental theorem of classical statistical mechanics (see, e.g. Landau and Lifshitz, 1969) relates the equilibrium fluctuations of any thermodynamic variable to the mean amount of this variable induced by a conjugate field. The present case affords the simplest possible example of this relation. It says (in units where the Boltzmann constant $k_B = 1$)

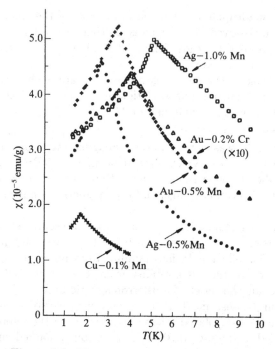

Figure 1.1: The ac susceptibility of Cu-0.1% Mn ($\times$), Ag-0.5% Mn ($\bullet$), Au-0.5% Mn (+), Au-0.2% Cr ($\triangle$), and Ag-1.0% Mn ($\square$) versus temperature for a magnetic field $H = 20$ Oe and 100 Hz (from Cannella and Mydosh, 1972, 1974).

$$T\chi_{ii} = \langle (S_i - \langle S_i \rangle)^2 \rangle = 1 - m_i^2 \qquad (1.2)$$

where the last step uses explicitly the fact that $S_i^2 = 1$. Averaging over all the sites in the system gives

$$\chi_{loc} \equiv \frac{1}{N} \sum_i \chi_{ii} = \frac{1 - N^{-1} \sum_i m_i^2}{T} \qquad (1.3)$$

That is, the reduction of the average local susceptibility χ_{loc} from the Curie law characteristic of free moments is a direct measure of the mean square local spontaneous magnetization in the frozen state. Although the experiments of Fig. 1.1 do not measure the local susceptibility, but rather the so-called uniform susceptibility (which we denote by χ without any subscript)

$$\chi = \frac{1}{N} \sum_{ij} \chi_{ij} = \frac{1}{N} \sum_{ij} \frac{\partial m_i}{\partial h_j} \qquad (1.4)$$

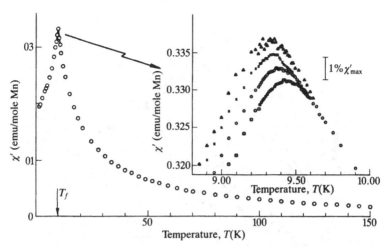

Figure 1.2: The ac susceptibility as a function of temperature for Cu-0.9% Mn for the frequencies 1.33 kHz ($\square$), 234 Hz (o), 10.4 Hz (x), and 2.6 Hz ($\triangle$) (from Mulder et al, 1981, 1982).

it can be shown (Fischer, 1976) that the off-diagonal elements of χ_{ij} vanish (in zero field) if the interactions between different spins are symmetrically distributed. More generally, the uniform χ will have a cusp if χ_{loc} does, so the experiments really do indicate the existence of a nonzero frozen spontaneous magnetization — a spin glass state.

The freezing temperature T_f, defined by the cusp in the ac susceptibility as seen in Fig. 1.1, actually turns out to depend on the frequency of the applied magnetic field. The 'true' T_f should therefore be defined by the limit of vanishing frequency. Furthermore,the cusp is not completely sharp, as shown in Fig. 1.2 for CuMn (which is one of the best investigated spin glass systems). In this more precise experiment one has to distinguish between the in-phase or real part $\chi'(\omega, T)$ and the out of phase or imaginary part $\chi''(\omega, T)$ of the complex susceptibility $\chi(\omega, T) = \chi'(\omega, T) + i\chi''(\omega, T)$.

There is a crude phenomenology for describing these slow dynamics (Lundgren et al, 1981; van Duyneveldt and Mulder, 1982) for frequencies in the range shown in Fig. 1.2. Below $T_f(\omega)$, the real part of $\chi(\omega)$ varies approximately logarithmically with frequency:

$$\chi'(\omega) = \chi_0 + a \ln\left(\frac{1}{|\omega|}\right) \tag{1.5}$$

Then the Kramers–Kronig relations imply a roughly frequency-independent χ'':

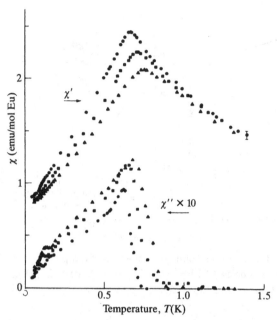

Figure 1.3: Temperature dependence of the real (solid symbols) and imaginary (open symbols) parts of the susceptibility for $Eu_{0.2}Sr_{0.8}S$ at an applied field $H \approx 0.1$ Oe (from Hüser et al, 1983).

$$\chi'' = \frac{\pi a}{2} \mathrm{sgn}\,\omega \qquad\qquad (1.6)$$

which is independently measurable (see Fig. 1.3).

The logarithmic dependence is not exact, just a rather good fit, and other functional forms such as a power law with a small exponent work as well. For much lower frequencies (i.e. times sufficiently longer than those characteristic of these experiments) the frequency dependence of χ' seems to disappear, indicating that a true equilibrium limit $\chi(0)$ has been reached. For much higher frequencies, the simple approximate frequency dependence of (1.5)–(1.6) breaks down, but the qualitative feature of frequency dependence extending over many orders of magnitude in frequency, from microscopic characteristic frequencies to the inverse of the longest experimental measuring times, has been found in a wide variety of experiments and in essentially all spin glass systems. This universal feature sets spin glasses apart from conventional magnets, where no significant frequency dependence is observed for frequencies much lower than the characteristic microscopic frequencies of the system.

The presence of this 'glassy' behaviour with such long characteristic

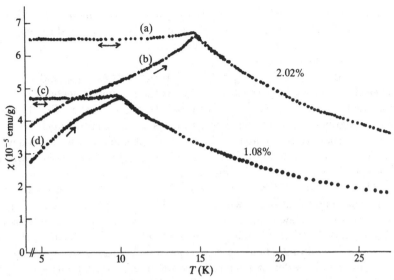

Figure 1.4: The static susceptibility of C̲uMn vs temperature for 1.08 and 2.02% Mn. After zero-field cooling ($H < 0.05$ Oe), initial susceptibilities (b) and (d) were taken for increasing temperature in a field of $H = 5.90$ Oe. The susceptibilities (a) and (c) were obtained in the field $H = 5.90$ Oe, which was applied above T_f before cooling the samples (from Nagata et al, 1979).

times suggests the possible presence of many metastable spin configurations with a distribution of energy barriers separating them. (Assuming that the typical time to cross a barrier depends exponentially on its height ΔE ($\tau \propto \exp(\beta \Delta E)$), we do not require too broad a distribution of barrier heights in order to get a very wide relaxation time distribution at low temperatures.)

Another important feature of all spin glasses is the onset of remanence effects below T_f. This is illustrated in Fig. 1.4 for the dc susceptibility of C̲uMn as measured in extremely small fields (0.05 Oe $\leq$ H $\leq$ 5.9 Oe). Even in these small fields χ_{dc} for $T < T_f$ depends strongly on the way the experiment is performed: $\chi_{dc}(T)$ is largest (and roughly temperature-independent) after 'field-cooling', i.e. if the field is applied above T_f and the sample subsequently cooled in this field to a temperature below T_f. This measurement is, to a very good approximation, reversible; that is, one can go up and down in temperature and measure the same magnetization, independent of history. This is in contrast to the 'zero-field-cooled' susceptibility χ_{zfc}, obtained by cooling the sample below T_f in zero field and then applying the field. After applying a field at a temperature below T_f, the magnetization jumps to a finite value, followed by a slow additional increase. This 'irreversible' contribution to χ_{zfc} decays only very slowly if the field is suddenly switched off.

The difference between χ_{ac} (Figs. 1.1 and 1.2) and χ_{dc} and the remanence effects in χ_{dc} have the same origin as the frequency dependence of χ_{ac} discussed above: the glass-like nature of the system below T_f. There are many roughly equivalant spin configurations, and the state which is reached depends crucially on details of the experiment such as the frequency and magnitude of the applied field, the speed with which one cools down, and whether one cools in zero or finite field.

There is also a difference at higher fields between the zero-field-cooled remanent magnetization, in which the field is applied at the measuring temperature and then switched off again ('isothermal' remanent magnetization (IRM)), and the 'thermoremanent' magnetization (TRM), that is, the magnetization remaining when the field is switched off after field-cooling. Fig. 1.5 shows the field dependence of these remanent magnetizations in AuFe. Again, both of them are time-dependent: Fig. 1.6a shows the decay of the IRM plotted against $\ln t$, which suggests a decay law

$$M_R(t) = M_0 - S_{RM} \ln t \tag{1.7}$$

This is in contrast to EuSrS (which again is a 'standard' spin glass), in which the power law

$$M_R(t) \propto t^{-a(T,H)} \tag{1.8}$$

indicated in Fig. 1.6b is found. Finally, one can also fit data by an exponential function of a power law:

$$M_R(t) \propto \exp[-(t/\tau)^\beta] \tag{1.9}$$

as indicated in Fig. 1.6c for AgMn, the exponent β being about 1/3 for T not too close to T_f. Any of these very slow decay laws is consistent with our qualitative ideas about the glassiness of the spin glass state, with many possible configurations separated by barriers of varying heights. However, this variety of fits in different systems (if it is meaningful) suggests that perhaps not all spin glass properties are universal and that the glass-like structure might vary in its details from system to system.

A similar non-universal property is the hysteresis of the magnetization. An example (CuMn) is shown in Fig. 1.7. As in ferromagnets, hysteresis effects are due to *anisotropy*, which might be extremely different in the various spin glass systems. The origin of this anisotropy will be discussed in Section 6.3.

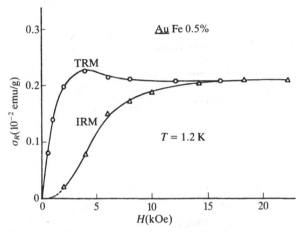

Figure 1.5: Field dependence of the thermoremanent magnetization (TRM) of AuFe 0.5% as obtained after cooling from $T > T_f$ to $T = 1.2$ K in a field H, and of the isothermal remanent magnetization (IRM) obtained when a field H applied at 1.2 K is suppressed (from Tholence and Tournier, 1974).

The behaviour of the magnetic specific heat was originally puzzling. The data of Fig. 1.8 for CuMn, showing a broad maximum above T_f but no indication of any anomaly right at T_f, are typical. If there is a phase transition at T_f (as we now believe to be the case in three dimensions in the presence of the kinds of anisotropy present in these systems), these experiments tell us that the corresponding singularity must so weak as to be undetectable in these experiments. Fortunately, the theoretically predicted singularity turns out to be quite weak, as we will see in Chapter 8, so there is no real puzzle.

In a ferromagnet, the approach to the ferromagnetic phase as one lowers the temperature toward the Curie point T_c is accompanied by a dramatic increase in the range of the spin correlations, which then diverges at T_c. Applying the fluctuation-response theorem again, in its more general form

$$\chi_{ij} = \frac{\partial m_i}{\partial h_j} = \beta \langle (S_i - \langle S_i \rangle)(S_j - \langle S_j \rangle) \rangle \tag{1.10}$$

we can see how this produces a divergent χ at the Curie point. A cor-

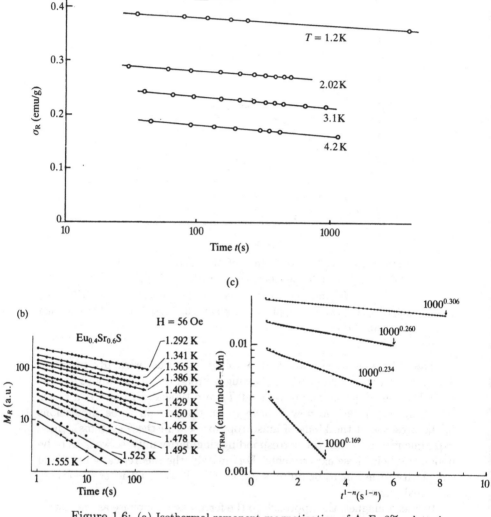

Figure 1.6: (a) Isothermal remanent magnetization of A̲uFe 8%, plotted vs the logarithm of time (from Holtzberg et al, 1977).
(b) Log–log plot of the saturated thermoremanent magnetization M_{TRM} vs time for $Eu_{0.4}$ $Sr_{0.6}S$ at various temperatures close to $T_f = 1.55$ K (from Ferré et al, 1981).
(c) Logarithm of the saturation value of the thermoremanent magnetization of A̲g-2.6% Mn plotted vs t^{1-n} at four temperatures, for a time range of about $1-10^3$ s. Fitted exponents $1-n$ are indicated in the figure (from Chamberlin et al, 1984).

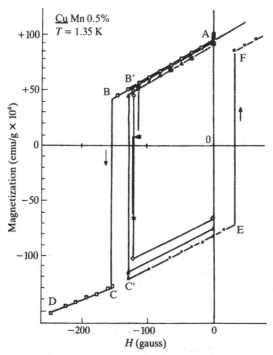

Figure 1.7: Hysteresis of Cu-0.3% Mn measured at 1.4 K in the sequence shown by letters (from Monod et al, 1979).

responding phenomenon also occurs in spin glasses, but it is not the spin correlation function $\langle S_i S_j \rangle$ that acquires long-range behaviour (we expect the randomness to be sufficiently strong that on average the correlation does not extend over more than a few spins), but rather its square. This then leads to the divergence, at the transition, of the quantity

$$\chi_{SG} = \frac{1}{N} \sum_{ij} \chi_{ij}^2 = \frac{\beta^2}{N} \sum_{ij} \langle S_i S_j \rangle^2 \qquad (T > T_f) \qquad (1.11)$$

It turns out that χ_{SG} can also be measured by analyzing the first deviations from linearity in the dependence of the magnetization M on the external field H, and the divergence of χ_{SG} on approaching T_f has indeed been observed in such measurements. We will return to this feature in Chapter 2. For now we just note that it, together with the measure of the degree of freezing given by the size of the cusp in χ, gives us the possibility to fit the spin glass transition into the standard theoretical framework of second order phase transitions characterized by critical exponents describing the

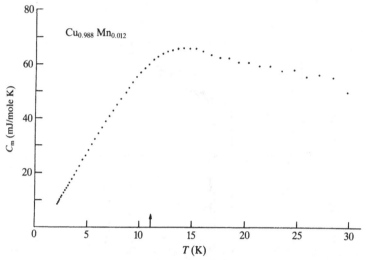

Figure 1.8: Magnetic contribution to the specific heat of Cu-1.2% Mn vs temperature. The arrow indicates T_f, as determined from the susceptibility (from Wenger and Keesom, 1975, 1976).

diverging correlations above the critical point and the onset of 'order' below.

Of course, it is not obvious in advance that this standard framework will be sufficient to describe the important physics of a class of systems with characteristic properties (like the time-dependent ones we have described here) which are very different from those of conventional materials. Indeed, we will see that at the mean field level (which one would think to be the natural starting point for understanding a new kind of system) completely new theoretical concepts and formalism turn out to be necessary. In fact, these are intimately connected with the basic physical mechanisms responsible for properties like remanence: They are the formal way of describing a system which can have many possible states, depending on its preparation. This problem is the essential theme of roughly the first half of this book (through Chapter 6).

Paradoxically, it appears as if a more conventional framework (Chapter 8) is sufficient when one goes beyond mean field theory to a description suitable for experimental systems. Spin glasses with short-range interactions in three dimensions probably do *not* have many possible different states, at least if one waits long enough for equilibrium to be established. However, 'waiting long enough' may mean an astronomically long time, so we also want to have a theory of what happens over experimental timescales. This theory is still in a primitive stage at present, so our story here is unfinished (Chapter 9).

Here is an outline of the rest of the book: Chapter 2 introduces the basic theoretical ideas, formal tools and experimental facts. In Chapter 3, in a sense the theoretical core of the book, we outline the statistical mechanics of Ising spin glasses in mean field theory (i.e. for infinite-range interactions). Chapter 4 introduces some basic material about dynamics, and the dynamics of the infinite-range Ising model are then discussed in Chapter 5. Chapter 6 extends this mean field theory to vector spin models. The next three chapters then deal with systems with short-range interactions: In Chapter 7 we focus on low-temperature properties in a description based on competing interactions or 'frustration', expressed mathematically by a kind of gauge theory. In Chapter 8 we treat the statistical mechanics of short-range systems using renormalization group ideas, and we examine their rather unusual dynamical properties in Chapter 9. Throughout all these chapters, the main experimental quantities we calculate are magnetic susceptibilities, but other kinds of measurements can also probe the spin glass state, if less directly, and we discuss them in Chapter 10. In Chapter 11 we examine systems where spin glass order competes with conventional magnetic order, and in Chapter 12 we see what kinds of things we can learn from one-dimensional models. The following two chapters deal with systems which are not spin glasses in the narrow sense, but which have something in common with them: materials with random external fields and random anisotropy instead of random exchange (Chapter 13), and some problems outside physics where concepts from spin glass theory play a central role (Chapter 14). Finally, Chapter 15 offers a quick tour through the history of spin glasses.

2

Models, order parameters, and systems

In order to study the interesting features of spin glass phenomena, we must employ models which capture the essential physics of the systems where the phenomena occur and theoretical concepts such as order parameters to describe the frozen state. In this chapter we describe the kinds of systems in which one finds spin glass behaviour and introduce a number of important and useful models. In the context of these models, which all involve randomness, we will then discuss how one averages over the randomness in order to calculate the relevant order parameters and related quantities that we can use to elucidate the phenomena. Finally, we will discuss how various known classes of spin glass systems fit into the models and the formal framework for calculating with them that we have set up.

2.1   Models: random sites and random bonds

The observed universality of spin glass phenomena suggests that we try to work with the simplest models we can that incorporate the apparent necessary features of disorder and competing interactions. One such kind of model one can construct in this spirit is based on an idealization of the classic spin glass systems consisting of transition metal impurities carrying local magnetic moments in noble metal hosts that we mentioned in the Introduction. In these systems, the interaction between impurity moments comes about because each of them polarizes the surrounding Fermi sea of the host conduction electrons; for a single spin $\mathbf{S}$ this polarization varies spatially like

$$\langle \mathbf{S}_{ind}(\mathbf{r}) \rangle \propto \sum_q \chi_0(\mathbf{q}) e^{i\mathbf{q}\cdot(\mathbf{r}-\mathbf{R}_i)} \mathbf{S} \qquad (2.1)$$

around an impurity at $\mathbf{R}_i$, where the function

$$\chi_0(\mathbf{q}) = -\frac{1}{N} \sum_k \frac{f_{\mathbf{k}} - f_{\mathbf{k}+\mathbf{q}}}{\epsilon_{\mathbf{k}} - \epsilon_{\mathbf{k}+\mathbf{q}}} \qquad (2.2)$$

is the spin susceptibility of the conduction electron gas. Here $f_{\mathbf{k}}$ is the Fermi function and $\epsilon_{\mathbf{k}}$ is the band energy of an electron of momentum $\mathbf{k}$. (See, e.g. Ashcroft and Mermin, 1976, Chapter 17, or any other elementary solid state textbook for the derivation.) In free-electron approximation ($\epsilon_{\mathbf{k}} = k^2/2m$), because of the discontinuity of $f_{\mathbf{k}}$ at the Fermi surface, $\chi_0(\mathbf{q})$ has a logarithmically divergent second derivative at $q = 2k_F$ (see Fig. 2.1a), which leads to oscillatory behaviour (Fig. 2.1b)

$$\chi_0(r) \propto \frac{\cos 2k_F r}{r^3}, \qquad 2k_F r \gg 1 \qquad (2.3)$$

in the induced conduction electron polarization. Since the interaction between impurity moments comes about when one impurity feels the polarization from the other, the effective interaction also has this oscillatory behaviour.

We can therefore consider as a simple model for spin glasses one in which spins $\mathbf{S}_i$ sit at random positions $\mathbf{R}_i$, interacting via an exchange proportional to (2.1) (or, more simply still, to its asymptotic form (2.3)). Though this model does not contain what turns out to be an important feature of these metallic spin glass systems — anisotropic interactions — it does apparently describe a good deal of their relevant physics, despite the fact that the host metals are not truly free-electron-like, so (2.3) is at best only a qualitatively correct form. It contains, in particular, the two ingredients we noted earlier were necessary for spin glass behaviour, namely, competing interactions (because of the oscillations in (2.3)) and randomness (because of the random positions of the spins). The earliest theoretical work on spin glasses (Klein and Brout, 1963) was based on this model.

The characteristic oscillatory form (2.3) was first studied in the context of nuclear magnetism by Ruderman and Kittel (1954); later work by Kasuya (1956) and Yosida (1957) applied it to the present context. It is therefore universally known today as the RKKY interaction, and the systems in this class are called RKKY glasses.

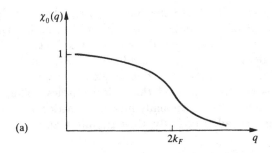

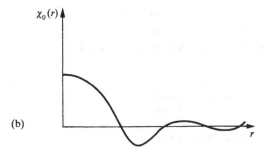

Figure 2.1: The static free-electron susceptibility, in (a) momentum space $(\chi_0(q))$, and (b) real space $(\chi_0(r))$.

Another possible kind of model is suggested by another well-known spin glass system, alloys of the form $Eu_xSr_{1-x}S$. In the Eu-rich limit, this is a well-studied ferromagnet with ferromagnetic nearest neighbour and antiferromagnetic next-nearest neighbour interactions. One can determine the exchange strengths from neutron scattering at low temperatures, and one also knows that all longer-range interactions are essentially negligible (Bohn et al, 1980). The Sr is magnetically dead, so substitution of Sr for Eu just effectively dilutes the latter. We can write a Hamiltonian of the form

$$H = -\tfrac{1}{2} \sum_{ij} J_{ij} c_i c_j \mathbf{S}_i \cdot \mathbf{S}_j \tag{2.4}$$

where $c_i = 1$ or 0 with probabilities x and $1 - x$. To fit the neutron scattering experiments, one takes $J_{ij} = 0.22K/k_B$ if i and j are nearest neighbours, and $J_{ij} = -0.10K/k_B$ if i and j are next-nearest neighbours. Like the RKKY model, then, this model has competing interactions and randomness.

However, the fact that spin glass behaviour is observed in a wide variety of systems in which the interactions are nothing like (2.3) suggests that we may try something even simpler, if it turns out to be theoretically convenient. Such a model was in fact introduced by Edwards and Anderson (1975) in the paper that, more than any other single one, marks the start of spin glass theory as an active area of theoretical physics. They took their spins to lie on a regular (translationally invariant) lattice, and let the interactions be random instead. Explicitly, their Hamiltonian is

$$H = -\tfrac{1}{2} \sum_{ij} J_{ij} \mathbf{S}_i \cdot \mathbf{S}_j \tag{2.5}$$

where the J_{ij} are taken to be independent random variables with a distribution which depends only on the lattice vector separation $\mathbf{R}_i - \mathbf{R}_j$. In particular, it is convenient to consider the cases of a symmetric Gaussian distribution

$$P(J_{ij}) = \frac{1}{(2\pi\Delta_{ij})^{\frac{1}{2}}} \exp\left[-\frac{J_{ij}^2}{2\Delta_{ij}}\right] \tag{2.6}$$

and the double delta-function distribution

$$P(J_{ij}) = \tfrac{1}{2}\delta(J_{ij} - \Delta_{ij}^{\frac{1}{2}}) + \tfrac{1}{2}\delta(J_{ij} + \Delta_{ij}^{\frac{1}{2}}) \tag{2.7}$$

In either case, the model is specified by the function

$$[J_{ij}^2]_{av} \equiv \Delta_{ij} \equiv \Delta(\mathbf{R}_i - \mathbf{R}_j) \tag{2.8}$$

(We introduce here our convention for averages over the distributions of the random variables in our theories: brackets with the subscript av.) We call these models Edwards–Anderson (or simply EA) models. They clearly also have randomness and competing interactions. More theoretical analysis has been done on these so-called random-bond models than on random-site models like those described above, because the averages are easier to perform.

An additional simplification arises if instead of the Heisenberg model (2.5) with vector spins $\mathbf{S}_i$ one considers an Ising model, in which one retains only a single component $S_{iz} = S_i$. It turns out that even this simplified model still describes nearly all essential spin glass properties. (Exceptions are mostly certain dynamical properties as discussed in Chapters 6 and 8.)

The cases where Δ_{ij} is nearest-neighbour or at least short-ranged have naturally attracted a great deal of attention, and the term 'EA model' is often taken to mean these special cases. However, long-range Δ's can also be considered, and we shall see that the infinite-range case, where Δ_{ij}

is independent of i and j, is extremely interesting, because infinite-range forces are a way of formally defining mean field theory, which is always the first step in understanding collective behaviour in a problem in statistical mechanics. The Ising version of such a model was first investigated by Sherrington and Kirkpatrick (1975) and is as a result universally called the SK model.

Another kind of infinite-range model is one where each spin interacts only with a finite ($O(1)$) number of other spins, but these may be anywhere in the system (Viana and Bray, 1985; Kanter and Sompolinsky, 1987). This kind of model is like short-range ones in its finite connectivity, but lacks any lattice topology.

We stress our belief (based on the experimental phenomenology) that there should not be a significant difference between random-site and random-bond models, provided of course that they have forces which fall off with distance in the same way and have spins of the same vector nature, etc. Thus, for example, one should be able to imitate an RKKY model of the type described here by an EA model in which $\Delta(r) \propto r^{-6}$, and a short-range EA model should be a good description of $Eu_xSr_{1-x}S$.

A number of people have also studied quantum spin models, where, in addition to all the effects we will be most concerned with, zero-point or quantum fluctuations can inhibit spin glass freezing. However, in this book we will only discuss classical models, because quantum effects are not very important in the materials which have been studied experimentally, nor have quantum models been a significant part of the conceptual theoretical developments.

2.2 Broken symmetry in pure systems

The theory of spin glasses relies heavily on concepts and techniques developed in the theory of phase transitions and broken symmetry in non-random systems. We therefore make a little digression to review the most important features of this simpler class of problems.

The most important basic notion is that of an *order parameter*. A broken symmetry state is characterized by a nonzero value of this parameter — in a ferromagnet, it is just the spontaneous magnetization (per site)

$$\mathbf{M} = \frac{1}{N} \sum_i \langle \mathbf{S}_i \rangle \qquad (2.9)$$

where $\mathbf{S}_i$ is the value of the spin at site i. In an antiferromagnet, the order parameter is the 'staggered magnetization'

$$\mathbf{M}_Q = \frac{1}{N} \sum_i \langle \mathbf{S}_i \rangle e^{i\mathbf{Q} \cdot \mathbf{r}_i} \tag{2.10}$$

where the wavevector $\mathbf{Q}$ specifies the spatial periodicity of of the antiferro-magnetic state.

The simplest way to try to describe a broken-symmetry state is with mean field theory. We illustrate it for the Ising ferromagnet (with $h = -g\mu_B B$, where B is the external magnetic field)

$$H = -\tfrac{1}{2} \sum_{ij} J_{ij} S_i S_j - h \sum_i S_i \qquad (S_i = \pm 1) \tag{2.11}$$

As the name of the theory implies, we approximate the internal field acting on spin i by its mean value:

$$h_{int} = h + \sum_j J_{ij} \langle S_j \rangle = h + J(0)M \tag{2.12}$$

where $J(0)$ means the $\mathbf{k} = 0$ component of the Fourier transform $J(\mathbf{k})$ of J_{ij}. We then calculate M self-consistently as the mean magnetization produced by this field:

$$M = \tanh[\beta(h + J(0)M)] \tag{2.13}$$

One finds nontrivial solutions, even with $h = 0$, below a temperature $T_c = J(0)$. (We take the Boltzmann constant $k_B = 1$.) Just below T_c, $M \propto (T - T_c)^{\frac{1}{2}}$, while in the low-temperature limit $M \to 1$.

The notion that the system can find itself in states which break the symmetry of the Hamiltonian is a very profound one. It means that the ergodic hypothesis (that in equilibrium the system should be found with Gibbs–Boltzmann probabilities $\propto e^{-\beta E}$ in any of its possible configurations) is violated. A ferromagnet with its net magnetization up will *never* be found later in a state with its net magnetization down (in the limit $N \to \infty$, of course). Its motion is restricted to the part of its configuration space with positive M. We call this kind of situation *broken ergodicity* (Palmer, 1983).

We note furthermore that we have to be careful about the meaning of thermal averages, such as those which appear in (2.9) and (2.10), in zero external field. They are *not* conventional Gibbs averages over all the spin configurations with the symmetric weight $\exp(-\beta H[S])$; such quantities vanish by symmetry. Rather, they are averages only over part of the con-figuration space — in the ferromagnetic case we can take this to be either the configurations with positive net magnetization or those with negative net magnetization, with a consequent sign difference in the resulting value of M. That is, the broken ergodicity has to be put in 'by hand' by re-stricting the trace used to define thermal averages to configurations near the chosen phase.

We can achieve this restricted trace formally in the ferromagnet by keeping a very small field term in the Hamiltonian and letting its magnitude go to zero after the thermodynamic limit. More explicitly, in a system of size N, the symmetry-breaking field need only be large compared to T/N in order to give, say, negative-magnetization states a negligible relative thermal weight. Thus it can still be very small in a large system, and the limit $h \to 0$ can formally be taken after $N \to \infty$. But notice that the order of the limits is crucial — if the field is taken to zero while N is still finite, we will just go back to the symmetric Gibbs average over all phases.

A way to restrict the trace to one phase without an external field is by means of a boundary condition. Thus, imposing that the boundary spins point up will put the whole system in the up-spin phase. More exotic phases can be induced by more complex boundary conditions: Making the spins on one end of the sample point up and those on the other end point down, for example, will produce a phase (or, actually, many phases) with a domain wall running through the middle of the sample, separating up- and down-magnetized regions.

And, of course, one can describe the breaking of ergodicity dynamically, interpreting the expectation values like $\langle S_i \rangle$ as averages over a time interval $[0, t_0]$ (taking the limit $t_0 \to \infty$). Broken ergodicity is simply the fact that these long-time averages do not vanish. To do this one has of course to go beyond the framework of equilibrium statistical mechanics and introduce specific dynamical models. On the one hand, we would like to be able to avoid doing this — equilibrium statistical mechanical quantities should be able to be calculated within the framework of equilibrium statistical mechanics, without recourse to a particular dynamical picture, and this is the motivation for introducing the symmetry-breaking fields or boundary conditions as we have described. On the other hand, dynamics is interesting in itself because many experiments are dynamical in nature, so we will also study the problem in dynamical formulations.

It is important to stress that broken symmetry and broken ergodicity can only occur, strictly speaking, in infinite systems. In a finite system, the entire configuration space is accessible: a finite ferromagnet in an up-spin state will eventually fluctuate over to the down-spin one (and back again, many times) at any nonzero temperature. This is why the order of limits ($h \to 0$ *after* $N \to \infty$) taken above is important.

People are normally quite sloppy about these points because they are nearly trivial in the ferromagnet and other conventional broken-symmetry systems. We belabour them here because we will find that we have to be much more careful and rethink the underlying assumptions in spin glasses. The notion of broken ergodicity will turn out to be an essential one for understanding spin glasses. The present ferromagnetic example is a rather trivial one, because here the broken ergodicity is simply broken symmetry. However, we will use it again when we come to study the more complicated sort of broken ergodicity, not expressible as broken symmetry, that we find in spin glasses.

A general framework for studying broken symmetry, both at the mean field level and beyond it, is provided by the phenomenological Landau theory (Landau and Lifshitz, 1969), which we now outline briefly for the simplest case: an Ising (one-component) ferromagnet. The basic idea in this approach is to focus on the energetics of slowly-varying local magnetization configurations. We therefore ignore the underlying discrete lattice structure of the material and describe it as a continuum, writing $M(\mathbf{r})$ for the magnetization at point $\mathbf{r}$. Near the transition point, the magnetization is small, so we try to expand the free energy in powers of $M(\mathbf{r})$. Except for the effect of the external field, which contributes a term linear in M, the problem is completely symmetric with respect to the sign of M, so only even powers of M appear. We therefore write

$$F = F_0(T) + \int d^d r [\tfrac{1}{2} a(T) M^2(\mathbf{r})$$

$$+ \tfrac{1}{4} u(T) M^4(\mathbf{r}) + \cdots + \tfrac{1}{2} c (\nabla M(\mathbf{r}))^2 + \cdots - h(\mathbf{r}) M(\mathbf{r})] \qquad (2.14)$$

Except for the uM^4 term, we have truncated the expansion in both the magnetization and its spatial variation at the first opportunity. The gradient term expresses the coupling between nearby spins, i.e. the fact that it costs energy to create a nonuniform magnetization configuration.

Looking then at uniform configurations, we see that the energetics of the system in zero field change dramatically when the quadratic coefficient a changes sign (Fig. 2.2): For positive a, F is minimised by $M = 0$, while for negative a, the minimum F is achieved for $M = (-a/u)^{\frac{1}{2}}$. Thus if we assume $a \propto (T - T_c)$ in the neighbourhood of T_c, we obtain

$$M \propto (T_c - T)^{\frac{1}{2}}, \qquad T < T_c \qquad (2.15)$$

just as we found from (2.13). The present approach suggests that this dependence on the distance to the critical temperature is not just a particular feature of the model (2.11), but a rather general one characteristic of all systems in which the free energy can be expanded as in (2.14).

Similarly, let us consider the response of the system for $T > T_c$ to the presence of a small, possibly spatially varying, external field $h(\mathbf{r})$. Then, writing F in terms of the Fourier transformed variables M_k and h_k (now for small enough magnetization the M^4 term plays no role), we have

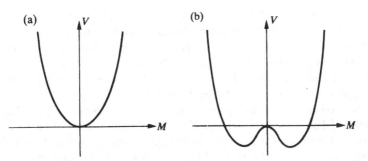

Figure 2.2: The Landau–Ginzburg 'effective potential' $\frac{1}{2}aM^2 + \frac{1}{4}uM^4$ for (a) positive a and (b) negative a.

$$F - F_0 = \sum_k [\tfrac{1}{2}(a + ck^2)|M_k|^2 - h_{-k}M_k], \qquad (2.16)$$

which is minimized by

$$M_k = \frac{h_k}{a + ck^2} \qquad (2.17)$$

This tells us, first of all, by considering the case where $h(\mathbf{r})$ is constant, that the uniform susceptibility χ diverges like

$$\chi = \frac{1}{a} \propto (T - T_c)^{-1} \qquad (2.18)$$

Furthermore, more generally, Fourier transforming (2.17) gives us the non-local susceptibility $\chi(r)$, or, using the fluctuation–response relation (1.10), the correlation function

$$C(r) = \langle M(0)M(r) \rangle = T\chi(r) \propto r^{2-d} \exp[-(a/c)^{\frac{1}{2}}r] \qquad (2.19)$$

in the large-r limit. Thus spin correlations have a range

$$\xi = (c/a)^{\frac{1}{2}} \propto (T - T_c)^{-\frac{1}{2}} \qquad (2.20)$$

which diverges as the critical point is approached.

The characteristic temperature dependences (2.15), (2.18) and (2.20) are the simplest examples of *universality* of properties near critical points. The exponents do not depend in detail on the microscopic parameters of the material, but follow quite generally from the overall symmetry of the system, as expressed in the expansion (2.14).

In systems with more complicated order parameters, more complicated terms occur in the corresponding expansion of the free energy. But always it is the symmetry of the system which dictates the nature of the singularities at the critical point, different kinds of symmetries leading to different so-called 'universality classes'. In general, to construct the counterpart of the free energy expansion (2.14), one has to consider all combinations of factors of different components of the order parameter which are invariant under the symmetry of the system. For example, for vector spins, the allowed quadratic and quartic terms have the form $a\mathbf{M}(\mathbf{r}) \cdot \mathbf{M}(\mathbf{r})$ and $u(\mathbf{M}(\mathbf{r}) \cdot \mathbf{M}(\mathbf{r}))^2$, respectively. In the case of vector spins subject to a cubic crystalline anisotropy, no other quadratic terms are allowed, but there is another possible quartic term of the form $v \sum_i M_i^4(\mathbf{r})$. In systems without symmetry under inversion of the order parameter (and we will see that spin glasses fall into this category), terms of cubic, fifth order, etc. also occur.

For simple Ising and rotationally invariant vector spin models, Landau theory is very simple, but for order parameters with a more complex matrix structure, it can be nontrivial. We will meet just such an example in the mean field theory of spin glasses in Chapter 3.

In this book we will focus nearly entirely on mean field theory until Chapter 8. Nevertheless, it is important to point out here that Landau theory is not in general correct; it contains implicitly an approximation like that made to obtain the Weiss mean field equation (2.13) — the neglect of thermal fluctuations in the internal field. It is true that Landau theory, through, e.g., the steps leading to (2.19), gives us a way to calculate fluctuations around the mean value of the magnetization. However, this procedure is not self-consistent, in that the effect of these fluctuations is not taken into efffect in calculating the mean magnetization. That they can be important can be seen dramatically in the following example, due to Landau and Peierls, for spins with two or more components. We suppose we are at low temperatures, so that the magnitude of the local magnetization $\mathbf{M}(\mathbf{r})$ is essentially fixed at its saturation value. The thermodynamically important fluctuations at low temperature are those in which the *direction* of $\mathbf{M}(\mathbf{r})$ varies slowly in space, since they have very low energy. This picture is summarized by the effective Hamiltonian

$$H = \tfrac{1}{2}J \int d^d x [\nabla \mathbf{M}(\mathbf{r})]^2 \tag{2.21}$$

Writing this in k-space and keeping only the $m - 1$ components of M perpendicular to the ordering direction, we have

$$H = \tfrac{1}{2}J \sum_k \sum_{\mu=2}^{m} k^2 |M_{k\mu}|^2 \tag{2.22}$$

For small values of these fluctuations, the different components are independent of one another, so (2.22) is a simple quadratic Hamiltonian (analogous to that of a classical gas) to which one can apply the classical equipartition theorem

$$\tfrac{1}{2}Jk^2\langle|M_{k\mu}|^2\rangle = \tfrac{1}{2}k_BT \tag{2.23}$$

Thus, for the order parameter components perpendicular to the ordering direction, the wavenumber-dependent susceptibility $\chi_\mu(k) = \beta\langle|M_{k\mu}|^2\rangle$ is proportional to k^{-2} for small k. This is simply a consequence of the rotational invariance of the system: a uniform ($k = 0$) rotation costs no energy, so the susceptibility perpendicular to the ordering direction is infinite. If we now use this result to calculate the mean square fluctuations of the magnetization perpendicular to the ordering direction, we find

$$\langle(\delta\mathbf{M}(\mathbf{r}))^2\rangle = \frac{(m-1)k_BT}{J}\int\frac{d^dk}{(2\pi)^d}\frac{1}{k^2} \tag{2.24}$$

which diverges for $d \leq 2$. Thus, for low enough dimensionality, the fluctuations ignored in Landau theory can actually be so strong as to destroy the phase transition entirely.

Even when these fluctuation effects are not sufficient to destroy the order, they may, depending on dimensionality and on the symmetry of the system, change the qualitative nature of the transition, e.g. the values of the universal exponents such as those in (2.15), (2.18) and (2.20). Nevertheless, the concept of universality classes for critical properties remains a valid and central one. We will return to these effects, in both conventional broken symmetry systems and spin glasses, in Chapter 8.

2.3 Averaging in disordered systems

Randomness introduces special features into theoretical problems in statistical mechanics. First we have the fundamental problem that we do not know all the parameters of the Hamiltonian of the sytem we are trying to study, only the parameters of a statistical ensemble of such Hamiltonians. Furthermore, we could not normally calculate properties of a single sample with a particular set of values of the random interactions J_{ij} even if we knew them. (Actually, sometimes we can do this on a computer, but such a calculation is more like an experiment than like theory in the sense we generally mean here: it can tell us what the value of some quantity is but cannot give us insight into the physics of the system.)

Fortunately, as is so often the case in statistical physics, statistical averaging comes to our aid in the limit of large systems. Just as in ordinary statistical mechanics we know that the relative fluctuations of the energy around its thermal mean value are of $O(N^{-\frac{1}{2}})$, we expect that its sample-to-sample fluctuations also go to zero in the limit of a large system. A quantity with this property is said to be *self-averaging*. If we know that quantities of interest to us are self-averaging, then not only can we expect the same results in experiments on different macroscopic samples, but we can also expect a theoretical calculation of the mean value of the quantity over the whole ensemble to give the same answer as the experiments. Therefore our general strategy is always to try to concentrate in our theory on self-averaging quantities and try to calculate their average values over the distribution of random interactions in the statistical ensemble.

Many quantities are *not* self-averaging. An example is the local internal field at a particular spin site i, which obviously depends sensitively on the local environment. Fortunately, many of the quantities we naturally think of measuring or calculating are sums or integrals over the entire volume of the sample, and for these so-called 'extensive' quantities statistical fluctuations become small for large samples. Slightly more carefully, the argument is as follows: We can divide our sample into many subsamples and write the energy (say) as a sum of energies from the individual subsamples plus corrections from the boundaries between subsamples (the energy difference between subsamples which are in direct contact with each other and subsamples which are widely separated). For large sample size L, the first of these terms should be $\propto L^d$ and the second $\propto L^{d-1}$, so we can ignore the latter. Now suppose we write the thermal mean energy of the (small) subsamples as $E_0 + \delta E$, where δE is the fluctuation due to the randomness of the system, i.e. the fact that the subsamples are not all identical. Then, combining all the little subsamples together into one big sample, the total energy will scale like the number of subsamples, while the fluctuations of the total energy will scale like the square root of this number. Thus we obtain the familiar $N^{-\frac{1}{2}}$-dependence of the relative fluctuations $\delta E/E$.

This argument implicitly assumes short-range forces (and we have not carefully investigated how fast the forces have to fall off with distance in order to qualify as short-range). This is a point we will have to be more careful with later on.

We will meet examples where the mean value of interesting quantities vanishes, as well, it turns out, as cases where variables we calculate will not be self-averaging, even though they are global quantities. In these cases, the theorist's job also includes calculating higher order moments of these quantities, or even their whole distribution.

Let us also observe explicitly that in the statistical mechanics of random systems, we have two distinct kinds of averages to perform: the usual 'thermal average' to be carried out in principle for each sample, and the average over the distribution of random parameters. This second average is sometimes called an 'impurity average' or a 'configuration average', because in random-site systems like the RKKY model of Section 2.1 it is just an average over all possible impurity configurations. In the random-bond models we will more commonly be studying, it is the average over the distribution of J_{ij}'s. Our notation for it, as already mentioned, is $[\ ]_{av}$.

The quantities which we want to average in this way can, according to the standard tricks of statistical mechanics, be expressed as or in terms of derivatives of a free energy with respect to auxiliary fields. (For example, the magnetization is the derivative with respect to a magnetic field.) So the general procedure is as follows: One starts in principle with a partition function which is a trace over the thermodynamic variables and a function of the fixed interaction strengths for that sample

$$F[J] = -T \ln Z[J] \tag{2.25}$$

Now $F[J]$ is an extensive variable, which we can expect to be self-averaging. The experimentally relevant quantity is therefore

$$F \equiv [F[J]]_{av} \equiv \int dP[J]F[J] = -T \int dP[J] \ln Z[J] \tag{2.26}$$

Note that it is $\ln Z$, not Z itself, which should be averaged. This is because Z is not an extensive quantity; self-averaging cannot be expected to apply to it, and so $[Z]_{av}$ is not a physically relevant quantity. (It is instructive to take a simple soluble example like the random ferromagnetic chain and verify explicitly the lack of self-averaging of Z.)

By obvious extension of this argument, the magnetization measured in a macroscopic sample is

$$[M]_{av} = T\frac{\partial[\ln Z[J,h]]_{av}}{\partial h} \tag{2.27}$$

and similarly for other extensive quantities. The calculation of correlation functions requires only a simple formal extension to site-dependent fields:

$$[\langle S_i S_j \rangle - \langle S_i \rangle \langle S_j \rangle]_{av} = T^2\frac{\partial[\ln Z[J,h]]_{av}}{\partial h_i \partial h_j} \tag{2.28}$$

How is one then to evaluate these averages? One way is of course to write down formal expressions for $F[J]$ or its derivatives (obtained, for example, by perturbation theory in J_{ij}/T) and average them, term by term, over the distribution of the J_{ij}'s. This procedure is in fact often practical, but it is also often very useful to be able to carry out the averaging formally from the beginning. This will leave us with a problem in which the disorder no longer appears explicitly; if the *ensemble* is translationally invariant, we will have a nonrandom, translationally invariant statistical mechanical problem. We have a lot of experience with problems like this, so we can hope to treat them by methods we have already developed elsewhere. Let us then see what kind of effective nonrandom problem we can get out of this approach.

We can note in passing that if it were Z rather than $\ln Z$ that had to be averaged, the calculation might be rather simple. Consider for example an Ising EA model with the Gaussian bond distribution (2.6). Then

$$[Z[J]]_{av} = Tr_S \int \prod_{\langle ij \rangle} \frac{dJ_{ij}}{(2\pi\Delta_{ij})^{\frac{1}{2}}} \exp\left(-\frac{J_{ij}^2}{2\Delta_{ij}} + \beta J_{ij}S_iS_j\right) \quad (2.29)$$

and each integral can be done by completing the square. However, it is also clear from this example that averaging Z is the wrong physics. The expression (2.29) has the form of a partition function for a nonrandom system in which both the spins and the J_{ij}'s are thermodynamic variables which are traced over on the same footing. That is, both the S_i's and the J_{ij}'s are free to take on values which tend to minimize the total effective energy

$$H_{annealed}[S, J] = \sum_{\langle ij \rangle} \left(\frac{TJ_{ij}^2}{2\Delta_{ij}} - J_{ij}S_iS_j\right) \quad (2.30)$$

But the real situation is that the J_{ij}'s are fixed for each sample; the S_i's can vary in response to them, but they cannot vary in response to the S_i's. Thinking again about our RKKY alloy systems, we could imagine heating the sample up so that the spins could diffuse around, making their interactions change their values like the J_{ij}'s in (2.30). This procedure — letting the J_{ij}'s come to equilibrium with the spins — is called annealing, and the kind of system described by (2.30) is called an annealed system. In the kind of materials we are interested in, the J_{ij}'s are frozen into their fixed values by rapid cooling when the sample is prepared. This kind of rapid cooling is called quenching, and one says that we have a quenched system and that the J_{ij}'s are 'quenched variables'. Thus the kind of averaging we have to do — of $\ln Z[J]$ rather than $Z[J]$ — is called a 'quenched average'.

It is also apparent that if we want to average $\ln Z$, we cannot generally carry out the integral as easily as in (2.29). There is a way around this difficulty, however. It is called the 'replica method' and is used very widely in the statistical mechanics of random systems. It makes use of the identity

$$\ln Z = \lim_{n \to 0} \frac{Z^n - 1}{n} \quad (2.31)$$

and the fact that the average $[Z^n]_{av}$ *can* be carried out for general integer n almost as simply as $[Z]_{av}$. We simply write Z^n as

$$Z^n[J] = Tr_{S_1, S_2, \cdots, S_n} \exp(-\beta \sum_{\alpha=1}^{n} H[S^\alpha, J]) \tag{2.32}$$

We say that we have hereby 'replicated' the system n times; hence the terminology 'replica method'. The indices α which appear attached to the spins are called replica indices. For the Ising EA model, the average of (2.32) over the Gaussian bond distribution is then simply

$$[Z^n]_{av} = Tr_{S_1 \ldots S_n} \exp \left(\tfrac{1}{4}\beta^2 \sum_{ij} \Delta_{ij} \sum_{\alpha\beta} S_i^\alpha S_i^\beta S_j^\alpha S_j^\beta \right)$$

$$\equiv Tr_S \exp(-\beta H_{eff}) \tag{2.33}$$

Thus we have converted this disordered problem into a non-random one involving 4-spin interactions between the n-component spins. More generally, for a general distribution of J_{ij}, we get

$$\beta H_{eff} = -\tfrac{1}{2} \sum_{ij} \sum_{p=1}^{\infty} \frac{1}{p!} [J^p]_c (\beta \sum_{\alpha=1}^{n} S_i^\alpha S_j^\alpha)^p \tag{2.34}$$

where $[J^p]_c$ is the pth cumulant of the distribution; that is, we have in general multispin interactions of all orders. After trying to solve these effective problems by whatever tools we can bring to bear on them, we have to take the limit $n \to 0$ in the result.

Notice the symmetry of these effective Hamiltonians. In addition to the standard symmetry under spin inversion of the original Ising model (or rotation invariance for Heisenberg-spin generalizations), there is a permutation symmetry in the replica index space. This is an obvious consequence of the way the theory was constructed: (2.32) is unchanged if two replica indices are permuted (the indices were just a formal tool in writing $Z^n[J]$), and this equivalence survives the averaging over the bond distribution.

In general, if we want to solve a model in such a way as to describe a broken-symmetry state, we want to set up a self-consistent (e.g. mean field) calculation for an order parameter. What sort of order parameters are possible for models like (2.33) and (2.34)? The simplest kind would appear to be averages like $\langle S_i^\alpha \rangle$. This is in fact the kind of order parameter which would be important in a ferromagnetically ordered random magnet, and we will use it when we study systems where spin glass ordering and ferromagnetism compete. But another kind of possible quantity which could describe a state of spontaneously broken symmetry is the matrix $\langle S_i^\alpha S_i^\beta \rangle$. We could call this a sort of 'quadrupolar' order parameter in the replica space. Going further, we could imagine 'octupolar' quantities $\langle S_i^\alpha S_i^\beta S_i^\gamma \rangle$, and so on. In this book we will be concerned mostly with $\langle S_i^\alpha \rangle$ and $\langle S_i^\alpha S_i^\beta \rangle$. The physical meaning of these formally defined matrices is not immediately obvious. We will see what it is in the next section.

One can well be skeptical about this bit of mathematical manipulation, and many of the troubles encountered in the early days of the mean field theory for spin glasses were often blamed on the replica method. However, none of these charges made against it has ever stuck. Furthermore, it is the only general method we have with which to deal systematically with the statistical mechanics of random systems. The main problem, of course, it the analytic continuation from integer n's to $n \to 0$. If we know the solution for a finite number of replicas this limit is not unique. However, in all physical problems where replicas have been introduced this procedure seems to work and leads to reasonable solutions.

2.4   Broken ergodicity and spin glass order parameters

If a spin glass state is qualitatively different from a disordered paramagnetic phase, we would like to characterize the difference by an order parameter, as we do for conventional broken symmetry systems. We will suppose that just as a pure ferromagnetic state is described by a nonzero value of $M = \langle S_i \rangle$ which is independent of i, that a spin glass state is characterized below a transition ('freezing') temperature T_f by

$$m_i \equiv \langle S_i \rangle \neq 0 \qquad (2.35)$$

without any long-range periodic order in their values.

As in the discussion of broken symmetry in the ferromagnet in Section 2.2, we must be careful about the precise meaning of the thermal average which appears in (2.35). If it is a complete average over *all* configuration space with Boltzmann–Gibbs weights, it vanishes identically in the absence of external fields. In order to make it nonzero, we can either restrict the thermal average in an *ad hoc* fashion to a suitable region of configuration space (like restricting the average to states with positive total magnetization in the ferromagnet), or impose a small external field which varies in space like m_i, which has the same effect (being careful, as in the ferromagnetic case, to take the limit of vanishing field strength *after* the thermodynamic limit $N \to \infty$). If the spin glass has a single pair of stable states (2.35) related by an overall spin flip, then the same effect can be achieved by a uniform external field: A given spin glass configuration m_i will have a net magnetization of $O(N^{\frac{1}{2}})$, so a uniform external field $h \gg TN^{-\frac{1}{2}}$ will be sufficient to make the thermal weight of one of the two states negligible relative to the other. (There is in fact nothing magical about a *uniform* field; any arbitrary field h_i uncorrelated with m_i will do equally well.)

However, in a spin glass we have at least to address the possibility of nontrivial broken ergodicity (Palmer, 1982), that is, the case where there are very many stable states. A helpful aid in visualizing this situation is the so-called many-valley picture illustrated in Fig. 2.3. This is the natural generalisation of the picture of Fig. 2.2b to a situation with many possible states. One imagines being able to impose on the system arbitrary local magnetizations m_i and, for each such set of values, calculating the free energy $F(m_1 \ldots m_N)$. We picture this function as a surface in the $(N+1)$-dimensional space whose axes are labeled by the m_i and F. Local minima of F, i.e. solutions of $\partial F/\partial m_i = 0$ with the eigenvalues of the matrix $\partial^2 F/\partial m_i \partial m_j$ all positive then correspond to locally stable magnetization configurations. If, in the limit of large N, some of the barriers between these mimima become infinite, we can partition the entire state space into mutually inaccessible 'valleys'. Each of these valleys corresponds to a thermodynamic phase or state like the up- or down-magnetized states of a ferromagnet. We label these states by an index a running from 1 to the number N_s of states and denote their local spontaneous magnetizations by m_i^a and their free energies by F_a. (Within such a state or valley, of course, there can be several sub-valleys which are local minima of F, but with *finite* barriers separating them. All but the lowest of these correspond to metastable magnetization configurations.) Microscopically, the configurations S_i which contribute to the partition function in a particular such phase a all lie in the region of spin configuration space near m_i^a, i.e. in a single valley in the free-energy landscape (possibly comprising several sub-valleys). Above T_f, of course, this landscape disappears, and there is just one valley centered at $m_i = 0$. (Also, the features of the landscape may be sensitive to T below T_f.)

When a system finds itself in one of these valleys, it will exhibit properties which in general are specific to that valley. They differ from true equilibrium properties, which involve averages over all valleys with appropriate relative thermal weights. For example, there might be valleys (or at least metastable sub-valleys) with very different magnetizations; this is the origin of the remanence observed in all spin glasses. In order to calculate the properties of the system in a single valley, the trace over configurations in the partition function must be restricted to the appropriate valley.

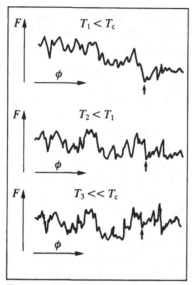

Figure 2.3: Schematic plot of the evolution of part of the free energy 'hypersurface' (shown here as a function of one phase space variable only) with decreasing temperature. Arrow denotes deepest minimum (from Reger et al, 1984).

If there are many possible states, the imposition of an infinitesimal external field uncorrelated with any m_i^a will no longer select out a single phase. Furthermore, we are handicapped in trying to use infinitesimal external fields h_i^a proportional to the m_i^a to try to generate appropriately restricted traces because we do not know these conjugate fields *a priori*.

Thus broken ergodicity makes the definition of relevant thermal averages in general, and that of order parameters in particular, a highly nontrivial one. Different ways of projecting out particular phases or sets of phases will lead to different results. With the clarity of hindsight we can see that a great deal of the difficulty that people experienced in constructing even a satisfactory mean field theory of spin glasses can be traced to this problem. A natural consequence of the existence of many phases is that a spin glass cannot be described by a single order parameter, but rather requires many of them.

So far the discussion deals with the characterization of the frozen state for a single sample. We will of course want to deal with quantities which are averages over the bond distribution. The only bond averaged quantity which is linear in the local magnetizations $\langle S_i \rangle$ is the mean magnetization $[\langle S_i \rangle]_{av}$. This is, of course, the order parameter of a ferromagnet and therefore not a suitable spin glass order parameter. Indeed, it vanishes in the limit of zero external field (at least for symmetric bond distributions, to which we restrict our attention for the moment). We clearly have to look at higher order moments. Let us consider, then, the possible kinds of second moments we obtain for the different ways of breaking ergodicity.

If we regard ergodicity breaking as essentially dynamical in nature, the most natural order parameter to consider is the one introduced originally by Edwards and Anderson (1975):

$$q_{EA} = \lim_{t\to\infty} \lim_{N\to\infty} [\langle S_i(t_0)S_i(t_0 + t)\rangle]_{av} \tag{2.36}$$

where the average is over a long (eventually infinitely so) set of reference times t_0. This will clearly be zero (in vanishing external field) if the system is ergodic, and will be nonzero if it is trapped in a single phase. One must take the $N \to \infty$ limit before the $t \to \infty$ one, since for a finite system the correlation will eventually die out as true equilibrium is reached. Since an infinite system can never escape the valley it is in, q_{EA} measures the mean square single-valley local spontaneous magnetization, averaged over all possible valleys. That is, in terms of thermal averages,

$$q_{EA} = \left[\sum_a P_a(m_i^a)^2\right]_{av} \tag{2.37}$$

where

$$P_a = \frac{e^{-\beta F_a}}{\sum_a e^{-\beta F_a}} \tag{2.38}$$

Assuming self-averaging, we can also write

$$q_{EA} = \frac{1}{N}\sum_a P_a \sum_i (m_i^a)^2 \tag{2.39}$$

(We have not proved the validity of self-averaging here, but it is possible to prove it within particular models and, in particular, in the mean field theory of the next chapter.)

q_{EA} is not, of course, the mean square local *equilibrium* magnetization. We call this the equilibrium or statistical mechanics order parameter and denote it simply by q. It is given by

$$q = [\langle S_i\rangle^2]_{av} = [m_i^2]_{av} = \left[\left(\sum_a P_a m_i^a\right)^2\right]_{av}$$

$$= \left[\sum_{ab} P_a P_b m_i^a m_i^b\right]_{av} \tag{2.40}$$

or, equivalently,

$$q = \frac{1}{N} \sum_i \left[\sum_{ab} P_a P_b m_i^a m_i^b \right]_{av} \tag{2.41}$$

We can also define a q for a single sample:

$$q_J = \frac{1}{N} \sum_i m_i^2 = \frac{1}{N} \sum_{iab} P_a P_b m_i^a m_i^b \tag{2.42}$$

We can see immediately that q differs from q_{EA} in having 'intervalley' contributions (if we have multiple phases). It will turn out that q_J is *not* self-averaging; $[q_J^2]_{av}$ exceeds q^2 by a quantity of $O(1)$ unless there is only one phase. We cannot prove this here directly from (2.40)–(2.42), however. The proof, which we will give in the next chapter, requires the use of the replica formalism as developed there.

Of course, these quantities vanish trivially in exactly zero external field. To make them nonzero, we need to impose a uniform or random (i.e. uncorrelated with the valley magnetizations) external field and then take the limit $h \to 0$ after $N \to \infty$.

It is sometimes convenient to consider the difference between q and q_{EA} as a new order parameter Δ that measures the degree of broken ergodicity:

$$\Delta = q_{EA} - q \tag{2.43}$$

It is a simple exercise to show that Δ is positive semidefinite and that the equality is attained only when there is just a single phase. One can picture the difference between q and q_{EA} dynamically by imagining N large but finite, so that transitions across the barriers separating different valleys are allowed. On a timescale long enough for the system to pass statistically many times through all valleys having significant thermal weight, true equilibrium is reached and all the intervalley terms in (2.40)–(2.42) contribute. On a short timescale, no transitions between valleys have time to occur, so q_{EA} is the physically relevant quantity. One can also imagine intermediate time scales, where certain groups of valleys are accessible to each other, but there is not time to climb over the higher barriers separating one group from another. Then a quantity between these two limiting cases is the physically relevant one. This picture of a gradual interpolation between complete broken ergodicity and full equilibrium is the physical basis of the dynamical formulation of mean field theory we will study in Chapter 5.

Measurement of the local susceptibility gives direct information about these order parameters, as we noted already in Chapter 1. Thus, from (1.3), for a system with vanishing average bond strength in a single phase, one measures

$$\tilde{\chi}_{loc} = \beta(1 - q_{EA}) \tag{2.44}$$

The average local equilibrium susceptibility, obtained from (1.3) with the equilibrium expression $m_i = \sum_a P_a m_i^a$ is

$$\chi_{loc} \equiv [\chi_J]_{av} = \beta(1 - q) \tag{2.45}$$

Notice that $\chi_J \equiv \beta(1 - q_J)$ will not be self-averaging if there is ergodicity breaking, since in that case, as we have remarked, $[q_J^2]_{av} > q^2$. As mentioned in Chapter 1, these averaged local susceptibilities are the same as the corresponding uniform susceptibilities (e.g. (1.4)) for symmetric bond distributions.

The equilibrium χ is not measurable in a finite-time experiment on a macroscopic sample if there is strict broken ergodicity, i.e. infinite barriers separating the phases. However, if there only finite barriers, separating metastable states, one will, by waiting long enough, observe a crossover from a single-valley χ at short times to an equilibrium one at the longest times.

If we have many phases, it is interesting to ask not only about the mean square magnetization in a single state, such as occurs in q_{EA}, but also about the correlation between states, i.e the 'overlap' (for a single sample)

$$q_{ab} = \frac{1}{N} \sum_i m_i^a m_i^b \tag{2.46}$$

(This occurs, for example, in the equilibrium expression (2.42).) As a and b range over the large number of phases, many possible values of q_{ab} (between -1 and $+1$) will in general be found. It is therefore useful to consider its distribution (Parisi, 1983). For a single sample, we have, formally,

$$P_J(q) \equiv \langle \delta(q - q_{ab}) \rangle \equiv \sum_{ab} P_a P_b \delta(q - q_{ab}) \tag{2.47}$$

and we can also define the bond-averaged quantity

$$P(q) \equiv [P_J(q)]_{av} \tag{2.48}$$

Just as q_J will turn out not to be self-averaging if there is broken ergodicity, neither, in general, is $P_J(q)$.

For a system with just two phases, $P(q)$ is just the sum of a pair of delta functions. (With the symmetry-breaking field it is just a single delta function.) If there is strong broken ergodicity, $P(q)$ may have a continuous part, indicating the possibility of a continuum of possible overlaps between various phases. Thus it is by the form of $P(q)$ that we can distinguish formally between systems with conventional broken symmetry and those with nontrivial broken ergodicity.

We can write the equilibrium q (2.40) in terms of $P(q)$ as

$$q = \int_{-1}^{1} P(q)q\,dq \qquad (2.49)$$

(and a corresponding expression for q_J in terms of $P_J(q)$). If $h = 0$, $P(q)$ is symmetric (since for every overlap q_{ab} which contributes to the distribution, one will also find $-q_{ab}$ by replacing either state a or state b by its equally probable spin-flipped counterpart). Therefore, as remarked above, the equilibrium q vanishes for $h = 0$. However, in the presence of a small field (which need only be large compared to $T/N^{\frac{1}{2}}$), only the states of one sign of the magnetization and therefore only positive overlaps contribute to $P(q)$. Then the lower limit on the integral in (2.49) is zero and q is finite.

On the argument that the correlation between phases cannot be greater than the mean square magnetization in a single phase, it is natural to identify q_{EA} as the largest value of q for which $P(q)$ has support.

The fact that q (or q_J) vanishes in exactly zero external field makes it convenient sometimes to consider instead the quantity (Binder, 1980a,b)

$$q^{(2)} = \lim_{|R_i - R_j| \to \infty} [\langle S_i S_j \rangle^2]_{av} \qquad (2.50)$$

This way of defining an order parameter is analogous to defining a ferromagnetic order parameter by

$$m^2 = \lim_{|R_i - R_j| \to \infty} \langle S_i S_j \rangle \qquad (2.51)$$

In the ferromagnet, this gives the same result as the more conventional way. But if there is ergodicity breaking, $q^{(2)} > q^2$. We can see this from the fact that in any single phase a

$$\lim_{|R_i - R_j| \to \infty} \langle S_i S_j \rangle_a = m_i^a m_j^a \qquad (2.52)$$

Therefore $q^{(2)}$ can be written

$$q^{(2)} = \frac{1}{N^2} \sum_{ij} \left[\sum_{ab} P_a P_b m_i^a m_i^b m_j^a m_j^b \right]_{av} = \int P(q)q^2\,dq \qquad (2.53)$$

which clearly exceeds $(\int P(q)q\,dq)^2 = q^2$.

In the preceding section, we saw that it was possible to represent random problems formally as translationally invariant systems by using the replica method. How do we express our spin glass order parameters in this formalism? Before we begin we consider the simpler case of a ferromagnetic order parameter

$$M = [m_i]_{av} = \left[\frac{Tr_S S_i e^{-\beta H[S,J]}}{Z[J]} \right]_{av} \qquad (2.54)$$

Inside the bond average, we now write $n-1$ factors of $Z[J]$ in both the numerator and denominator. In the limit $n \to 0$, the denominator then goes to unity, so the bond average need only be carried out over the numerator. Introducing replica indices as in (2.32), we find

$$M = \left[Tr_{S_1 \ldots S_n} S_i^\alpha \exp \left(-\beta \sum_{\beta=1}^{n} H[S^\beta, J] \right) \right]_{av} \tag{2.55}$$

where α is any one of the replica indices. The bond average can then be carried out exactly as in obtaining (2.33) or (2.34), and the result can be written

$$M = Tr_S[S_i^\alpha \exp(-\beta H_{eff})] \tag{2.56}$$

Again making use of the fact that

$$Tr_S \exp(-\beta H_{eff}) \equiv [Z^n]_{av} \to 1 \tag{2.57}$$

as $n \to 0$ we can write this simply as

$$M = \langle S_i^\alpha \rangle \tag{2.58}$$

where the average is a 'thermal' one in a Gibbs ensemble with effective Hamiltonian H_{eff}. (Note that this and other averages which appear in the replica approach are over a translationally invariant effective thermal ensemble, since the disorder has already been averaged over, as described in Section 2.3. Therefore no further averages over disorder, such as appear in (2.40), are necessary.)

This answer must of course be independent of which replica index α is singled out. If, in the solution of the problem using the replica Hamiltonian, all the $\langle S_i^\alpha \rangle$ turn out to be equal, this happens automatically. But what if they do not all turn out to be equal? Suppose, for example, that there was a solution of the problem in the replica formulation in which only one of the $\langle S^\alpha \rangle$ was nonzero. Such a situation is conventionally called *replica symmetry breaking*. A hint about how to resolve this apparently paradoxical situation can be got by imagining that the replica indices are real, physically meaningful indices labeling different directions in space. Then Hamiltonians like (2.34) would represent systems with a hypercubic anisotropy in this space, i.e. it is energetically favourable to align the magnetization along the n crystal axes, rather than in some oblique direction. In the absence of an external symmetry-breaking field, a ferromagnetic state in such a crystal could be aligned equally well along any of these axes. For a full equilibrium quantity, we must average over all these states, which will in general give different results from those for just one state. Now going back to our replica theory, where the Hamiltonian is fully replica symmetric, we conclude that if solutions are possible in which replica symmetry is broken, the equilibrium result is obtained by averaging over all these different solutions (De Dominicis and Young, 1983).

We now go through the same arguments for the equilibrium spin glass order parameter q. We simply write the expression for m_i inside the bond average in (2.54) twice and then multiply numerator and denominator by $Z^{n-2}[J]$. Carrying the bond average out then leads straightforwardly to

$$q = q^{\alpha\beta} \equiv \langle S_i^\alpha S_i^\beta \rangle \tag{2.59}$$

for any arbitrary pair of replica indices α and β (as long as $\alpha \neq \beta$, since $(S_i^\alpha)^2 \equiv 1$ for Ising spins). Again, if it turns out that the solution of the replica-theoretic model breaks replica symmetry, one must average over all the possible ways one could break it:

$$q = \lim_{n \to 0} \frac{1}{n(n-1)} \sum_{\alpha \neq \beta} q^{\alpha\beta} \tag{2.60}$$

Similarly, one finds

$$q^{(2)} = \lim_{n \to 0} \frac{1}{n(n-1)} \sum_{\alpha \neq \beta} (q^{\alpha\beta})^2 \tag{2.61}$$

Exactly the same reasoning gives a replica-theoretic expression for $P(q)$:

$$P(q) = \lim_{n \to 0} \frac{1}{n(n-1)} \sum_{\alpha \neq \beta} \delta(q - q^{\alpha\beta}) \tag{2.62}$$

That is, comparing (2.62) with (2.47), the distribution of values of the matrix elements $q^{\alpha\beta}$ in a replica-symmetry-breaking solution must be the same as the distribution of overlaps between different states when there are many states. Thus there is an intimate connection between broken ergodicity and broken replica symmetry.

Finally, we can then identify q_{EA} with the largest $q^{\alpha\beta}$ in a broken replica symmetry solution:

$$q_{EA} = \max_{\alpha\beta} q^{\alpha\beta} \tag{2.63}$$

In the Introduction we also mentioned a quantity χ_{SG} which plays the role in a spin glass that the uniform susceptibility does in a ferromagnet. Formally, we define the spin glass susceptibility as

$$\chi_{SG}(\mathbf{R}_{ij}) = [\chi_{ij}^2]_{av} = \beta^2 [(\langle S_i S_j \rangle - \langle S_i \rangle \langle S_j \rangle)^2]_{av} \tag{2.64}$$

When we write χ_{SG} without an argument, we mean the Fourier transform $\chi_{SG}(\mathbf{k})$ evaluated at $\mathbf{k} = 0$. Above T_f this reduces simply to (1.11).

In a ferromagnet, the ordinary susceptibility is the magnetization induced per unit external field h. In a spin glass, we can induce a nonzero q even above T_f by turning on a random external field h_i. We then have

$$\langle S_i \rangle = \sum_j \chi_{ij} h_j \tag{2.65}$$

and squaring and averaging gives

$$q = \sum_{ij} [\chi_{ij}^2]_{av} \sigma^2 = \chi_{SG} \sigma^2 \tag{2.66}$$

(assuming the h_i at different sites are uncorrelated), where σ^2 is the variance of the random field. Thus σ^2 acts as a conjugate field to the order parameter q.

One can see that a uniform external field h will also induce a $q \propto h^2$ in the same way. However, it also of course induces a net magnetization, so it is more natural to identify the variance of a random field as the conjugate field.

Below T_f, q is finite even when h or $\sigma^2 \to 0$. Thus we expect by analogy with the ferromagnetic case that as one approaches T_f from above, χ_{SG} diverges. We will see in the following chapters that this actually happens in soluble theories and in experiments. Furthermore, the correlation length of the function $\chi_{SG}(\mathbf{r})$ (2.64) also diverges in just the same way (2.20) that the correlation length of the ordinary spin correlation function does in the ferromagnet.

χ_{SG} is also measurable, through the quantity called the 'nonlinear susceptibility' (Chalupa, 1977a,b; Suzuki, 1977). It is defined as the coefficient of $-h^3$ in the expansion of the magnetization in powers of the external field:

$$m = \chi h - \chi_{nl} h^3 + \cdots \tag{2.67}$$

From standard linear reponse theory, just as χ is proportional to the thermal variance $\langle (S - \langle S \rangle)^2 \rangle$, one finds

$$\chi_{nl} = -\frac{\beta^3}{3} N \left\langle \left(\sum_i S_i \right)^4 \right\rangle_c \tag{2.68}$$

where the subscript c means a cumulant average. Above T_f, the cumulant average in (2.68) is simply

$$\left\langle \left(\sum_i S_i \right)^4 \right\rangle_c = \sum_{ijkl} (\langle S_i S_j S_k S_l \rangle - 3 \langle S_i S_j \rangle \langle S_k S_l \rangle)$$

$$= 4N - 6 \sum_{ij} \langle S_i S_j \rangle^2 \tag{2.69}$$

Thus we can identify

$$\chi_{nl} = \beta(\chi_{SG} - \tfrac{2}{3}\beta^2) \tag{2.70}$$

Measurements of χ_{nl} therefore give us important information about the nature of the spin glass transition.

So far, we have just discussed Ising models. For Heisenberg spin glasses (by which we mean any rotationally invariant system with more than one spin component), q and q_{EA} become tensors:

$$q_{\mu\nu} = [\langle S_{i\mu} \rangle \langle S_{i\nu} \rangle]_{av} = [m_{i\mu} m_{i\nu}]_{av} \tag{2.71}$$

where $S_{i\mu}$ is the μth component of $\mathbf{S}_i$. Similarly, the overlap between a pair of states a and b becomes a tensor

$$q_{ab,\mu\nu} = \frac{1}{N} \sum_i m_{i\mu}^a m_{i\nu}^b \tag{2.72}$$

Now when we want to discuss the overlap distribution function, we do not want to count as separate states configurations which only differ from each other by an overall spin rotation. So we define

$$q_{ab}(\mathsf{R}) = \frac{1}{N} \sum_{i\mu\nu} m_{i\mu}^a R_{\mu\nu} m_{i\nu}^b = \sum_{\mu\nu} R_{\nu\mu} q_{ab,\mu\nu} \tag{2.73}$$

where R is a rotation matrix. The relevant overlap to use in $P(q)$ is then

$$q_{ab} = \max_{\mathsf{R}} q_{ab}(\mathsf{R}) \tag{2.74}$$

One can then define $P_J(q)$ as before (2.47). This procedure separates true broken ergodicity from conventional broken symmetry. (Note that one has to make a corresponding maximization over rotations in a Heisenberg *ferromagnet* if one wants to have a delta-function form for $P(q)$.) We could of course have performed a similar maximization over overall spin flips in the Ising case; this would just have restricted the domain of support of $P(q)$ to $q \geq 0$, just as imposing a symmetry-breaking field does.

In replica formalism we can accordingly define

$$q^{\alpha\beta} = \max_{\mathsf{R}} \sum_{\mu\nu} R_{\nu\mu} \langle S_{i\mu}^\alpha S_{i\nu}^\beta \rangle \tag{2.75}$$

so $P(q)$ and the moments of q are given again as in (2.60)–(2.62).

Broken ergodicity and broken replica symmetry as outlined here will be a key ingredient of the mean field theory of spin glasses as described in

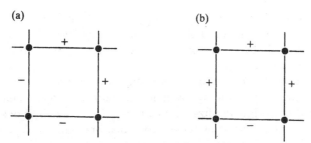

Figure 2.4:   Frustration in a square lattice:
(a) an unfrustrated plaquette.
(b) a frustrated plaquette.

Chapter 3. It seems unlikely as far as we know, however, that the broken symmetry in real three-dimensional spin glasses is of this exotic sort (Chapter 8). Nevertheless, qualitative features which are reminiscent of nontrivial broken ergodicity do persist in real systems, especially in their dynamics. That is, even though there are not infinitely many thermodynamically stable phases, there do appear to be many *metastable* states. Thus broken ergodicity *on a given timescale* remains an important concept in real spin glasses, even if strict thermodynamic broken ergodicity is not.

2.5   Frustration

We have been stressing that what sets spin glasses apart from conventional broken symmetry systems is broken ergodicity. But what is it that causes broken ergodicity? In this section we argue that the important underlying feature is 'frustration' or competing interactions (Toulouse, 1977), combined with the intrinsic disorder in the system.

To illustrate the physics of frustration, consider first a two-dimensional EA Ising model on a square lattice with nearest-neighbour couplings J_{ij} which can take on (independently) only the values $\pm J$. (This is conventionally referred to as the '$\pm J$ model'.) We examine its energetics at the level of a single square of 4 spins and their 4 mutual couplings (Fig. 2.4). Such an elementary unit of a lattice is called a 'plaquette'. Since the bonds are random and independent, it is equally likely to find an even or an odd number of (say) negative bonds around this plaquette.

If the number of negative bonds is even (Fig. 2.4a), then it is always possible to find a pair of spin configurations (related by an overall flip) which satisfy all the bonds. One simply chooses one of the spins to be, say, up and then moves, say, clockwise around the loop, determining the value of the next spin to be that of the previous one multiplied by the sign of the bond connecting them. Then there will be no conflicting instructions coming from the originally-fixed spin when we get all the way around the loop to the starting point again.

But if the number of negative bonds is odd (Fig. 2.4b), there will be a conflict when one gets all the way around the loop; the bond connecting the last spin and the originally-fixed spin will not be satisfied. If one tries to satisfy it by flipping either of these two spins, one will break another bond instead. Thus this plaquette has an extra ground state degeneracy, beyond that which simply follows from symmetry under an overall spin flip, corresponding to the freedom to break any one of the bonds, i.e. to move the defect or 'kink' marking the broken bond anywhere around the loop. (In a loop topology, it is convenient to describe the state of the system in 'bond variables' $A_{ij} = S_i S_j \mathrm{sgn} J_{ij}$ instead of the S_i's; $A_{ij} = -1$ then describes a kink on bond ij, and flipping a single spin moves the kink from one bond to a neighbouring one.)

The term 'frustration' refers to this inability to satisfy all the bonds simultaneously. This picture is obviously more general than this example of an isolated plaquette of 4 spins. For example, the foregoing argument applies equally well to a triangle or to any closed loop of spins.

If the magnitude of the bonds is also random (as, for example, in the Gaussian EA model), the exact degeneracy will be broken, but it is clear that frustration still gives a possibility of low-lying metastable configurations that would be absent in the unfrustrated case. For example, suppose in the cluster of Fig. 2.4b that two bonds on opposite sides of the plaquette have the smallest $|J_{ij}|$'s. Then the ground state is the configuration in which the weakest bond is broken, while there is a metastable state in which the next-weakest bond, across the plaquette, is broken. It is metastable because in order for the kink to move around to the weakest bond, it must move through a stronger bond, which is a higher-energy configuration. Four spins obviously do not make a spin glass, but this simple example does show in the most elementary fashion how frustration and randomness can combine to give extra metastable states, which is the basis of broken ergodicity.

These simple examples show that frustration requires more than a mixture of positive and negative interactions. The following example serves to illustrate the point, in a negative way. It is a model first introduced by Mattis (1976) in which one takes the bonds as

$$J_{ij} = J\xi_i\xi_j \tag{2.76}$$

where the ξ_i are independent and take on the values ± 1 with equal probabilities. Half the bonds are indeed positive and half negative, but they are not independent: if we take the product of J_{ij}'s around any plaquette

$$J_{12}J_{23}J_{34}J_{41} = J^4\xi_1\xi_2\xi_2\xi_3\xi_3\xi_4\xi_4\xi_1 = J^4 \qquad (2.77)$$

the result is always positive. That is, all plaquettes in the Mattis model are unfrustrated, in contrast to the independent-bond $\pm J$ model described above. If we think spin glass behaviour has something to do with frustration, the Mattis model is not a spin glass.

Indeed, by making a change in how we define 'up' and 'down' locally, i.e. defining new spins

$$S_i' = \xi_i S_i \qquad (2.78)$$

the Hamiltonian of the Mattis model reduces (in zero external field) simply to that of a uniform ferromagnet. Accordingly, its free energy, or any other field-independent quantity, is the same as that of a ferromagnet. Furthermore, even with an external field, the transformed problem is that of a ferromagnet in a random external field, which is also known to be rather different from a spin glass (see Chapter 13).

How can we extend the systematic analysis of frustration from the single plaquettes of Fig. 2.4 to a full lattice? There is a cute geometric formulation which, though restricted, at least in its simplest form, to nearest neighbour models, gives a geometric construction for determining the ground states of the system. This is illustrated for two dimensions in Fig. 2.5. We mark each frustrated plaquette of the lattice with an x. In the symmetric $\pm J$ model, the x's are distributed randomly over the plaquettes. If there is a bias toward ferromagnetism (say), the frustrated plaquettes are not independently distributed, but tend to occur in adjacent pairs. (To see this, consider the limit of dilute, randomly distributed antiferromagnetic bonds in an otherwise ferromagnetic lattice. Then the frustrated plaquettes occur only in such pairs.) In any case, for a given sample, the positions of these frustrated plaquettes are fixed.

We indicate the state of the spins on the lattice by the defect variables A_{ij} instead of the S_i's themselves. By our previous arguments we know that each frustrated plaquette must have an odd number of the A_{ij}'s around it equal to -1 (i.e. an odd number of broken bonds), while each unfrustrated plaquette must have an even number. So if we draw (dotted) lines passing through each of the broken bonds, these lines must both enter and exit from every unfrustrated plaquette, while they must originate or terminate in frustrated plaquettes.

Any spin configuration can be represented this way; its energy is given simply (relative to the ground state of the unfrustrated system) by $2J$ times the total length of these 'strings'.

One can of course make the same description of the configurations of an unfrustrated system. Then all the configurations are closed loops of defects.

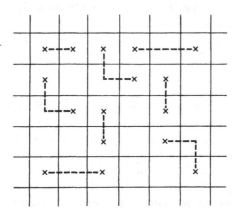

Figure 2.5: Frustrated plaquettes and strings.

In the low-T limit, the thermodynamics of a ferromagnet is simply that of a dilute gas of these loops. What is different about the frustrated case is that the strings cannot all simply evaporate as $T \to 0$. Even in a ground state there remain strings coupling the frustrated plaquettes in pairs; a ground state is distinguished by the fact that its total string length is the minimum possible value.

A similar construction applies in higher dimensions (Fradkin et al, 1978). Again one starts with the frustrated plaquettes and constructs strings which run through these plaquettes (perpendicular to them), forming closed loops (Fig. 2.6). If, as in the figure, this loop lies in a plane, it is clear that the lowest-energy spin configuration is one in which all the bonds lying inside the loop and oriented perpendicular to it are broken. More generally, the construction is of the minimum-area surface of broken bonds spanning the loop running through the frustrated plaquettes (like the configuration of a soap film spanning a wire network).

We remark that this construction presupposes knowing only which plaquettes are frustrated, not all the values of the J_{ij}'s. That is, we can divide the disorder into two kinds: the kind that gives rise to frustration, as in this construction, and the rest, such as occurs in the Mattis model, which is irrelevant to possible spin glass behaviour. One can formulate this separation mathematically by going over to bond variables in the partition function. For the $\pm J$ model, then, the energy is just a sum over bonds of

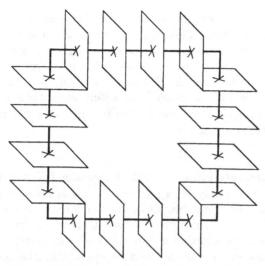

Figure 2.6: A closed tube of frustrated plaquettes. The broken line represents the loop of dual links involved in a loop integral (from Fradkin et al, 1978).

$-|J|$ times the A_{ij} for that bond. What makes the problem nontrivial are the constraints that there must be an even (odd) number of broken bonds around each unfrustrated (frustrated) plaquette. So we write

$$Z[\Phi] = Tr_A \exp\left(\beta \sum_{\langle ij \rangle} A_{ij}\right) \prod_{\langle ijkl \rangle} \delta_{A_{ij}A_{jk}A_{kl}A_{li}, \Phi_{ijkl}} \qquad (2.79)$$

where we have set $|J| = 1$, $\langle ijkl \rangle$ labels the plaquettes and Φ_{ijkl} is $+1$ or -1, according to whether plaquette $ijkl$ is unfrustrated or frustrated, respectively. The constraints can be put into a form which perhaps makes the partition function look less cumbersome (Kirkpatrick, 1977):

$$Z[\Phi] = \lim_{\beta_p \to \infty} Tr_A \exp(-\beta H_{eff}[A; \beta, \beta_p]) \qquad (2.80)$$

where

$$H_{eff} = -\beta \sum_{\langle ij \rangle} A_{ij} - \beta_p \sum_{\langle ijkl \rangle} (\Phi_{ijkl} A_{ij} A_{jk} A_{kl} A_{li} - 1) \qquad (2.81)$$

We write $Z[\Phi]$ to emphasize that Z depends only on the frustration variables Φ_{ijkl}. In this formulation the J's have disappeared except insofar as they determine the Φ's; different sets of bonds which have the same Φ's manifestly have the same Z.

Hamiltonians similar to (2.81) appear in lattice formulations of gauge theories. (See, e.g. Kogut, 1979.) These are theories in which one not only has an overall global symmetry, like the up–down one in our problem, but also a local symmetry. That is, one can make different symmetry transformations at different points, and the Hamiltonian is still invariant. The simplest way to construct such a theory is to define variables V_{ij} which belong to a particular group (in our case Z_2) on the links of a lattice (like our A_{ij}'s) and define them to transform as

$$V'_{ij} = U_i^{-1} V_{ij} U_j \qquad (2.82)$$

under an arbitrary, in general position–dependent operation U_i in the group. (For Z_2, U_i, like V_{ij}, can simply be $+1$ or -1.) This is called a gauge transformation. A gauge-invariant Hamiltonian, therefore, must be made up of combinations of the V's which are invariant under this transformation. The simplest such combination is a product of 4 elements around an elementary plaquette of the lattice, such as appears in the β_p-term in (2.81). In our problem, the gauge transformation (2.82) on the A_{ij}'s corresponds to a transformation like (2.78) on the S_i's in the original formulation of the problem, with an arbitrary U_i replacing ξ_i. Note however that the first term in our effective Hamiltonian is *not* gauge invariant.

For the $\pm J$ model, all the disorder in (2.81) is contained in the gauge-invariant β_p-term. If we also allow variations in the bond magnitude, that will appear in the first term. That is, this formulation separates frustrating disorder from other, presumably less important disorder: frustrating disorder is gauge-invariant disorder.

We have argued here that frustration is necessary for spin glass behaviour. But is it sufficient? The answer is apparently no. There has also been a great deal of study of periodic frustrated systems. The simplest example is a two-dimensional nearest-neighbour Ising model on a triangular lattice with all bonds antiferromagnetic. Here we know from work by Wannier (1950) that there is a large ground state degeneracy (i.e. a ground state entropy proportional to N), but no phase transition at any nonzero temperature. Similar behaviour has been found in some other fully frustrated models (Villain, 1977a,b; Wolff and Zittartz, 1982,1983a,b). In these cases, apparently, the effective free energy landscape is too smooth. There is degeneracy, but there are not high enough barriers between the different configurations to produce broken ergodicity. Other frustrated models, including the nearest-neighbour Ising antiferromagnet on an fcc lattice (Mackenzie and Young, 1981), which is a three-dimensional counterpart of the two-dimensional triangular antiferromagnet, exhibit periodic order below a nonzero transition temperature, but so far no one has found such a

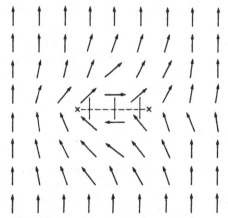

Figure 2.7: Positive and negative half-vortices for XY spins on a square lattice.

model with anything resembling a spin glass state in its equilibrium thermodynamics.

One can also study frustration and construct a formal description in similar terms for XY (2-component spins) and Heisenberg spin glasses. Here we just outline the fundamental elements for the XY case, following Villain (1977), starting again from the frustrated plaquette of Fig. 2.4b. If we put the spins in a configuration they would have in the Ising case, we find the system to be unstable; the spins rotate to try to relieve the frustration. One finds easily that the stable configuration has every spin rotated 45° relative to its neighbour if the bond between them is positive, and 225° if the bond is negative. There are two distinct locally stable configurations (Fig. 2.7): one corresponding to these increases in the orientation angles as we go around the plaquette clockwise; the other, counterclockwise. (Of course, each such configuration can be rotated uniformly without cost in energy, but such a rotation will not take one kind of configuration into the other.)

If we add up the 'extra' twist angles as we go around the plaquette, we get π and $-\pi$ in the two cases. Thus the defects which are centered at a frustrated plaquette in a lattice act like half-vortices of either positive or negative vorticity. Recalling that an unfrustrated two-dimensional XY model may be recast into the form of a gas of logarithmically-interacting vortices (Kosterlitz and Thouless, 1973), we see that the problem of its frustrated counterpart is one in which those vortices also interact with a set of half-vortices frozen on the frustrated plaquettes. These half-vortices can have either positive or negative vorticity, so the frustrated problem has an Ising-like discrete degree of freedom which is absent in the unfrustrated case.

These examples show how using the idea of frustration to formulate models of spin glasses can clarify the consequences of the disorder in the system and make new insight into the physics of the problems possible. We will exploit it again in different ways in Chapter 7.

2.6   Spin glass systems

Most of this chapter has dealt with quite formal questions. In this last section, we make a brief survey of the classes of systems that exhibit spin glass behaviour and try to indicate their relationship to the models and theoretical concepts we have introduced in the preceding sections. This survey is not intended to be a comprehensive review. The aim is just to illustrate how a wide variety of physical systems can combine disorder and frustration in such a way as to produce the universal properties we sketched in Chapter 1, despite the fact that the theoretical models we have introduced seldom represent more than their grossest qualitative features.

We have already mentioned a few examples of spin glass systems, including the first class we will discuss here, the so-called RKKY spin glasses. In addition to the classical systems of transition metal impurities in noble metal hosts ($\underline{Cu}$Mn, $\underline{Ag}$Mn, $\underline{Cu}$Fe, etc.), we include in this general category some systems (like $\underline{Pt}$Mn) where the host metal is itself nearly magnetically ordered, alloys with rare earth constituents ($\underline{Y}$Dy, $\underline{Sc}$Tb, $\underline{Y}$Er and others), as well as a number of ternary systems (e.g. $La_{1-x}Gd_xAl_2$). An extreme example of the first is $\underline{Pd}$Fe, where the Fe atoms induce parallel moments in neighbouring Pd atoms, and the resulting 'giant moments' interact to form a spin glass state at very low Fe concentrations. Thus even in systems with such strong enhancement of the ferromagnetic moments, there are still sufficient negative interactions (remnants of the oscillatory tail of the RKKY interaction) to produce spin glass behaviour. The only restriction we make is to crystalline metallic systems with reasonably dilute magnetic impurities. In the idealized picture presented in Section 2.1, these systems were described as randomly situated localized spins associated with the magnetic impurities in a free-electron host. The indirect exchange between the impurities via the host electrons then has the oscillatory, decaying asymptotic form (2.3), so this picture contains the basic necessary ingredients of disorder and frustration. The real microscopic physics of these systems is of course much more complicated.

To start with, even the notion of a well-defined impurity moment is nontrivial: (See Fischer (1978) for a review.) We know that for a magnetic transition metal impurity in a free-electron-like host there is a characteristic temperature called the Kondo temperature. Near and below this temperature, the impurity moment is quenched by the formation of a collective singlet state coupling the impurity spin with the spins of the nearby conduction electrons (Wilson, 1975). The problem of the competition between this effect and spin glass ordering is itself an interesting one, but we will not discuss it in this book. We will assume for convenience that we are concerned with temperatures and concentrations where spin glass effects dominate any possible Kondo effects.

Furthermore, the host conduction electrons feel a strong periodic potential from the crystal lattice, as well as the random potential scattering from the impurities and other defects. Thus the free-electron expression (2.2) should be generalized to include band structure effects, and the finite mean free path effects should be put in. These are complicated calculations in microscopic solid state physics, and to our knowledge no one has ever put all these effects together in a calculation for a single material. These problems are important if one wants to calculate the RKKY interaction for real metals. Of course, it is clear in principle that if we are dealing with impurities which can couple to a degenerate electron gas, the existence of a Fermi surface is likely to give rise to oscillatory spatial dependence of interaction energies, so competing effective interactions are implicit in the microscopic electronic structure.

We must also keep in mind that the positions of the impurities are not really as random as we might have imagined them to be. Even if the sample could be prepared by infinitely rapid quenching from the liquid state, there would be correlations in the impurity positions. Experimentally, one finds by neutron scattering that there can be significant correlations — clustering or, more commonly, 'anticlustering'. These correlations can also be changed by aging or plastic deformation. An example of the influence of annealing and cold working on the cusp in the ac susceptibility is shown in Fig. 2.8.

If one knows the impurity position correlations, one can in principle extract the exchange interactions $J(\mathbf{R}_i - \mathbf{R}_j)$ from the temperature and concentration dependence of the susceptibility at high temperatures. Fig. 2.9 shows the separation dependence of the interactions found in this way by Morgownik and Mydosh (1983) for several of the classic RKKY systems. It is clear that the free-electron RKKY function (2.3) has very little to do with the true interactions, which also vary a great deal from system to system. Of course, the true interactions are positive at some separations and negative at others, so the fundamental requirements of frustration and disorder are met in these systems.

The lesson we draw from these remarks and conclusions is that while the RKKY model is certainly not correct in quantitative detail, it contains

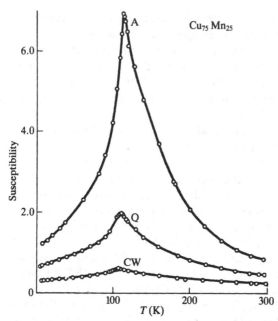

Figure 2.8: The ac susceptibility in arbitrary units for $Cu_{75}Mn_{25}$ for aged (A), quenched (Q), and cold-worked (CW) samples vs temperature (from Tustison, 1976).

the basic physics common to this class of systems; in fact, the theoretically simpler EA model ought to be equally good in this respect because it, too, contains frustration and disorder. One must, however, take corresponding care in making comparisons of theory and experiment on these systems. For example, if there is strong chemical clustering and ferromagnetic interaction for nearest neighbours, or a tendency toward ferromagnetism in the host metal, the 'spins' of the formal model ought to be taken to describe ferromagnetic clusters rather than the physical impurity spins themselves.

On the other hand, we also know of situations where the simple (Heisenberg) RKKY or EA models can mislead us qualitatively. This is not because of anything like the omission of band structure effects, however, but because they contain no anisotropic interactions. In the alloys we are discussing here, these interactions are mostly a spin–orbit effect, so one expects them to be small. And indeed they are, but we now know that the fact that they are nonzero plays a crucial role in the stabilization of the spin glass state in these systems. This is similar to ferromagnets, where crystalline anisotropy plays an analogous role. We will return to these forces in some detail in Chapters 6–8. For now we will just anticipate the result that, insofar as

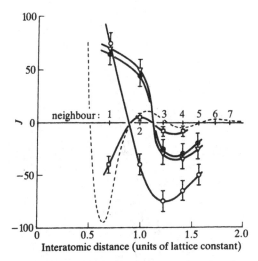

Figure 2.9: The exchange parameters J as function of neighbour distance R in units of lattice parameter for ○ AuFe, ▽ CuMn, ● AuMn, ▢ PtMn. The broken line represents the RKKY conduction electron polarization (from Morgownik and Mydosh, 1983).

they affect whether there is a stable spin glass state, they essentially turn the problem into an Ising-like one. Thus the Ising models on which we will lavish so much theoretical attention are actually quite relevant to real systems which are superficially much more Heisenberg-like.

Another phenomenon that most of these systems exhibit is competition between spin glass and ferromagnetism or periodic magnetic order. At sufficiently high concentration of the magnetic impurities, one can find a transition from a spin glass state to, for example, a ferromagnetic state. A particularly interesting aspect of these systems is the finding that, for a certain concentration range, one sees, with decreasing temperature, first a transition from a paramagnetic state to a ferromagnetic one and then a second transition from the ferromagnetic state to a spin glass. Fig. 2.10a shows this illustrated in a phase diagram of AuFe. The competition between spin glass and ferromagnetic order is the subject of Chapter 11.

The class of insulating spin glasses exemplified by $Eu_xSr_{1-x}S$ (also mentioned in Section 2.1) is better understood at the microscopic level, so the Hamiltonian (2.4) stands on a reasonably firm footing, except in principle for possible chemical clustering effects and the omission of other kinds of interaction than simple exchange. (Chemical clustering of the sort described above in the RKKY systems does not, however, appear to play a significant

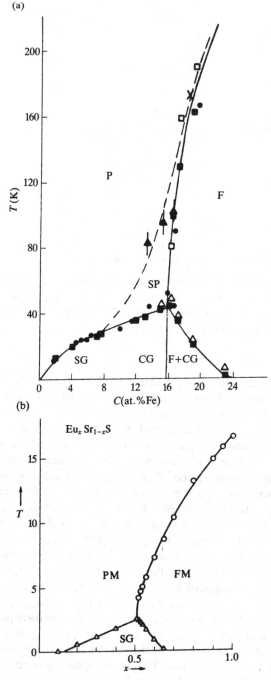

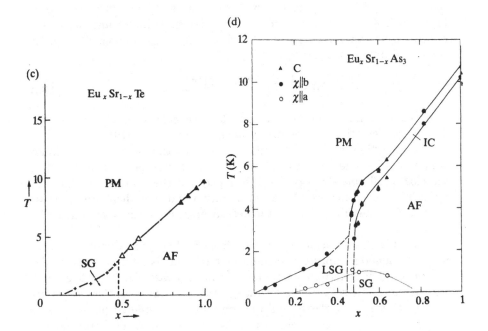

Figure 2.10: (a) Magnetic phase diagram of A̲uFe. ● Mössbauer anomalies; ■ susceptibility anomalies (above 15% upper points Curie temperatures, lower points T_f); □ critical peaks in the neutron scattering; × specific heat peak; ▲ peak in the ESR linewidth; △ minimum in the ESR linewidth; P paramagnetic; SP superparamagnetic; SG spin glass; CG cluster glass; F ferromagnetic (from Coles et al, 1978).
(b) Magnetic phase diagram of $Eu_xSr_{1-x}S$ (from Maletta and Convert, 1979).
(c) Magnetic phase diagram of $Eu_xSr_{1-x}Te$ (from Börgermann et al, 1986).
(d) Magnetic phase diagram of $Eu_xSr_{1-x}As_3$. PM paramagnetic; AF antiferromagnet; IC incommensurate antiferromagnet; LSG longitudinal spin glass; SG spin glass; ▲ from specific heat; other symbols from susceptibility (from Schröder et al, 1989).

role in these materials: For example, the critical Eu concentration below which there is no spin glass state agrees quite well with the percolation threshold $x_{nnn} \approx 0.136$ for next-nearest neighbour percolation in the fcc lattice.) Furthermore, the observed phase diagram (Fig. 2.10b) of this system agrees quite well in general with the results of computer simulations based on (2.4) (Kinzel and Binder, 1981). The most important interaction omitted from the model (2.4) in this class of systems is apparently simple classical dipole–dipole forces:

$$E_{dip} = \frac{\mathbf{m}_1 \cdot \mathbf{m}_2 - 3(\mathbf{m}_1 \cdot \mathbf{r})(\mathbf{m}_2 \cdot \mathbf{r})/r^2}{r^3} \tag{2.83}$$

They play an important role at very low Eu concentrations, where the exchange interactions are relatively weak (Eiselt et al, 1979).

For analytic theory, on the other hand, even the model (2.4) is rather cumbersome. If one wants to address conceptual questions in these systems, the most profitable strategy is still to argue that the microscopic Hamiltonian clearly contains disorder and frustration and that therefore the most appropriate starting point for theoretical analysis is the short-range EA model.

An interesting cousin of $Eu_xSr_{1-x}S$ is $Eu_xSr_{1-x}Te$, where the next-nearest-neighbour antiferromagnetic interaction is stronger than the nearest-neighbour ferromagnetic one, so in this case the Eu-rich limit is antiferromagnetic rather than ferromagnetic. In both cases, there are transitions between spin glasses and the conventionally ordered phases as one increases x. The question of the possible reentrance of the ferromagnetic phase boundary (as described above for AuFe) has also been studied extensively in systems in this family. In $Eu_xSr_{1-x}As_3$, a similar situation to that in $Eu_xSr_{1-x}Te$ occurs, with the additional complication that crystalline anisotropy leads to a splitting of the spin glass transition into two: first, freezing of the spin components parallel to the b axis, then, at a lower T, that of the components in the other two directions. Fig. 2.10b–d shows the phase diagrams of all three of these systems in the $T - x$ plane.

There are many other insulating spin glasses with similar behaviour, some of them rather similar to these examples and some more complicated. In the former category we include cases where Mn is the magnetic ion, and Cd, Zn or Sn take the place of Sr, as well as other possibilities. Another is the system $Eu_xSr_{1-x}S_ySe_{1-y}$. Like EuTe, EuSe is antiferromagnetic, so varying y can be thought of roughly as varying the ratio of the nearest-neighbour to the next-nearest neighbour interaction in the model (2.4). Of course, there is extra disorder as well, but that does not change the basic picture. Examples with different basic structures include spinel systems such as $CdIn_{2-2x}Cr_{2x}S_4$ or analogous structures with Al, Ga or Sn in place of the In or O in place of the S, transition metal compounds like Fe_2TiO_3, and many more too diverse and numerous to list here.

We have mentioned the importance of anisotropy. In addition to the dipolar forces and the anisotropy arising from spin–orbit interactions in metallic spin glasses, there are crystal field anisotropies, in particular uniaxial ones, which allow one to study the m-dependence of spin glass ordering experimentally. Of the systems we have mentioned, the rare earth metallic spin glass $\underline{Y}Er$ and the insulators $Eu_xSr_{1-x}As_3$ and Fe_2TiO_3, as well as semimetallic $(Ti_{1-x}V_x)_2O_3$ are Ising-like. Examples of XY spin glasses include $\underline{Y}Dy$, $\underline{Y}Tb$, $\underline{Sc}Dy$, and $\underline{Sc}Tb$. A convenient set of systems to study is $\underline{Zn}Mn$ (Ising), $\underline{Cd}Mn$ (XY), and $\underline{Mg}Mn$ (Heisenberg).

It should be no surprise that many amorphous materials are also found to be spin glasses. Some such systems are $Al_{1-x}Gd_x$ for x between about 0.3 and 0.4 and a number of amorphous transition metal and rare earth systems like Fe_xNi_{1-x}, Tb_xFe_{1-x}, and La_xGd_{1-x}. Spin glass behaviour is also seen in aluminosilicate glasses like $MnO.Al_2O_3.SiO_2$ and its Co counterpart, and in metallic glasses such as $(Fe_xNi_{1-x})_{75}P_{16}B_6Al_3$. Again, this is only a small part of the full list of such materials that have been studied. In these, we generally have even less microscopic knowledge of the microscopic Hamiltonian and exactly how it gives rise to the frustration that causes the spin glass state than we do in the crystalline metallic alloys discussed above, and certainly less than we do in $Eu_xSr_{1-x}S$ and its close relatives. Nevertheless, the fundamental universal phenomena we described in Chapter 1 are also observed in these materials. In these systems, the spin glass state might be due to either random exchange (as in conventional spin glasses) or random uniaxial anisotropy. This will be discussed in detail in Chapter 13.

As we have already mentioned, spin-glass-like behaviour has also been seen in many materials where the 'spins' are electric dipoles or quadrupoles. In these materials, the microscopic physics tends to be more complicated, and the distance to simple EA-like models correspondingly greater than in magnetic systems. One feature, in particular, that complicates the analysis is the fact that electrical moments couple much more strongly to lattice distortions than magnetic ones do. Thus the spin-glass-like transitions we would like to study are difficult to separate from structural ones.

Dipole–dipole and quadrupole–quadrupole interactions are already rather frustrated in pure isotropic systems. For example, consider the dipole–dipole interaction (2.83). A ferroelectric configuration $m_1 = m_2 = m\hat{z}$ lowers the interaction of those pairs of dipoles whose separation vector r lies along or near the z axis, but raises it for those whose r lies in or near the $x-y$ plane. The isotropic angular average of E_{dip} vanishes. The same is true (by definition) of quadrupole–quadrupole and higher-order multipole-

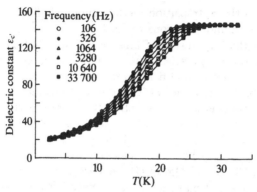

Figure 2.11: Real part of the dielectric constant of RDP–ADP mixtures (from Courtens, 1984).

multipole forces. When this frustration is combined with substitutional disorder, we have the necessary conditions for spin glass behaviour, and many of the universal features of spin glass experiments are observed in the corresponding properties in these systems.

Examples of dipolar glasses include mixtures of RbH_2PO_4 (colloquially called RDP) and $NH_4H_2PO_4$ (ADP). The former is ferroelectric and the latter antiferroelectric. For suitable concentration and temperature, the electric polarizability of the mixtures shows glassy features strongly reminiscent of the frequency-dependence of the magnetic susceptibility in spin glasses (Fig 2.11). Another likely dipole glass is $(KDP)_{1-x}(ADP)_x$ (KDP is like RDP but with K in place of the Rb), which apparently exhibits the electric counterpart of the reentrant ferromagnetic phase boundary seen in $Eu_xSr_{1-x}S$. There is also another small family of dilute mixtures like $K_{1-x}Li_xTaO_3$ (or Na in place of Li) and $KTa_{1-x}Nb_xO_3$.

Spin glasses where the 'spin' is quadrupolar include such systems as $(KCN)_x(KBr)_{1-x}$, $(N_2)_{1-x}Ar_x$, and solid mixtures of ortho- and para-hydrogen. The first of these is the most extensively studied. The ordering involves the CN units, which, in pure KCN, line up parallel to each other. (They also have a dipole moment, and the order is antiferroelectric.) For large dilution, the CN's apparently freeze into a randomly oriented configuration. Dielectric susceptibility measurements exhibit the same kind of frequency dependence indicative of a broad spectrum of relaxation times and barrier heights that we have described in spin glasses and dipolar glasses (Fig. 2.12). (It is the existence of the CN dipole which makes it possible to probe the freezing through the electrical response. However, $(KCN)_x(KBr)_{1-x}$ is *not* an electric dipole glass. The dipole energies are

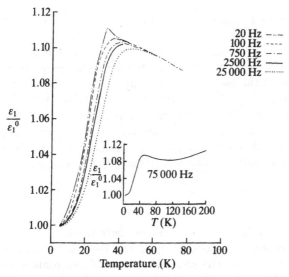

Figure 2.12: Real part of the dielectric constant of $(KCN)_x(KBr)_{1-x}$ (from Bhattacharya et al, 1982).

apparently small compared to the quadrupolar ones, and the freezing is not sensitive to the electric field the way spin glass ordering is in a magnetic system.)

The other two systems are harder to probe because there is no dipole moment for an external field to couple to, and a uniform field *gradient*, which does couple to a quadrupole, is difficult to impose experimentally. Nevertheless, neutron scattering shows evidence of orientational freezing of the N_2 in $(N_2)_{1-x}Ar_x$. In the ortho–para H_2 mixtures, the periodic quadrupolar order which occurs in pure orthohydrogen is frustrated by a sufficiently large concentration of the effectively inert para molecules. The experimental evidence comes from NMR, which is a rather indirect probe; there is still not a consensus on the location or nature of the transition.

In none of these systems, dipolar or quadrupolar, is there any good evidence for a stable equilibrium glassy state, in contrast to the apparent situation in many magnetic spin glasses. It would clearly be desirable to see whether more such systems with more favourable conditions for a finite–T transition can be made. This is especially interesting for the quadrupolar case because, as we will see in the next chapter, at least at the mean field level, quadrupole glasses are expected to order rather differently from dipolar ones. They are expected to exhibit broken ergodicity *without* the broken symmetry that accompanies it in the dipole case. In this sense they are a more general case, and it would be very desirable to be able to study this phenomenon better experimentally.

Finally, we mention superconducting analogues of spin glass systems. The important feature of the superconducting order parameter for the present discussion is the fact that it is complex, i.e. it has an amplitude and a phase. Here, we will be concerned only with systems where the amplitude is effectively fixed (or its fluctuations are small and uninteresting). For example, imagine a matrix of weakly-coupled superconducting grains, each well below its bulk transition temperature. Letting ϕ_i be the phase of grain i, the coupling between grains then takes the Josephson form

$$H = -\tfrac{1}{2} \sum_{ij} J_{ij} \cos(\phi_i - \phi_j) \tag{2.84}$$

which is just the Hamiltonian of an XY spin ferromagnet.

It is the imposition of an external magnetic field that makes the problem interesting for us here. The gauge invariance of electromagnetism tells us that one can change the phase of the superconducting order parameter arbitrarily by an angle $\theta(\mathbf{r})$, in general different at different points in space, provided the vector potential is changed at the same time by $\nabla\theta(\mathbf{r})$. In order that this be true, the Hamiltonian must be invariant under this transformation. In the present example, this means that the energy must have the 'minimal coupling' form

$$H = -\tfrac{1}{2} \sum_{ij} J_{ij} \cos(\phi_i - \phi_j - A_{ij}) \tag{2.85}$$

where

$$A_{ij} = \frac{2\pi}{\Phi_0} \int_{\mathbf{r}_i}^{\mathbf{r}_j} \mathbf{A} \cdot d\mathbf{l} \tag{2.86}$$

and Φ_0 is the elementary flux quantum $hc/2e$ associated with the Cooper pairs (charge $2e$). Thus a magnetic field can act as a source of frustration. An A_{ij} which is an odd multiple of π effectively makes the bond ij negative.

Of course, only gauge-invariant quantities are thermodynamically relevant. In this case, this means that the values of the A_{ij}'s themselves are not relevant as such; only the magnetic field flux through each plaquette (i.e. the sum of the A_{ij}'s around the plaquette) matters. If this sum is a multiple of 2π, the plaquette is unfrustrated, otherwise it is frustrated. The spin systems we described in the preceding section effectively have plaquette fluxes which are integral multiples of $\tfrac{1}{2}\Phi_0$. For example, the fully frustrated two-dimensional XY model is realized by a regular network of these superconducting grains with half a flux quantum per plaquette. The superconducting systems we are considering here are more general, however, in that the flux through a plaquette is not constrained to values $\frac{1}{2}n\Phi_0$.

If one can add disorder to the frustration, one then expects to be able to see spin-glass-like behaviour. This can be done by making the system amorphous, so that the amount of flux through a plaquette is random, or, on a lattice, by dilution. Glass-like behaviour has been reported in the high-T_c superconducting ceramic system BaLaCuO (Müller et al, 1987).

There is even the possibility (though it runs contrary to our general principle as stated hitherto) of spin-glass-like behaviour without disorder if the flux per plaquette is *irrational*. This is a kind of 'incommensurate' frustration, which one expects can be more serious than the commensurate kind discussed in the preceding section. Simulations do in fact suggest a spin glass phase in this case (Halsey, 1985).

In addition to all these physical systems that show phenomena identical or similar to spin glass behaviour, there are a number of nonphysical problems, including combinatorial optimization problems and so-called neural networks, which share many features with spin glasses. We will discuss them in Chapter 14.

Throughout our discussions so far, we have stressed the universality of spin glass phenomena. At the same time, it is by now clear, however, that within the large class of systems that we call spin-glass-like, there emerge distinct universality classes, distinguished by dimensionality, spin dimensionality, interaction range (short or long), and so forth. Much of spin glass research deals with sorting out these subclasses, both theoretically and experimentally.

3
Mean field theory I: Ising model, equilibrium theory

In this chapter we will put the concepts we have developed in Chapter 2 to work in constructing the mean field theory for an Ising spin glass. The term 'mean field theory' (henceforth frequently abbreviated 'MFT') can be interpreted in many ways. Here we will take it to mean the exact solution of a model in which the forces are of infinitely long range, so that each spin interacts equally strongly with every other one. For spin glasses, this is the Sherrington–Kirkpatrick (SK) model (1975), which we solve heuristically (though unfortunately incorrectly) in Section 3.1. The error is a somewhat subtle one, as is evident from the fact that we obtain exactly the same result in the more systematic calculations of Sections 3.2 (a direct summation of the leading terms in N^{-1} in high-temperature perturbation theory) and 3.3 (which uses the replica formalism). In both these approaches, we can also see how the theory itself reveals that it is wrong, since they lead to negative values of quantities which are necessarily positive.

We then study the correct mean field theory, with replica symmetry breaking, in Section 3.4, including the remarkable nature of the broken ergodicity it implies and an examination of its stability. Further physical insight into the problem is obtained in Section 3.5 from the mean field equations first introduced by Thouless, Anderson and Palmer (TAP) (1977). Finally, we turn to some rather simpler models which are also soluble in the mean field limit: the spherical and random energy models (Section 3.6), and the Potts glass (Section 3.7), which displays even richer properties than the SK Ising spin glass.

3.1   The SK model and a heuristic solution

The SK model is simply an Ising EA model (see Section 2.1)

$$H = -\tfrac{1}{2} \sum_{ij} J_{ij} S_i S_j - h \sum_i S_i \tag{3.1}$$

in which the variance of the interactions is independent of the distance between between the spins:

$$[J_{ij}^2]_{av} \equiv \Delta_{ij} = \frac{J^2}{N} \tag{3.2}$$

The scaling with N^{-1} is necessary in order that thermodynamic quantities be finite in the large N limit. (Consider, for example, the energy at $T = 0$ to see why (3.2) has to have this N-dependence.)

The heuristic derivation (Southern, 1976) is done within Weiss mean field theory (which is correct for a ferromagnet), where spin i feels an internal field

$$h_i^{eff} = \sum_j J_{ij} m_j + h \tag{3.3}$$

and the magnetization m_i is just the magnetization induced by h_i^{eff}:

$$m_i = \tanh \beta h_i^{eff} \tag{3.4}$$

In the SK model, this field is the sum of $N - 1$ terms from the other sites, and we argue hopefully that they may be regarded as independent, so that we can apply the central limit theorem. Then h_i^{eff} may be taken to be a Gaussian random variable:

$$P(h_i^{eff}) = \frac{1}{(2\pi\sigma^2)^{\frac{1}{2}}} \exp[-\tfrac{1}{2}(h_i^{eff} - h)^2/\sigma^2] \tag{3.5}$$

From (3.3) we find that h_i^{eff} has a variance

$$\sigma^2 = \left[\left(\sum_j J_{ij} m_j \right)^2 \right]_{av} = \left[\sum_j J_{ij}^2 \right]_{av} [m_j^2]_{av} = J^2 q \tag{3.6}$$

assuming further that the site magnetizations are independent of each other and of the J_{ij}'s. But q is by definition the mean square value of the local magnetization (3.4) over the distribution (3.5). Thus (with the change of variable $h_i^{eff} = Jq^{\frac{1}{2}}z$) we obtain a self-consistent equation for q:

$$q = \int \frac{dz}{\sqrt{2\pi}} e^{-\frac{1}{2}z^2} \tanh^2[\beta(h + Jq^{\frac{1}{2}}z)] \tag{3.7}$$

(3.7) is known as the SK equation.

Setting the external field h equal to zero, we can solve for q. We find that there is a critical temperature T_f above which the only solution is $q = 0$, but below which there is also a solution with positive q. Explicitly, by expanding the right hand side to order q^2, we find

$$q = (\beta J)^2 q - 2(\beta J)^4 q^2 + \cdots \tag{3.8}$$

from which we find that $T_f = J$ and $q = (T_f - T)/T_f + O((T_f - T)^2)$ for T just below T_f. At low T we can expand the integral in T instead of q; the result is

$$q = 1 - \left(\frac{2}{\pi}\right)^{\frac{1}{2}} \frac{T}{T_f} + \cdots \tag{3.9}$$

The full result is shown in Fig. 3.1a, where the root mean square local magnetization $q^{\frac{1}{2}}$ is plotted as a function of T. This result is appealing, because $q^{\frac{1}{2}}$ behaves roughly the way the order parameter does in an ordinary broken symmetry state. The theory thus appears to fit nicely into the standard theory of phase transitions, at least at the mean field level.

Fig. 3.1b shows the susceptibility obtained from this q, using the relation (2.44). Again, the existence of a cusp in χ is appealing in view of the experimental findings and encourages us to proceed further along this direction.

Despite the natural appeal of these simple physical arguments and results, however, both the derivation and the results are wrong. In fact, if they could be justified, this book (if it were written at all) would be much shorter, for then spin glass order would not be all that different from conventional broken symmetry. It is just the failure of this simple picture that makes spin glasses interesting and novel.

The obvious flaws in the derivation are apparent: The fields from different sites are not independent, so the central limit theorem is not justified. The site magnetizations, moreover, are not independent of each other or of the J_{ij}'s, so the factorization of the average in (3.6) is not valid. And finally, the Weiss mean field equations (3.3) and (3.4) are incorrect. Though

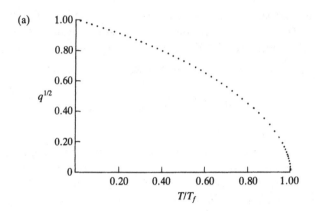

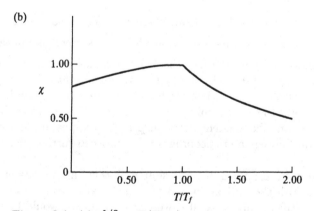

Figure 3.1: (a) $q^{1/2}$ vs T/T_f .
(b) Susceptibility $\chi(T)/\chi(T_f)$ vs T/T_K (from Sherrington and Kirkpatrick, 1975).

they are valid for the MFT of a ferromagnet, there are corrections to them which do not vanish in the large N limit of the SK spin glass. This is because the *average* field, which is the dominant term in the ferromagnet, vanishes in the spin glass. The dominant terms in the free energy of the spin glass, which are weaker by a factor N^{-1}, are of two kinds: contributions from the fluctuations (in the averaging over the bond distribution) of the mean fields, which are what we have considered above, and the first thermodynamic fluctuation corrections to these fields, i.e. the lowest-order corrections to the Weiss equations. A consistent MFT must include both these contributions. We will try to correct these deficiencies systematically in the treatments of the next few sections. Remarkably, the same (incorrect) result (3.7) will emerge from these derivations. The correct mean field description of a spin glass turns out to be rather elusive! (And even this correct MFT will not be a correct description of real spin glasses with short-range interactions.)

3.2 SK solution from perturbation theory

In this section we try to solve the SK model by perturbation theory in the inverse temperature β, i.e. in βJ. The perturbation theory can be formulated on a lattice of arbitrary dimensionality; we obtain the SK solution by keeping only the terms at each order in β which are of leading order in the reciprocal coordination number z of the lattice. (For a hypercubic lattice in d dimensions, $z = 2d$.) Alternatively, we can formulate the perturbation theory for an SK model of finite size N and obtain our solution by taking the $N \to \infty$ limit. Either way, we get the same answer.

There are two stages to the perturbation theory: first, the formal expression of the quantities of interest for a single sample, then the averaging of these quantities over the bond distribution. Here we carry out the first part for a general Ising model (Brout, 1974); the generalization to systems with more components is straightforward. It is convenient to represent the theory diagrammatically, and to derive the diagrammatic rules it is convenient to have the partition function expressed in a field-theory-like form. This is all facilitated by applying the identity

$$\exp(\tfrac{1}{2}\lambda a^2) = \left(\frac{\lambda}{2\pi}\right)^{\frac{1}{2}} \int_{-\infty}^{\infty} dx \exp(-\tfrac{1}{2}\lambda x^2 + a\lambda x) \qquad (3.10)$$

N times to the partition function

$$Z = Tr_S \exp\left(\tfrac{1}{2}\beta \sum_{ij} J_{ij} S_i S_j + \beta \sum_i h_i S_i\right) \qquad (3.11)$$

This replaces the exponential of the quadratic form in the S's by a Gaussian-weighted integral of the exponential of a quantity *linear* in the S's. That is, the partition function becomes equal to the average, over a Gaussian distribution of external fields, of the partition function of N free spins in an inhomogeneous external field. The latter is formally trivial to evaluate, with the result

$$Z = \int \prod_i \frac{d\psi_i}{(2\pi)^{\frac{1}{2}}} [\det(\beta J)]^{-\frac{1}{2}}$$

$$\times \exp\left\{ -\frac{1}{2} \sum_{ij} \psi_i [(\beta J)^{-1}]_{ij} \psi_j + \sum_i \ln 2 \cosh(\psi_i + \beta h_i) \right\}$$

$$\equiv \int D\psi \exp(-A[\psi, h]) \tag{3.12}$$

This now has the form of the partition function of a lattice field theory, with a real scalar field ψ_i. We can therefore directly identify the ingredients of the diagrammatic expansion from the form of the effective action $A[\psi, h]$ in (3.12). Thus the zero order correlation function can be read off as the inverse of the quadratic coefficient $(\beta J)^{-1}$, i.e. $\langle \psi_i \psi_j \rangle_0 = \beta J_{ij}$, and the bare interaction vertices are obtained as the coefficients of the powers of ψ_i in the expansion of the ln cosh term. Thus, in zero external field, we have

$$\lambda_2^0 = \left(\frac{\partial^2}{\partial \psi^2} \ln \cosh \psi \right)_{\psi=0} = \quad \quad (= 1) \tag{3.13}$$

$$\lambda_4^0 = \left(\frac{\partial^4}{\partial \psi^4} \ln \cosh \psi \right)_{\psi=0} = \quad \tag{3.14}$$

$$\lambda_6^0 = \left(\frac{\partial^6}{\partial \psi^6} \ln \cosh \psi \right)_{\psi=0} = \quad \tag{3.15}$$

etc. In this field theory, one can then write diagrams for the free energy or for correlation functions by tying together lines (which stand for $\langle \psi_i \psi_j \rangle_0 = \beta J_{ij}$) with all these vertices. The generalization to an external field h_i is simple: for every power of the field attached at a vertex at site i, another derivative of ln cosh ψ is taken.

Now what we really want are the correlation functions of the original Ising spins S_i; the fields ψ_i were only introduced as a formal construct to enable us to use standard field-theoretic perturbation theory. So we have to express the correlation functions of the S_i's in terms of those of the ψ_i's. Fortunately, there is a fairly simple relation between them. To find this, it is simplest to shift variables in (3.12) so that a shift by βh_i appears in the Gaussian weight term rather than the ln cosh one. Then we find that

$$\langle S_i \rangle = \frac{\partial \ln Z}{\partial \beta h_i} = \sum_i [(\beta J)^{-1}]_{ij} \langle \psi_j \rangle \tag{3.16}$$

This says that graphs for $\langle S_i \rangle$ are obtained from graphs for $\langle \psi_i \rangle$ simply by removing the first βJ_{ij} bond. Similarly, taking one more derivative, we find

$$\langle \psi_i \psi_j \rangle = \beta J_{ij} + \sum_{kl} \beta J_{ik} \langle S_k S_l \rangle \beta J_{lj} \tag{3.17}$$

This just tells us that if we take the graphs for $\langle \psi_i \psi_j \rangle$, excluding the zeroth-order term βJ_{ij}, and remove the bonds from both ends, we get the graphs for $\langle S_i S_j \rangle$.

The diagrams which we generate by the above formalism are of course functions of the random variables J_{ij}. They must therefore be averaged over the bond distribution. We can describe how to do this very simply in the case of zero-mean Gaussian-distributed bonds as follows: In any diagram constructed as described above, connect the bond lines together in pairs in all possible ways and assign the value

$$\beta^2 [J_{ij} J_{kl}]_{av} = \beta^2 \Delta_{ij} (\delta_{ik}\delta_{jl} + \delta_{il}\delta_{jk}) = \qquad \qquad \tag{3.18}$$

to each such pair of bonds. For the SK model, the factor Δ_{ij} just becomes J^2/N. (This rule is just a simple consequence of the theorem for multivariate Gaussian variables that says that the average of the product of such variables is equal to the sum of the products of the averages of pairs, summed over all ways of arranging the variables into pairs.) If we had a nonGaussian distribution of bond strengths, we would also need to draw graphs with groups of 3 or more bonds linked together and given values proportional to the higher cumulants of the distribution. But even if we had a model with these higher-order cumulants, they would turn out to be irrelevant in the infinite-range limit, so we ignore them here.

Before we proceed to derive the SK equation with this formalism (Feigelman and Tsvelick, 1979; Khurana and Hertz, 1980), we will do a somewhat simpler calculation — that of the spin glass susceptibility χ_{SG} (2.64), the quantity which diverges as the spin glass transition is approached from above. What we want here is thus the average of a pair of spin correlation functions $\langle S_i S_j \rangle$ over the distribution of J_{ij}'s. We encourage the reader to play around with the relevant diagrams to convince herself that in the infinite-range or infinite-dimensionality limit, the crucial feature of the dominant diagrams is the fact that they can be arranged so that the pairs of bonds representing the averages in (3.18) lie next to each other forming double chains, as shown in Fig. 3.2. The result can be expressed

(a)

(b)

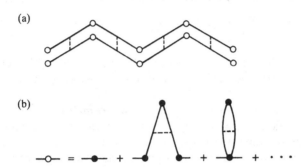

Figure 3.2: (a) Diagrams for χ_{SG}.
(b) some diagrams which go into the definition of the full vertex λ_2 indicated by the open circles in (a).

concisely as a simple sum of double chains (Fig. 3.2a) in which a 'renormalized' single-site vertex λ_2 occurs (indicated by the open circle in the figure). λ_2 contains contributions from all possible side excursions off the main chain of Fig. 3.2a. Some such contributions are shown in Fig. 3.2b.

Fortunately, we do not have to sum these all up here, because by the terms of our diagrammatic rules, λ_2 contains exactly all diagrams for the local spin correlation function $[\langle S_i^2 \rangle]_{av}$. More carefully, it is a diagram for a second partial derivative of Z with respect to the auxiliary external field βh_i. Thus it is, except for a factor of β, the local susceptibility (1.2), i.e. the so-called *connected* local spin correlation function $[\langle (S_i - \langle S_i \rangle)^2 \rangle]_{av}$. But we are above T_f and in the limit of zero external field, so this is just $[\langle S_i^2 \rangle]_{av}$, and for Ising spins this is identically unity (Feigelman and Tsvelick, 1979). Thus the double-chain diagrams of Fig 3.2a give simply

$$[\langle S_i S_j \rangle^2]_{av} = \delta_{ij} + \beta^2 \Delta_{ij} + \sum_k \beta^2 \Delta_{ik} \beta^2 \Delta_{kj} + \cdots \qquad (3.19)$$

Since we are now dealing with averaged quantities (the Δ_{ij}'s), which are translationally invariant, we can sum this geometric series analytically in k-space:

$$\chi_{SG}(k) = \frac{\beta^2}{1 - \beta^2 \Delta(k)} \qquad (3.20)$$

This result is analogous to the Curie–Weiss law for a ferromagnet:

$$\chi(k) = \frac{\beta}{1 - \beta J(k)} \qquad (3.21)$$

where $J(k)$ is the Fourier component of the exchange $J(\mathbf{R}_i - \mathbf{R}_j)$. This expression has the Curie or free response β, enhanced by the coupling between sites. In the spin glass case we have an enhancement of the free result β^2 by the fluctuations in the couplings.

Now we are ready to go below T_f, where we will both derive the SK equation for q and recalculate χ_{SG} in the presence of the spin glass order. To calculate q, we need to average all pairs of diagrams for the magnetization $m_i = \langle S_i \rangle$ over the bond distribution. We do this in the presence of an infinitesimal external symmetry-breaking field $h_i = h$ (which we recall is necessary in order to have a nonzero q). In the diagrams, the factors of this external field which can be attached to a spin vertex are represented by wavy lines. Thus if all interactions between spins are set equal to zero, we have only the diagrams of Fig. 3.3a, which just represent a familiar result:

$$\langle S_i \rangle = \sum_n \frac{1}{n!} \left(\frac{\partial^n}{\partial \psi^n} \tanh \psi \right)_{\psi=0} (\beta h)^n = \tanh \beta h \qquad (3.22)$$

With finite J_{ij}, we get graphs like Fig. 3.3b as well, with vertices connected by βJ_{ij} bonds, as in the graphs for $\langle S_i S_j \rangle$ which occurred in χ_{SG} above. The difference is that now unconnected tree-like structures which eventually connect to the external field lines can grow out of the vertices of order higher than 2. These parts of the diagrams, excluding the bonds J_{ij} by which they are rooted to the vertex, are just themselves (by definition) the diagrams for the local magnetizations $m_j = \langle S_j \rangle$. That is, the diagrams for the magnetization occur as parts of themselves. In this way a self-consistent equation for these magnetizations can be obtained. This is actually quite simple in the so-called 'tree approximation', where we exclude any closed loops. Then, in addition to the factors of the external field h in (3.22), we have simply any number of factors of the *internal* field $\sum_j \beta J_{ij} \langle S_j \rangle$ attached to a vertex at site i. This is the diagrammatic derivation of the Weiss equations (3.4). (However, as we remarked, the Weiss equations are not sufficient in the mean field limit of the spin glass.)

When we now turn to averaging pairs of these diagrams over the bond distribution, the argument determining the dominant graphs in the large z limit is the same as in the calculation of χ_{SG} — one needs pairs of bonds connecting the same pair of sites lying next to each other (Fig. 3.4a). The sum of all graphs which can be added on to the end at any of the pairs of bonds tied together in the averaging procedure is just q itself, so we get a self-consistent equation for q. As in the calculation of χ_{SG} above, the full vertices λ_n (indicated in Fig. 3.4a as open circles) contain a great deal of

(a)

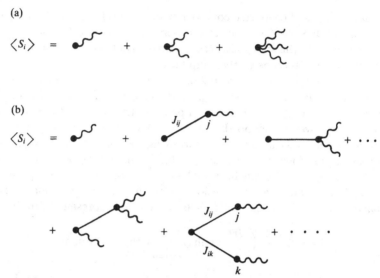

(b)

Figure 3.3: (a) Diagrams for local magnetization $\langle S_i \rangle$ for an isolated spin in the presence of a field h. The field is represented by a wavy line. (b) Diagrams for $\langle S_i \rangle$ with interactions J_{ij} (solid lines).

possible internal structure analogous to that shown in Fig. 3.2b for λ_2. But again, in the same way we argued before, these vertices are *local* quantities which are proportional to derivatives of the free energy with respect to a local source field and, therefore, to cumulants $[\langle S_i^n \rangle_c]_{av}$ of the single spin S_i. Furthermore, as these graphs contain in themselves no factors of the external field, they are just the cumulants of the free Ising spin distribution $P(S_i) = \delta(S_i^2 - 1)$, i.e. the bare vertices (3.13)–(3.15). Adding up terms of all orders in the graphs for q, we get

$$q = \sum_{mn} \frac{1}{m!} \left(\frac{\partial^m}{\partial \psi^m} \tanh \psi \right)_{\psi=0} \frac{1}{n!} \left(\frac{\partial^n}{\partial \psi^n} \tanh \psi \right)_{\psi=0}$$

$$\times (\beta^2 J^2 q)^{\frac{1}{2}(n+m)} (n+m-1)!! \qquad (3.23)$$

since $(n+m-1)!! \equiv (n+m-1)(n+m-3) \cdots 3 \cdot 1$ is the number of ways of pairing the $n+m$ bonds. But this is just the power series expansion of

$$q = \int \frac{dz}{\sqrt{2\pi}} e^{-\frac{1}{2}z^2} \tanh^2(\beta J q^{\frac{1}{2}} z) \qquad (3.24)$$

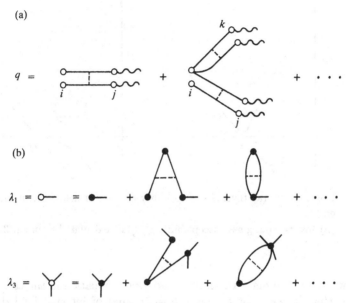

Figure 3.4: (a) Diagrams for $q = [\langle S_i \rangle]^2_{av}$: pairs of magnetization diagrams like Fig. 3.3b, tied together as in eq.(3.18) in all possible ways.
Only diagrams which survive in the $N \to \infty$ limit of the SK model are
shown.
(b) Examples of contributions to the 'dressed' vertices λ_1 and λ_3.

because $(p - 1)!!$ is the pth moment of a Gaussian distribution of unit
variance. Thus we have just recovered the zero-field limit of the SK equation
(3.7). The errors due to several uncontrolled approximations we made in the
preceding section turn out to cancel. We see from the derivation that these
cancellations (e.g. between the open-loop and closed-loop contributions
to λ_2 in Fig. 3.2b) occur because the local vertices λ_n have fixed values,
independent of the interactions in the problem.

What about χ_{SG} below T_f? In the ferromagnet below T_c we can simply
modify the paramagnetic calculation by replacing the factor β in (3.21),
which represents the local Curie susceptibility, by

$$\chi_{loc} = \beta(1 - M^2(T)) \tag{3.25}$$

where $M(T)$ is the self-consistent solution of the ferromagnetic Weiss mean
field equation. This restores stability to the system, as shown in Fig. 3.5a.

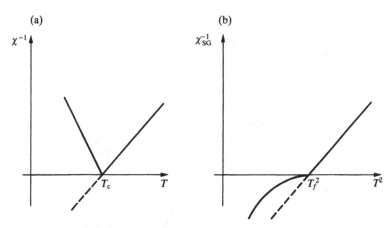

Figure 3.5: (a) Inverse susceptibility of a mean field ferromagnet, from eq.(3.25).
(b) Inverse spin glass susceptibility $\chi_{SG}^{-1}(T)$ calculated from eq.(3.26).

Now we want to make the corresponding modification in the calculation of χ_{SG}, that is, we want to put the same kind of internal field dressing on the λ_2 vertices that we put on the local magnetization vertices (λ_1) in calculating q. The calculation is then almost exactly analogous to that for q, except that each vertex has one more derivative (i.e. two external lines instead of one). Thus we find

$$\chi_{SG} = \frac{\beta^2 \int \frac{dz}{\sqrt{2\pi}} e^{-\frac{1}{2}z^2} \operatorname{sech}^4(\beta J q^{\frac{1}{2}} z)}{1 - (\beta J)^2 \int \frac{dz}{\sqrt{2\pi}} e^{-\frac{1}{2}z^2} \operatorname{sech}^4(\beta J q^{\frac{1}{2}} z)} \qquad (3.26)$$

This equation has a simple physical interpretation: We simply replace the factor of β^2 in (3.20) by an average of $\beta^2(1 - M^2(h))^2$ (where $M(h) = \tanh \beta h$ is the magnetization in a field h) over the distribution of internal fields. That is, we take into account the effect of the molecular field on the susceptibility just as we do in the ferromagnet, but now we have a distribution of molecular fields to average over.

The shock is that this does not stabilize χ_{SG} the way the corresponding procedure did in the ferromagnet; the disastrous result is shown in Fig. 3.5b. It is instructive to expand the sech^4 in the denominator of (3.26) in powers of q. Then one finds that to first order in q, i.e. to first order in $T_f - T$, the denominator is exactly zero. To this order, the reduction in the local susceptibility by the internal fields exactly compensates the divergence which comes from the ladder sum. Therefore the curve showing χ_{SG}^{-1} in Fig. 3.5b starts out flat just below T_f before curving below the axis. If the curve stayed flat at $\chi_{SG}^{-1} = 0$ for all $T < T_f$, we could live with the result. This kind of *marginal stability* is familiar in, e.g. Heisenberg ferromagnets, where, as we emphasized in Section 2.2, the susceptibility in a direction perpendicular to that of the magnetization is infinite at infinite wavelength. But we obviously cannot live with a negative χ_{SG}, since it was defined in (2.64) as the average of the sum of squares of real numbers.

It is a simple matter to generalize these calculations to include a uniform external field h. The result is just that the field simply enters in the argument of the tanh in (3.24) or the sech in (3.26) the way it did in the tanh in the heuristic theory leading to (3.7). In particular, the condition for the stability of the SK solution from the positivity of χ_{SG} is

$$1 \geq (\beta J)^2 \int \frac{dz}{\sqrt{2\pi}} e^{-\frac{1}{2}z^2} \mathrm{sech}^4[\beta(Jq^{\frac{1}{2}}z + h)] \qquad (3.27)$$

This equation defines a line $T_f(h)$ in the $h - T$ plane marking the boundary of the region where the SK solution is stable, as shown in Fig. 3.6. Expanding it near $T = T_f(0) = J$ and $h = 0$, we obtain the behaviour

$$\frac{T_f(0) - T_f(h))}{T_f(0)} = \left(\frac{3}{4}\right)^{1/3} \left(\frac{h}{J}\right)^{2/3} \qquad (3.28)$$

This result was first found (in a very different way which we will examine in the next section) by de Almeida and Thouless (1978). The line $T_f(h)$ is now universally known as the AT line, and the finding that the theory gives a negative χ_{SG} below this line is called the AT instability. It clearly indicates that there is something seriously wrong with the SK solution in this region.

But the foregoing derivation seemed to be a straightforward exercise in summing perturbation theory to all orders. So where is the error? Are there mysterious nonperturbative effects that cause the problem? To answer these questions, we have to go back and look more closely at the meaning of the self-consistent equations we have written down in the light of our careful formulations of broken symmetry and broken ergodicity in Section 2.4. Just which order parameter is it that we are supposed to be evaluating by these arguments? The two candidates are q and q_{EA}, so let us see if it makes sense to interpret our derivation in either of these ways.

We consider the equilibrium order parameter q first. Now in the diagrams before averaging over the bonds, every tree rooted at a vertex at site i should represent the equilibrium $m_i = \sum_a P_a m_i^a$. Now we assumed

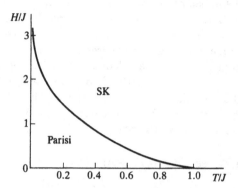

Figure 3.6: Plot of the Almeida–Thouless (AT) line for the SK model. To the right of the line, the SK solution with a single order parameter is correct, while to the left of the line the Parisi solution (see Section 3.4) is believed exact. The Parisi solution represents the many–valley structure of phase space and nonergodic behaviour. The AT line, therefore, signals the onset of irreversibility.

that when we carried out the bond average, trees rooted at different sites would decouple, so each pair of trees rooted at the same site gave a factor q. However, If there is nontrivial ergodicity breaking, this decoupling is not valid. It presumes, e.g. that the quantity $[\langle S_i S_j \rangle^2]_{av}$ is equal to q^2. But we showed in (2.53) that this is not true; this is just the fact that $q^{(2)} > q^2$. Therefore our derivation cannot apply to the equilibrium order parameter.

What about q_{EA}, the order parameter for a single state? One can certainly start by interpreting our graphs for the local magnetizations (before bond averaging) as describing m_i^a, the magnetization at site i in a single pure state a. To obtain q_{EA} (2.37), we must average the quantity $(m_i^a)^2$ (i.e. all pairs of these local magnetization graphs) over all possible states a with their proper weights P_a (2.38), and then average the result over the bond distribution. The problem with our derivation, however, is that it ignores the problem of the averaging over the different states; it assumes effectively that there is just one state a with $P_a = 1$. Thus it is incorrect if there is nontrivial broken ergodicity, so our theory does not describe q_{EA} either. Of course, if there were only one equilibrium phase, there would be nothing wrong with the present treatment. It is nontrivial broken ergodicity that is the source of the problem. It will be incorporated into the correct solution, which we study in Section 3.4. That solution uses the replica formalism we introduced in Section 2.3, so as an introductory step we will next derive the incorrect solution a third time (and the AT instability a second), using replicas. This will give us the grounding in replica manipulation we need to do the correct solution.

3.3 Replica-symmetric theory

We derived in Sections 3.1 and 3.2 a simple mean field equation (the SK equation (3.7) or (3.24)) for the spin glass order parameter $q(T, h)$ defined in (2.40). We also found that below the AT line (3.27) the solution of the SK equation leads to a negative spin glass susceptibility χ_{SG} as defined in (2.64), which is clearly unphysical. A stable solution was first found by Parisi (1979), using a very ingenious scheme based on the replica method introduced in Section 2.3. In this section we apply the replica method to the SK model and solve it within an ansatz that once again gives us the SK equations (3.7) or (3.24) (Sherrington and Kirkpatrick, 1975; Kirkpatrick and Sherrington, 1978). While this leaves us at this point with the same severe problems that we found in the preceding section, it lays necessary groundwork for the Parisi theory in the next section.

We want to calculate the averaged free energy (2.26) by means of the replica trick (2.31) and (2.32). Instead of the symmetric distribution of bonds (2.6) with (3.2) we consider here a more general case with a nonzero mean J_0:

$$P(J_{ij}) = \left[\frac{N}{2\pi J^2}\right]^{\frac{1}{2}} \exp\left[-\frac{N(J_{ij} - J_0/N)^2}{2J^2}\right] \tag{3.29}$$

This distribution again is assumed to be the same for all pairs of spins, with

$$[J_{ij}]_{av} = \frac{J_0}{N} \tag{3.30}$$

$$[J_{ij}^2]_{av} - [J_{ij}]_{av}^2 = \frac{J^2}{N} \tag{3.31}$$

Again, the factors of $1/N$ are necessary in order to obtain finite thermodynamic quantities for $N \to \infty$. The parameter J_0 describes a tendency toward ferromagnetic order. For $J_0 \gg J$ one has only ferromagnetic bonds and (3.29) should describe a more or less ideal ferromagnet.

We proceeed as in Section 2.3 and obtain for the Hamiltonian (3.1)

$$[Z^n]_{av} = Tr_S \exp\left[\frac{1}{N}\sum_{ij}(\tfrac{1}{4}(\beta J)^2 \sum_{\alpha\beta} S_i^\alpha S_j^\alpha S_i^\beta S_j^\beta\right.$$

$$\left. + \beta J_0 \sum_\alpha S_i^\alpha S_j^\alpha) + \beta h \sum_{i\alpha} S_i^\alpha\right] \tag{3.32}$$

We extract in the double sum over α and β the terms with $\alpha = \beta$ and omit terms of order unity in the exponent. This leads with $(S_i^\alpha)^2 = 1$ to

$$[Z^n]_{av} = Tr_S \exp\left[\frac{(\beta J)^2}{4N}\sum_{(\alpha\beta)}[(\sum_i S_i^\alpha S_i^\beta)^2 + nN^2]\right.$$

$$\left. + \frac{\beta J_0}{2N}\sum_\alpha(\sum_i S_i^\alpha)^2 + \beta h \sum_{\alpha i} S_i^\alpha\right] \tag{3.33}$$

where $(\alpha\beta)$ means summation over $\alpha \neq \beta$. Equation (3.33) can be simplified by means of the identity (3.10) with $a = \sum_i S_i^\alpha S_i^\beta$, $\lambda = \tfrac{1}{2}\beta J$ and $a = \sum_i S_i^\alpha$, $\lambda = \tfrac{1}{2}\beta J_0$ for the two quadratic terms in the exponent, respectively, giving

$$[Z^n]_{av} = \exp[nN(\tfrac{1}{2}\beta J)^2]\int_{-\infty}^{\infty}\prod_{(\alpha\beta)}\frac{\beta J N^{\frac{1}{2}}}{\sqrt{2\pi}}dy^{\alpha\beta}\prod_\alpha\left(\frac{\beta J_0 N}{2\pi}\right)^{\frac{1}{2}}dx^\alpha$$

$$\times \exp\left[-\tfrac{1}{2}N(\beta J)^2\sum_{(\alpha\beta)}(y^{\alpha\beta})^2 - \tfrac{1}{2}N\beta J_0\sum_\alpha(x^\alpha)^2\right]$$

$$\times Tr_S \exp\left[(\beta J)^2\sum_{i\alpha\beta}y^{\alpha\beta}S_i^\alpha S_i^\beta + \beta\sum_{i\alpha}(J_0 x^\alpha + h)S_i^\alpha\right] \tag{3.34}$$

Thus, just as in (3.12), the many-spin problem has been reduced to an average over simpler problems in which different sites are decoupled. In return for this simplification, we have to deal with inter-replica couplings in these single-spin problems. The single-spin property can be used to write (where g is an arbitrary function)

$$Tr_S \exp[\sum_i g(S_i^\alpha)] = \exp[N \ln tr_S \exp g(S^\alpha)] \tag{3.35}$$

so we have

$$[Z^n]_{av} = \exp[nN(\tfrac{1}{2}\beta J)^2] \times$$

$$\int_{-\infty}^{\infty} \prod_{(\alpha\beta)} \frac{\beta J N^{\frac{1}{2}}}{\sqrt{2\pi}} dy^{\alpha\beta} \prod_{\alpha} \left(\frac{\beta J_0 N}{2\pi}\right)^{\frac{1}{2}} dx^{\alpha} \exp(-NG) \qquad (3.36)$$

with

$$G = \tfrac{1}{2}(\beta J)^2 \sum_{(\alpha\beta)} (y^{\alpha\beta})^2 + \tfrac{1}{2}\beta J_0 \sum_{\alpha} (x^{\alpha})^2$$

$$- \ln tr_S \exp[\tfrac{1}{2}(\beta J)^2 \sum_{(\alpha\beta)} y^{\alpha\beta} S^{\alpha} S^{\beta} + \beta \sum_{\alpha} (J_0 x^{\alpha} + h) S^{\alpha}] \qquad (3.37)$$

and where the trace tr_S extends over all states of a single replicated spin S^{α}. For $N \to \infty$ the integral (3.36) with (3.37) can be done by the method of steepest descents: schematically,

$$\int dy \exp[-NG(y)] =$$

$$\int dy \exp[-NG(y_0) - \tfrac{1}{2}NG''(y_0)(y - y_0)^2 + \cdots] \qquad (3.38)$$

where $G'(y_0) = 0$ defines the saddle point y_0. Note that the second term can be ignored for $N \to \infty$, provided $G''(y_0) > 0$; otherwise the resulting Gaussian integral diverges and the saddle point procedure is senseless. The conditions $\partial G/\partial y^{\alpha\beta} = 0$ and $\partial G/\partial x^{\alpha} = 0$ lead to

$$y_0^{\alpha\beta} = \langle S^{\alpha} S^{\beta} \rangle \equiv \tilde{Z}^{-1} tr_S[S^{\alpha} S^{\beta} \exp H_{eff}] \qquad (3.39)$$

$$x_0^{\alpha} = \langle S^{\alpha} \rangle \equiv \tilde{Z}^{-1} tr_S[S^{\alpha} \exp H_{eff}] \qquad (3.40)$$

$$\tilde{Z} \equiv tr_S \exp H_{eff} \qquad (3.41)$$

with the effective (dimensionless) single-spin Hamiltonian

$$H_{eff} = \tfrac{1}{2}(\beta J)^2 \sum_{(\alpha\beta)} y_0^{\alpha\beta} S^{\alpha} S^{\beta} + \beta \sum_{\alpha} (J_0 x_0^{\alpha} + h) S^{\alpha} \qquad (3.42)$$

In deriving (3.39) to (3.42) we have assumed $N \to \infty$ but n finite, that is, we have interchanged the limits $n \to 0$ and $N \to \infty$. There is no way to justify this assumption *a priori*, but the resulting equation for the order parameter will agree with (3.7) and (3.24).

We can now express the free energy in terms of the saddle-point values $y_0^{\alpha\beta}$ and x_0^{α} of the integration variables $y^{\alpha\beta}$ and x^{α}. We have from the definition (2.26) with (2.31), (3.36), (3.37) and the saddle point condition ($f = F/N$ is the free energy per spin)

$$-\beta f = \lim_{n \to 0} \left[(\tfrac{1}{2}\beta J)^2 \left(1 - \frac{1}{n} \sum_{(\alpha\beta)} (y_0^{\alpha\beta})^2 \right) \right.$$

$$\left. - \frac{\beta J_0}{2n} \sum_{\alpha} (x_0^{\alpha})^2 + \frac{1}{n} \ln tr_S \exp H_{eff} \right] \tag{3.43}$$

with

$$\frac{\partial f}{\partial y_0^{\alpha\beta}} = \frac{\partial f}{\partial x_0^{\alpha}} = 0 \qquad (\text{all } \alpha \neq \beta) \tag{3.44}$$

The second derivative of this free energy with respect to the field h leads for $J_0 = 0$, $h \to 0$ to the local susceptibility

$$\chi_{loc} = \beta \lim_{n \to 0} \left(1 - \frac{1}{n} \sum_{(\alpha\beta)} y_0^{\alpha\beta} \right) \tag{3.45}$$

This expression agrees for $n \to 0$ with (2.45) and (2.60) if we identify the parameters $y_0^{\alpha\beta}$ with the parameters $q^{\alpha\beta}$ defined in (2.59) and

$$q = -\lim_{n \to 0} \frac{1}{n} \sum_{(\alpha\beta)} y_0^{\alpha\beta} \tag{3.46}$$

with the spin glass order parameter (2.40). Similarly, the quantity

$$M = \lim_{n \to 0} \frac{1}{n} \sum_{\alpha} x_0^{\alpha} \tag{3.47}$$

can be identified as the local magnetization. The expressions (3.45)–(3.47) are completely general within the replica formalism; they are not restricted to the region of validity of the ansatz we make below. We will also use them in the Parisi theory in the next section.

The simplest approach for solving the self-consistent equations (3.39) to (3.42) (i.e. (3.44)) consists of making the so-called replica-symmetric ansatz, for which all parameters $y_0^{\alpha\beta} = q^{\alpha\beta}$ and x_0^{α} are independent of their replica indices (Edwards and Anderson, 1975; Sherrington and Kirkpatrick, 1975)

$$y_0^{\alpha\beta} = q \quad (\alpha \neq \beta); \quad x_0^{\alpha} = M \tag{3.48}$$

The free energy becomes, with (3.48), $\sum_{(\alpha\beta)} 1 = n(n-1)$, and

$$tr_S \exp(A \sum_{\alpha} S^{\alpha}) \approx 1 + n \ln(2 \cosh A) \quad (n \to 0) \tag{3.49}$$

$$-\beta f = (\tfrac{1}{2}\beta J)^2 (1-q)^2 - \tfrac{1}{2}\beta J_0 M^2$$

$$+(2\pi)^{-\frac{1}{2}} \int_{-\infty}^{\infty} dz e^{-\frac{1}{2}z^2} \ln \cosh \eta(z) \tag{3.50}$$

where

$$\eta(z) = \beta(Jq^{\frac{1}{2}}z + J_0 M + h) \tag{3.51}$$

The variation of the free energy (3.50) with respect to q and M then leads to the SK equations for the equilibrium values of these quantities:

$$M(T,h) = (2\pi)^{-\frac{1}{2}} \int_{-\infty}^{\infty} dz e^{-\frac{1}{2}z^2} \tanh \eta(z) \tag{3.52}$$

$$q(T,h) = (2\pi)^{-\frac{1}{2}} \int_{-\infty}^{\infty} dz e^{-\frac{1}{2}z^2} \tanh^2 \eta(z) \tag{3.53}$$

For $J_0 = 0$, of course (3.53) just reduces to (3.7), and we have analyzed the solution for this case in the preceding sections.

The parameter $\eta(z)$ can be interpreted as a random local field with mean $J_0 M + h$ and variance $J^2 q$. The equations (3.51) and (3.53) indicate that one obtains exactly the same solution for $q(T,h)$ and $q(T,J_0M)$, where the magnetization M is determined by (3.52). We found already in Section 3.2 that the solution (3.53) with $J_0 = 0$ becomes unstable below the AT line. The same also holds now for $J_0 M \neq 0$, $h = 0$ at sufficiently low temperatures.

The various critical lines are indicated in Fig. 3.7. The lines 1 and 5 are AT instability lines, below which the susceptibility χ_{SG} (1.6) becomes negative (as in (3.26)). Below line 2 one has a spin glass state with spontaneous magnetization $M \neq 0$ and with the enhanced susceptibility (Sherrington and Kirkpatrick, 1975)

$$\chi(T) = \frac{1-q}{T - J_0(1-q)} \tag{3.54}$$

In the region above line 2 in zero external field, where q vanishes, this susceptibility reduces to an ordinary Curie–Weiss law, which diverges as $T \to J_0 = T_c$, indicating the onset of ferromagnetic order. Thus one has for $J_0 > J$ below the Curie temperature T_c (line 3) a ferromagnetic state with a finite spontaneous magnetization M, as well as a finite q.

The boundary between the state with $M = 0$ and that with finite M is given by the vanishing of the denominator in (3.54) in the spin glass phase, using the SK equation (3.53) with (3.51) for q with $M = 0$. This yields the

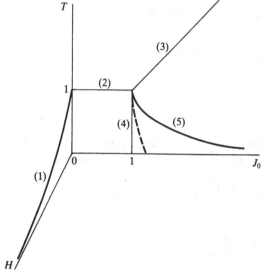

Figure 3.7: Phase diagram of the Ising SK model with infinite-range interactions, as a function of temperature, average ferromagnetic interaction J_0/J and the magnetic field in units of $J = k_B T_f$. Lines 1 and 5 are determined by the AT instability, line 2 by $q \to 0$, line 3 by the vanishing of the spontaneous magnetization $M_s \to 0$, $q = 0$, and line 4 by $q \neq 0$, $M_s \to 0$. The broken line represents the (incorrect) solution of SK.

broken line in the figure. However, the SK equations are not valid (because of the AT instability, as indicated above) anywhere below lines 1, 2, and 5. The true boundary is the vertical line 4; the argument goes as follows (Toulouse, 1980): In mean field theory, the only change introduced in the thermodynamic potential in the presence of a nonzero J_0 is the addition of the ferromagnetic condensation energy:

$$A(T, M, J_0) = A(T, M) - \tfrac{1}{2} J_0 M^2 \qquad (3.55)$$

where $A(T, M) \equiv A(T, M, J_0 = 0)$. In what follows we assume this also to be true below lines 2 and 5, i.e. in the whole $J_0 - T$ plane. Now the M–dependence of $A(T, M)$ for small M is quite generally

$$A(T, M) = A_0(T) + \tfrac{1}{2} \chi^{-1} M^2 \qquad (3.56)$$

with $A_0(T) \equiv A(T, M = 0)$, where χ is the zero-field susceptibility with $J_0 = 0$. Combining (3.55) and (3.56), we see that the total coefficient of M^2 goes to zero, indicating the ferromagnetic instability, when $J_0 = \chi^{-1}$. Indeed, application of this criterion, using the (incorrect) value of χ obtained from the SK equations (3.51)–(3.53), just gives the previously mentioned (incorrect) broken boundary line. If we now anticipate the result for χ obtained in the Parisi theory in the next section, $\chi = J^{-1}$ (independent of temperature), the condition for the boundary of the ferromagnetic region becomes instead just $J = J_0$, i.e. the vertical line 4.

This argument applies only to the zero-field limit. In nonzero fields, the transition lines 3 and 4 disappear. Apparently, the states below line 1 with $q \neq 0$, $M = 0$ and line 5 with $q \neq 0, M \neq 0$ are different. As we saw in Section 2.6, a double transition superficially resembling lines 3 and 5 has been observed in many experiments, and we shall come back to this point in Chapter 11.

The magnetization $M(h)$ for $J_0 = 0$ turns out to be nonanalytic at T_f. One has from the SK equations (3.51)–(3.53)

$$M = \frac{h}{T}\left[1 - \frac{h^2}{3T^2}\frac{T^2 + 2T_f^2}{T^2 - T_f^2} + \cdots\right], \qquad T > T_f \qquad (3.57)$$

$$M = \frac{h}{T_f}\left[1 - \frac{|h|}{2^{\frac{1}{2}}T_f} + \frac{23}{24}\left(\frac{h}{T_f}\right)^2 + \cdots\right], \qquad T = T_f \qquad (3.58)$$

The nonlinear susceptibility χ_{nl} (2.64) diverges for $T \to T_f$ in a mean-field fashion, proportional to $(T - T_f)^{-1}$. The situation beyond mean field theory and in experiments, where the exponent is considerably larger, will be discussed in Chapter 8.

The free energy (3.50) leads to a cusp in the specific heat at T_f and to T^{-2} behaviour above T_f. Neither is observed, as we noted in Chapter 1. The discrepancy is simply a consequence of the fact that mean field theory fails to take any account of short-range spin correlations. The improved theoretical treatment of short-range-interaction spin glasses described in Chapter 8 does better.

These results have all been obtained from the replica-symmetric SK saddle point (3.48). But the validity of this procedure depends on whether the first corrections to the results in an expansion in $1/N$ can be ignored. These corrections are obtained by expanding the exponent (3.37) of the integral (3.36) for $[Z^n]_{av}$ to quadratic order in the deviations from the saddle point value (as indicated schematically by (3.38)) and carrying out the remaining $\frac{1}{2}n(n-1)$-dimensional Gaussian integration. Whether the theory based on this saddle point is sensible, then, depends on whether this integral converges, and this will be true only if the eigenvalues of the Hessian matrix

$$A^{\alpha\beta,\gamma\delta} \equiv \frac{\partial^2 G}{\partial y^{\alpha\beta} \partial y^{\gamma\delta}}$$

$$= \beta^2 J^2 \left[\delta_{\alpha\beta,\gamma\delta} - \beta^2 J^2 (\langle S^\alpha S^\beta S^\gamma S^\delta \rangle - \langle S^\alpha S^\beta \rangle \langle S^\gamma S^\delta \rangle) \right] \qquad (3.59)$$

evaluated at the SK saddle point, are all positive. (We restrict ourselves in equation (3.59) and the rest of the formal expressions in this section to the simpler $J_0 = 0$ case; more generally, the relevant matrix of second derivatives also contains elements of the forms $\partial G/\partial x^\alpha \partial x^\beta$ and $\partial G/\partial x^\alpha \partial y^{\beta\gamma}$. The generalization of the calculation to include these is straightforward if slightly tedious.)

Exploiting the permutation symmetry of the replicated effective Hamiltonian, de Almeida and Thouless (1978) were able to find all the eigenvalues of (3.59). We outline their calculation here. The elements of the Hessian matrix are of three kinds: The diagonal terms are given by

$$(\beta J)^{-2} A^{\alpha\beta,\alpha\beta} = 1 - (\beta J)^2 (1 - q^2) \qquad (3.60)$$

The elements with one index in common between $\alpha\beta$ and $\gamma\delta$ are given by

$$(\beta J)^{-2} A^{\alpha\beta,\alpha\gamma} = -(\beta J)^2 (q - q^2) \qquad (3.61)$$

Finally, there are those with no elements in common:

$$(\beta J)^{-2} A^{\alpha\beta,\gamma\delta} = -(\beta J)^2 (r - q^2) \qquad (3.62)$$

where q is the SK order parameter (3.53) and

$$r = (2\pi)^{-\frac{1}{2}} \int_{-\infty}^{\infty} dz\, e^{-\frac{1}{2}z^2} \tanh^4 \eta(z) \qquad (3.63)$$

The eigenvectors of the Hessian matrix are relatively simple. They fall into three distinct classes. The first is the fully symmetric eigenvector, which is found (for $n \to 0$) to have the eigenvalue λ_1 given by

$$(\beta J)^{-2} \lambda_1 = 1 - (\beta J)^2 (1 - 4q + 3r) \qquad (3.64)$$

This mode consists simply of varying the magnitude of the SK q without changing the structure of the matrix in replica space. This is analogous to varying the magnitude but not the direction of the magnetization in a Heisenberg ferromagnet, so it is known colloquially as the 'longitudinal' mode. The next class of eigenvectors consists of those which are symmetric under interchange of all but one index. These modes are known as 'anomalous', and in the limit $n \to 0$ their common eigenvalue λ_2 is also given by (3.64). Finally, there are eigenvectors, called 'replicons', which are symmetric under interchange of all but two indices, with eigenvalue λ_3:

$$(\beta J)^{-2}\lambda_3 = 1 - (\beta J)^2 (1 - 2q + r)$$

$$= 1 - (\beta J)^2 (2\pi)^{-\frac{1}{2}} \int_{-\infty}^{\infty} dz e^{-\frac{1}{2}z^2} \operatorname{sech}^4 \eta(z) \tag{3.65}$$

The eigenvalues $\lambda_{1,2}$ are always positive, but the condition that λ_3 not be negative is simply the condition (3.27) we found that χ_{SG} not be negative. This, as we observed, is violated in the SK solution below the AT line, so the SK saddle point is not a stable one in this part of the $T - h$ plane.

We note a curious feature of this stability analysis: Suppose we have a stable solution, i.e. one in which all the eigenvalues of the Hessian matrix (3.59) are positive (such as the SK solution *above* the AT line). That means that whenever we move a little away from the saddle point in the space of the $q^{\alpha\beta}$'s, the free energy functional G will increase; we are sitting in a local minimum of G with respect to variation in any direction in this space. This is just what we would expect in any system where we examine the free energetics of fluctuations around a mean field solution. But the present situation has an odd twist to it: Because the dimensionality $\frac{1}{2}n(n-1)$ of the replica space is *negative* in the limit $n \to 0$, the correction to the free energy obtained by summing the fluctuation contributions from the Gaussian integrals in all the principal directions in the space is negative. That is, in this sense, a stable saddle point solution is a local free energy *maximum*, rather than a minimum.

This makes it easier to understand why the free energy obtained in the SK solution is actually *higher* than the extrapolation of the free energy of the paramagnetic state to $T < T_f$. Similarly, the free energy of the Parisi solution which we will obtain in the next section will be even higher than that of the SK solution. One might say that in replica formalism we have to maximize rather than minimize the free energy, but a better way to look at it is in terms of the fluctuations around the saddle point as described above: The relevant criterion is the positivity of the eigenvalues of the free energy fluctuation matrix (3.59).

Another serious objection to the SK solution comes from its low-temperature properties. The free energy (3.50) leads to a *negative* entropy for $T \to 0$, which is clearly unphysical. Both this defect and the AT instability will be removed in the Parisi solution.

In summary, the replica formulation of the SK model, as we have solved it here under the assumption of replica symmetry, leads to exactly the same problems that we found in our previous attempts to solve it in the preceding sections. But this exercise has not been a total waste of our time: with this warm-up experience in using the replica formalism, we are now ready to go on to handle replica symmetry breaking and the Parisi solution in the following section.

3.4   Replica symmetry breaking and broken ergodicity: the Parisi solution

Our discussion at the end of Section 3.2 pointed to the possibility of broken ergodicity as a reason for the failure of the SK theory below the AT line. Given the connection we noted in Chapter 2 between broken ergodicity and replica symmetry breaking, the instability we have just found for the replica-symmetric ansatz points even more clearly in this direction. The replica formalism, moreover, offers the possibility of a way to describe the broken ergodicity formally, by finding a stationary point of the exponent (3.35) in the averaged, replicated partition function (3.34) which is not replica-symmetric.

For simplicity, we restrict ourselves in this section to the case of symmetrically distributed bonds (3.2); we have already indicated how to generalize to nonzero J_0. As the solution is still quite involved at general temperatures, we further restrict ourselves, following Parisi (1979, 1980), to the neighbourhood of T_f. Then we can expand (3.35) in powers of $y^{\alpha\beta}$, keeping only the first few (hopefully important) terms. As in any Landau expansion (cf. Section 2.2), the kinds of terms one can get in the exponent are all invariants under the symmetry of the problem. (In order to conform with the notation which has become standard in the literature, we now change our notation slightly and write $q^{\alpha\beta}$ instead of $y^{\alpha\beta}$.)

The only quadratic invariant is $tr(\mathbf{q}^2) = \sum_{\alpha\beta}(q^{\alpha\beta})^2$, and the only cubic invariant is $tr(\mathbf{q}^3) = \sum_{\alpha\beta\gamma} q^{\alpha\beta}q^{\beta\gamma}q^{\gamma\alpha}$. One might also expect a term of the form $\sum_{\alpha\beta}(q^{\alpha\beta})^3$, but this is not allowed. True, the permutation symmetry of the replica indices does not forbid it, but one has to remember that these terms come from the expansion of (3.35) in powers of y. Thus any term that contains a single replica index α a given number of times must have come from a trace which contained the α-th replica spin the same number of times. Since the original spin Hamiltonian was even in the S^α's, it follows that terms such as $\sum_{\alpha\beta}(q^{\alpha\beta})^3$ are forbidden. (They do appear, however, in problems such as those we will examine in Section 3.6b and 3.7.)

There are several quartic terms, including $tr(\mathbf{q}^4)$, but in this simplified treatment, again following Parisi, we will keep only one of them, namely $\sum_{\alpha\beta}(q^{\alpha\beta})^4$. Retention of the others would not make any qualitative difference (Bray and Moore, 1978, Pytte and Rudnick, 1979, Thouless et al 1980). We ignore sixth- and higher-order terms, which would only make any difference at all at higher order in $T - T_f$ than we are concerned with here. Thus we have an effective model

$$G[\mathsf{q}] = \lim_{n \to 0} \frac{1}{2n} \left[\theta tr\, \mathsf{q}^2 - \frac{1}{3} tr\, \mathsf{q}^3 - \frac{1}{6} \sum_{\alpha\beta} (q^{\alpha\beta})^4 \right] \qquad (3.66)$$

where $\theta = (T - T_f)/T_f$. It is necessary to go to fourth order in q because if we stopped at third order, a replica-symmetric solution would turn out to solve the model. The crucial physics of the problem first appears in the particular quartic term which we have retained in (3.66).

Even with these simplifications, the search for the optimal $q^{\alpha\beta}$ lies in a very large space. No general direct way to find it exists; we are forced to try ansätze with some variational parameters in them. Parisi made such an ansatz which seems to be successful, after all the *a posteriori* tests that have been made (which we discuss later). It has the following hierarchical form: We start with the SK form, in which all elements of the q-matrix have the same value q_0 (except the diagonal ones, which are taken to be zero). We then break the big $n \times n$ matrix into $n/m_1 \times n/m_1$ blocks of size $m_1 \times m_1$. In the off-diagonal blocks, nothing is changed, but in the diagonal blocks we replace q_0 by q_1:

$$\mathsf{q}^{(1)} = \begin{bmatrix} 0 & q_1 & q_1 & q_0 & q_0 & q_0 \\ q_1 & 0 & q_1 & q_0 & q_0 & q_0 \\ q_1 & q_1 & 0 & q_0 & q_0 & q_0 \\ q_0 & q_0 & q_0 & 0 & q_1 & q_1 \\ q_0 & q_0 & q_0 & q_1 & 0 & q_1 \\ q_0 & q_0 & q_0 & q_1 & q_1 & 0 \end{bmatrix} \qquad (3.67)$$

We then do the same thing within each of the blocks along the diagonal: they are broken into $m_1/m_2 \times m_1/m_2$ subblocks, each of size $m_2 \times m_2$, and in each of these along the diagonal we replace q_1 by q_2. This procedure is then repeated infinitely many times.

If we imagine doing this for positive-integral-dimensional matrices, then each m_i has to be evenly divisible by m_{i+1}, which requires

$$n > m_1 > m_2 > \cdots > 1 \qquad (3.68)$$

But in the $n \to 0$ limit, we must turn this around and take

$$0 < m_1 < m_2 < \cdots < 1 \qquad (3.69)$$

instead. In the limit where the procedure is done infinitely many times, the m_i become continuous: $m_i \to x$, $0 < x < 1$. The information in the set of q_i's and m_i's is then contained in a function $q(x)$ on the unit interval.

It takes a little algebra to make this step (see, e.g. Thouless et al, 1980). It is not hard for the quadratic and quartic terms: We can see quite simply that

$$\sum_{\alpha\beta}(q^{\alpha\beta})^2 = n\sum_j(m_j - m_{j+1})q_j^2 \rightarrow = -n\int_0^1 q^2(x)dx \qquad (3.70)$$

and

$$\sum_{\alpha\beta}(q^{\alpha\beta})^4 = n\sum_j(m_j - m_{j+1})q_j^4 \rightarrow = -n\int_0^1 q^4(x)dx \qquad (3.71)$$

The cubic term is harder; it comes out to be

$$\mathrm{tr}q^3 = n\int_0^1 dx\left[xq^3(x) + 3q(x)\int_0^x q^2(y)dy\right] \qquad (3.72)$$

Thus the free energy functional (3.66) can be written

$$G[\mathbf{q}] = \tfrac{1}{2}\int_0^1 dx\left[|\theta|q^2(x) + \frac{1}{6}q^4(x) - \frac{1}{3}xq^3(x)\right.$$

$$\left. -q(x)\int_0^x q^2(y)dy\right] \qquad (3.73)$$

We vary it with respect to $q(x)$:

$$2|\theta|q(x)-xq^2(x)-\int_0^x q^2(y)dy-2q(x)\int_x^1 q(y)dy+\frac{2}{3}q^3(x) = 0 \qquad (3.74)$$

Differentiating this equation gives

$$|\theta| - xq(x) - \int_x^1 q(y)dy + q^2(x) = 0 \quad \text{or} \quad \frac{dq}{dx} = 0 \qquad (3.75)$$

and differentiating it again yields

$$q(x) = \tfrac{1}{2}x \quad \text{or} \quad \frac{dq}{dx} = 0 \qquad (3.76)$$

Assuming $q(x)$ to be continuous, we find the solution by taking the first of these for small q and a constant $q(x) = q(1)$ for $x > 2q(1)$. Substituting this form into (3.75) then allows us to solve for the plateau value $q(1) = |\theta|+O(|\theta|^2)$. Physically, $q(1)$, being the largest overlap, must be the single–phase order parameter q_{EA}. The solution is shown in Fig. 3.8, including the behaviour for finite field h.

It is an instructive exercise to derive the existence of the low-x plateau in an external field. One finds the plateau value (Thouless et al, 1980)

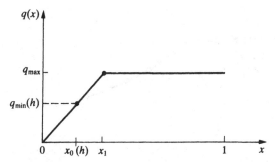

Figure 3.8: The Parisi solution for $q(x)$ close to $T = T_f$. The solid line is for $h = 0$ and the broken line is for small h. The solution is independent of h, except for the appearance of the second plateau. For $h = 0$, one has $q_{min} = x_0 = 0$. There is a 'plateau' region $q = q_{max}$ for $x \geq x_1$ and a region $x_0 < x < x_1$ where $q(x)$ increases monotonically (from Parisi, 1980).

$$q(0) = \frac{3}{4} \left[\frac{h^2}{J^2} \right]^{\frac{2}{3}} \tag{3.77}$$

As the field increases, this plateau rises, and once it reaches the height of the second plateau, the only solution is x-independent. That is, the replica symmetry breaking disappears and one is back to the SK solution. We recognize the point at which this happens as the AT transition line discussed in the preceding sections.

For the Parisi form of q-matrix, the equilibrium susceptibility (from (2.45) and (2.60)) becomes

$$\chi = \lim_{n \to 0} \beta \left[1 - \frac{1}{n(n-1)} \sum_{\alpha \neq \beta} q^{\alpha\beta} \right] = \beta \left[1 - \int_0^1 q(x) dx \right] \tag{3.78}$$

To the present order of accuracy, i.e. the Landau expansion to $O(q^4)$, our solution then implies a temperature-independent equilibrium $\chi = 1/J$ below $T_f = J$. In fact a fuller treatment which we do not present here turns out to give this result exactly for all $T > T_f$. This result will also come out in a simple way in the dynamical formulation of the theory in the next chapter.

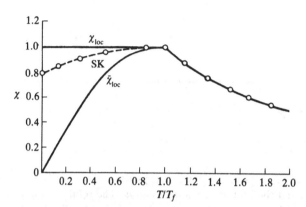

Figure 3.9: Susceptibilities $\chi_{loc} = \beta(1-q)$, $\tilde{\chi}_{loc} = \beta(1-q(1))$, and the solution of SK as obtained for the SK model with infinite-range interactions (from Thouless et al, 1977).

It is clear from Fig. 3.8 that $q(1) > \int_0^1 q(x)dx$; thus we have $q(1) = |\theta| + O(|\theta|^2)$, but the coefficient of $|\theta|^2$ is not determined at the level of accuracy of the model (3.66). A consistent treatment to order $|\theta|^3$ (Thouless et al, 1980) gives $q(1) = |\theta| + |\theta|^2 - |\theta|^3 + O(|\theta|^4)$. $q(1)$ is not known for general $T < T_f$, but we will discuss arguments in the next section which predict $q(1) = 1 - \alpha(T/T_f)^2$ at low T.

Fig. 3.9 shows the temperature dependences of the resulting susceptibilities $\chi_{loc} = \beta(1-q)$ and $\tilde{\chi}_{loc} = \beta(1-q(1))$ together with the corresponding SK result for comparison.

The principal reason we were driven to something so exotic as replica symmetry breaking was the AT instability in the SK solution. We therefore have to ask whether the Parisi solution is stable, i.e. whether the eigenvalues of the Hessian fluctuation matrix (3.56), *evaluated at the Parisi saddle point*, are all positive. The diagonalization is a rather large technical task, but it was thoroughly carried out by De Dominicis and Kondor (1983, 1984, 1985a,b). They found two families of eigenvectors, corresponding roughly to the two classes of eigenvectors found by de Almeida and Thouless in their analysis of the fluctuations around the SK saddle point, each now with a spectrum of eigenvalues (Fig. 3.10). Some of the eigenvalue spectra have support only on the positive real axis (these can be thought of crudely

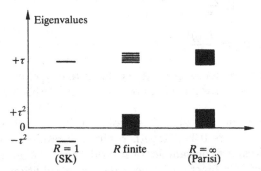

Figure 3.10: Eigenvalue spectrum of the Hessian fluctuation matrix, for different numbers of levels R in the Parisi replica-symmetry-breaking scheme. $R = 1$ is the SK solution and $R = \infty$ the full Parisi solution (from De Dominicis and Kondor, 1985a).

as analogous to the positive eigenvalue λ_1 of the AT analysis), while other spectra extend down to zero. In addition there are some isolated zero eigenvalues. But the important thing is that *there are no negative eigenvalues.* Thus the AT instability has been cured!

The presence of zero eigenvalues and of parts of the eigenvalue spectrum extending down to zero deserves a little comment. This says that the system is *marginally stable*, like a system at the critical point. The system is apparently very 'soft'. This softness could be due either to a softness or marginal stability of the different locally stable phases, or to an extra softening that comes in from 'intervalley' contributions (as in 2.40) when different phases are coupled together in the thermodynamic averaging that gives the full equilibrium response. The detailed investigations of De Dominicis and Kondor indicate that there are contributions from both sources. We will be able to get more insight into the marginal stability from the dynamical theory of the next chapter.

The problem of the negative low-temperature entropy is also cured by this solution.

Now given that this broken-replica-symmetry solution seems to be correct, what does it tell us about the nature of the broken ergodicity we are looking for? A great deal, it turns out, and some of it striking and unexpected. We start with the more straightforward aspects: These are seen from the overlap distribution function $P(q)$, which was given in general in the replica formalism by (2.62). Now, with the Parisi parametrization of the $q^{\alpha\beta}$ we have

$$P(q) = \lim_{n\to 0} \frac{1}{n(n-1)} \sum_{\alpha\neq\beta} \delta(q - q^{\alpha\beta})$$

$$= \int_0^1 dx\delta[q - q(x)] = \frac{dx(q)}{dq} \tag{3.79}$$

where $x(q)$ is the inverse function of $q(x)$. It is evident that $q(x)$ must be a nondecreasing function, as we assumed it to be, in order that this inversion can be performed.

Fig. 3.11 shows the form of $P(q)$ derived from the Parisi $q(x)$ in zero external field at a temperature $0.4T_f$, together with the results of Monte Carlo simulation measurements on samples of several sizes. The important point about the form of this function is the presence, in addition to the delta-function spike at $q = q(1)$, of a continuous part. There therefore must exist very many states resembling each other in all possible degrees. On the other hand, the presence of the spike tells us that only a few of these states dominate the thermodynamic sum.

Numerical simulations (Mackenzie and Young, 1982, 1983) can tell us about how the size of the barriers between the different phases scale with the size N of the system. One finds that the log of the longest relaxation time τ (i.e. the largest energy barrier) in the system is proportional to $N^{\frac{1}{4}}$. (This result is obtained if the total magnetization (which is of order $N^{\frac{1}{2}}$) does not change sign in the time of the simulation. There is a still longer time, the so-called ergodic time τ_e, within which the total magnetization typically changes its sign, for which $\ln \tau_e \propto N^{\frac{1}{2}}$.) Thus all barriers between phases are infinite in the thermodynamic limit, but some barriers (i.e. those separating phases with different signs of the total magnetization) are more infinite than others (Orwell, 1946).

Another surprising feature of the theory is the lack of self-averaging of certain quantities (Mézard et al, 1984a,b; Young et al, 1984). For example, consider the single-sample equilibrium order parameter q_J (2.42). By straightforward generalization of the arguments we used to express q, $q^{(2)}$, and $P(q)$ ((2.60)–(2.62)) in replica language, we can write

$$(\Delta q)^2 \equiv [q_J^2]_{av} - q^2 = \lim_{n\to 0} \left[\frac{1}{n(n-1)(n-2)(n-3)} \sum_{(\alpha\beta\gamma\delta)} q^{\alpha\beta} q^{\gamma\delta} \right.$$

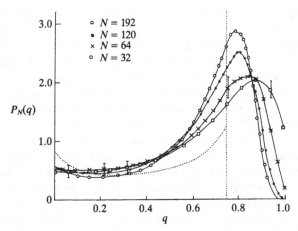

Figure 3.11: Data for $P_N(q)$ for the SK model for several sizes at $T = 0.4J$, $h = 0$. The distribution has been symmetrized, so only $q \geq 0$ is shown. The dotted line is the prediction of an approximate solution of Parisi's equations and consists of a delta function of weight $\frac{4}{7}$ at $q = q_{max} = 0.774$ and a continuous part with a finite weight down to $q = 0$ (from Young, 1983b).

$$- \left(\frac{1}{n(n-1)} \sum_{(\alpha\beta)} q^{\alpha\beta} \right)^2 \Bigg] \tag{3.80}$$

where the notation $(\alpha\beta\gamma\delta)$ in the sum means that all indices are unequal. So far this is quite general and does not depend on the particular form of replica symmetry breaking. We can now evaluate (3.80) for the Parisi parametrization of $q^{\alpha\beta}$, with the result

$$(\Delta q)^2 = \frac{1}{3} \left[\int_0^1 q^2(x) dx - \left(\int_0^1 q(x) dx \right)^2 \right] \tag{3.81}$$

Thus in the Parisi scheme, it is manifest that self-averaging of q (or χ) fails whenever $q(x)$ is not a delta-function. More generally, the same argument applies to $P_J(q)$ and other quantities derived from it.

On the other hand, one can see very simply that quantities such as the equilibrium energy, free energy, or magnetization, which do not involve intervalley correlations, will be self-averaging. One may well then ask the question: How can the equilibrium magnetization be self-averaging and the susceptibility χ_J, which is just its derivative, not be? The answer is that the different states and the free energy landscape are very sensitive to field. In our formal expression for the susceptibility (2.45) with (2.40) we have *not* let the valleys change with the external field. On the other hand, another susceptibility, defined as the limit of the difference of two equilibrium magnetization measurements at slightly different fields, divided by the field difference, *does* implicitly take changes in the free energy landscape into account, so these two susceptibilities are not the same. The former is not self-averaging; the latter is. Furthermore, the directly measurable quantity is the latter, so lack of self-averaging is difficult to observe.

Thus we have to be careful in all our quantities involving overlaps between different phases about whether the phases are taken with differing or the same fields (or temperatures, by the same argument). For example, our old expression for the average overlap, which occurs in the susceptibility, is taken with the fields and temperatures of the phases compared with each other exactly equal. But, on the other hand, one can show (Sompolinsky, 1984; Binder and Young, 1986) that

$$q' \equiv \lim_{h_1 \to h_2} \left[\frac{1}{N} \sum_{iab} P_a(h_1) P_b(h_2) m_i^a(h_1) m_i^b(h_2) \right]_{av} = q(0), \quad (3.82)$$

the minimum overlap, as suggested by the argument about extreme sensitivity of the free energy landscape to field. We will use this difference in the formulation of the dynamic theory in the next chapter.

An even more dramatic and celebrated feature than the lack of self-averaging which one discovers in the Parisi solution is *ultrametricity* (Mézard et al, 1984a,b; Rammal et al, 1986). This is the terminology for a very special hierarchical structure in the overlaps between the various phases. It is revealed by computing the joint distribution of the mutual overlaps between three randomly chosen states:

$$P_J(q_1, q_2, q_3) = \sum_{abc} P_a P_b P_c \delta(q_1 - q_{ab}) \delta(q_2 - q_{bc}) \delta(q_3 - q_{ca}) \quad (3.83)$$

In replica formalism the bond average of this quantity is

$$P(q_1, q_2, q_3) \equiv [P_J(q_1, q_2, q_3)]_{av}$$

$$= \frac{1}{n(n-1)(n-2)} \sum_{(\alpha\beta\gamma)} \delta(q_1 - q^{\alpha\beta}) \delta(q_2 - q^{\beta\gamma}) \delta(q_3 - q^{\gamma\alpha}) \quad (3.84)$$

and in the Parisi parametrization it becomes

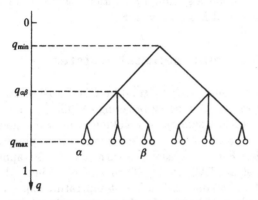

Figure 3.12: Sketch of the tree structure of ultrametric overlap. The overlap between a pair of states (represented by 'leaves' of the tree) depends only on how many levels one must go up to find a 'common ancestor'.

$$P(q_1, q_2, q_3) = \tfrac{1}{2}P(q_1)x(q_1)\delta(q_1 - q_2)\delta(q_3 - q_1)$$

$$+\tfrac{1}{2}[P(q_1)P(q_2)\theta(q_1 - q_2)\delta(q_2 - q_3) + \text{ permutations }] \qquad (3.85)$$

This result is very interesting. Suppose we take the 'distance' between two configurations in the N-dimensional spin configuration space to be the fraction of spins which are different in the two states, i.e. $d_{ab} = \tfrac{1}{2}(1 - q_{ab})$. Then it says that if we take any three states, either they are all mutually equidistant or two of the separations are equal and greater than the third. Upon a little reflection, this implies that the space of states has a sort of 'family tree' structure (Fig. 3.12) in which the degree of relatedness in the family is associated with the degree of overlap. A set with this kind of connectivity is called 'ultrametric'. It is quite remarkable that such a highly disordered system as the SK spin glass should possess such a rich structure in its states.

It is clear that the ultrametricity of the phases comes out in this theory because we put it in in the hierarchical structure (3.67) of the $q^{\alpha\beta}$ matrix. It would of course be desirable to establish this striking feature on an independent basis, and some evidence has been obtained in numerical simulations (Parga et al, 1984, Bhatt and Young, 1986a).

All our discussion in this chapter has been confined to the SK model; we have not considered the Viana–Bray infinite-range model with *very dilute* but strong bonds ($J_{ij} = O(1)$ instead of $O(N^{-\frac{1}{2}})$). However, the theoretical situation for this other kind of mean field model is more difficult than in the SK model. While it is known that the replica-symmetric solution is unstable, no stable broken-replica-symmetry solution has been found yet (de Almeida, De Dominicis and Mottishaw, 1988).

3.5   TAP equations and metastable states

In the preceding section we found a stable solution (the Parisi solution) of the SK model based on replica symmetry breaking. While we know that the physical meaning of replica symmetry breaking is broken ergodicity, the replica formalism is rather abstract, and some more direct physical insight would be desirable. For this reason we turn now to the approach introduced by Thouless et al (TAP, 1977). Their starting point, which is related to what we tried in Section 3.2, is a high-temperature expansion of the free energy for a *fixed* set of bonds J_{ij}. In Section 3.2, we developed a formal diagrammatic representation for equations for the local magnetizations $m_i = \langle S_i \rangle$ (for arbitrary-range interactions) and tried to study the behaviour of the bond average of m_i^2 in the infinite-range limit. In the TAP approach, one works instead with the infinite-range limit of these equations for a single sample, averaging over the different solutions and the distribution $P[J]$ only at a later stage of the calculation. A rigorous derivation of these self-consistent equations is not trivial, and we give only the result

$$m_i = \tanh[\beta(\sum_j J_{ij} m_j + h_i - \beta \sum_j J_{ij}^2 (1 - m_j^2) m_i)] \qquad (3.86)$$

and discuss the physical content. Equations (3.86) are N coupled nonlinear equations for the local magnetizations m_i, and their exact solution is a hopeless task. Their physical interpretation is rather simple: The first two terms on the right hand side describe the (site-dependent) conventional mean field as in (3.3)–(3.4). The third term is called the 'Onsager' or 'reaction' term and describes the contribution to the internal field from the spin S_i itself: The magnetization m_i at site i produces a mean field $m_i J_{ij}$ at site j, which induces a magnetization $\chi_{jj} m_i J_{ij}$ at site j and hence a mean field $J_{ij}^2 \chi_{jj} m_i$ at site i. Following the argument first made by Onsager (1936), the ordering of spin S_i is induced by the internal fields of the rest of the spins S_j *in the absence of S_i*, so this term has to be subtracted from the full mean field $\sum_j J_{ij} m_j$ in computing m_i. Use of the linear response relation (1.2) between the spin correlations and the susceptibility then leads to (3.86). The reaction term is quadratic in J_{ij}

and has a nonzero average value of $O(1)$, while the ordinary molecular field (the first term) has zero average (for a symmetric bond distribution) and an rms value also of $O(1)$. Thus in the SK model the reaction field is of the same order as the ordinary molecular field, and a correct mean field treatment must include both terms on the same footing, as we remarked at the end of Section 3.1. (In ferromagnets, in contrast, the reaction field is smaller than the (nonzero) mean molecular field by a factor $1/N$ (or $1/z$ for finite range lattice models).)

The Onsager correction to the naive molecular field can be seen as the first term in a systematic development in $1/N$ or $1/z$ (Nakanishi, 1981). It is exact in the thermodynamic limit of the SK model.

The TAP equations can be derived by variation ($\partial F / \partial m_i = 0$) of a free energy functional

$$
\begin{aligned}
F\{m_i\} = \quad & - \tfrac{1}{2} \sum_{ij} J_{ij} m_i m_j - \sum_i h_i m_i \\
& - \tfrac{1}{4}\beta \sum_{ij} J_{ij}^2 (1 - m_i^2)(1 - m_j^2)
\end{aligned}
$$

$$
+ \tfrac{1}{2} T \sum_i \{(1 + m_i) \ln[\tfrac{1}{2}(1 + m_i)] + (1 - m_i) \ln[\tfrac{1}{2}(1 - m_i)]\} \quad (3.87)
$$

which is just the Weiss mean field free energy, augmented by the last term in the first line. This term contains the reaction field effects.

An analytic solution of the TAP equations can be obtained for $h_i \to 0$ and $T > T_f$, where one can linearize. This leads to

$$
m_i = \beta \sum_j J_{ij} m_j + \beta h_i - (\beta J)^2 m_i \tag{3.88}
$$

with $\sum_j J_{ij}^2 = N[J_{ij}^2]_{av} = J^2$, replacing the sum over all sites by an average over the bond distribution $P[J]$. Equation (3.88) is solved by diagonalization of the symmetric matrix J_{ij}

$$
J_{ij} = \sum_{\lambda=1}^{N} \langle i | \lambda \rangle \langle \lambda | j \rangle \tag{3.89}
$$

with the eigenvalues J_λ and the real orthogonal eigenfunctions $\langle \lambda | i \rangle$:

$$
\sum_i \langle \lambda | i \rangle \langle i | \lambda' \rangle = \delta_{\lambda \lambda'}, \quad \sum_\lambda \langle i | \lambda \rangle \langle \lambda | j \rangle = \delta_{ij} \tag{3.90}
$$

One has the staggered magnetizations and fields

$$
m_\lambda = \sum_i \langle \lambda \mid i \rangle m_i, \quad h_\lambda = \sum_i \langle \lambda \mid i \rangle h_i \tag{3.91}
$$

and from (3.88)

$$m_\lambda[1 - \beta J_\lambda + (\beta J)^2] = \beta h_\lambda \tag{3.92}$$

For $N \to \infty$ the random matrix J_{ij} has a distribution of eigenvalues $\rho(J_\lambda)$ which obeys a semicircular law with the largest eigenvalue $J_\lambda^{max} = 2J$ (Mehta, 1967; Edwards and Jones, 1978):

$$\rho(J_\lambda) = \frac{1}{(2\pi J^2)}(4J^2 - J_\lambda^2)^{\frac{1}{2}} \tag{3.93}$$

The staggered susceptibility

$$\chi_\lambda = \frac{dm_\lambda}{dh_\lambda} = \frac{\beta}{1 - \beta J_\lambda + (\beta J)^2} \tag{3.94}$$

diverges for the largest eigenvalue J_λ^{max} at the freezing temperature $T_f = J$. Hence the TAP equations (3.86) lead to a phase transition at the same temperature as the SK theory. Note that without the Onsager term the divergence would be at $T = J_\lambda^{max} = 2J$, so the freezing temperature would have been overestimated by a factor 2.

The TAP equations can be considerably simplified in general if one replaces the Onsager term $\sum_J J_{ij}^2(1 - m_j^2)$ by

$$[J_{ij}^2]_{av} \sum_j (1 - m_j^2) = J^2(1 - \tilde{q}) \tag{3.95}$$

with

$$\tilde{q} = \frac{1}{N} \sum_j m_j^2 \tag{3.96}$$

We will later be able to identify $\tilde{q}$ with the single-phase spin glass order parameter q_{EA} (2.37). A simple estimate of the low-temperature limit of $\tilde{q}(T)$ can be obtained from the entropy $S(T)$. One has from the free energy (3.87) for $h_i = 0$ with $\partial F/\partial m_i = 0$

$$S(T) = -\frac{\partial F}{\partial T} = -\tfrac{1}{2}\beta^2 \sum_{ij} J_{ij}^2(1 - m_i^2)(1 - m_j^2)$$

$$-\tfrac{1}{2}\sum_i \{(1 + m_i)\ln[\tfrac{1}{2}(1 + m_i)] + (1 - m_i)\ln[\tfrac{1}{2}(1 - m_i)]\} \tag{3.97}$$

In the limit $T \to 0$ one has $m_i \to \pm 1$ and the second term in (3.97) vanishes. The first term is approximated by (3.95). This leads to

$$S(T = 0) = -\tfrac{1}{2}NJ^2 \lim_{T \to 0} \beta^2(1 - \tilde{q})^2 \tag{3.98}$$

The condition of a vanishing entropy at $T = 0$ is therefore that $1 - \tilde{q}$ vanish faster than T as $T \to 0$. Monte Carlo simulations indeed indicate

$$1 - \tilde{q}(T) = \alpha \left(\frac{T}{T_f}\right)^2, \qquad (T \to 0) \qquad (3.99)$$

If $\tilde{q}(T)$ is identified with q_{EA} ($= q(1)$ in Parisi's replica approach), this leads to the linear temperature dependence of the susceptibility $\tilde{\chi}(T)$ as $T \to 0$ that we saw in Fig. 3.9.

The constant α is related to the distribution $p(\tilde{h}_i)$ of (total) internal fields $\tilde{h}_i = \sum_j J_{ij} m_j$. TAP (1977) and Palmer and Pond (1979) found numerically that $p(\tilde{h})$ is *linear* in $\tilde{h}$:

$$p(\tilde{h}) = \frac{\tilde{h}}{h_0^2} \qquad (\tilde{h} \to 0, \ T \to 0) \qquad (3.100)$$

A relation between the coefficients α and h_0^{-2} can then be obtained as follows: If one defines $\tilde{q}$ by

$$\tilde{q} = \int_0^\infty d\tilde{h} \, m^2(\tilde{h}) p(\tilde{h}) \qquad (3.101)$$

one has at low T

$$1 - \tilde{q}(T) =$$

$$\int_0^\infty p(\tilde{h})[1 - m^2(\tilde{h})]dh \approx h_0^{-2} \int_0^1 (1 - m^2)\tilde{h}(m) \frac{d\tilde{h}}{dm} dm \qquad (3.102)$$

Thus, using the low-T limit (3.99) in the approximation (3.95) for the reaction term and writing the TAP equations (3.86) in the form

$$\beta \tilde{h}_i = \alpha m_i + \tanh^{-1} m_i \qquad (3.103)$$

we obtain

$$\alpha \left(\frac{T}{T_f}\right)^2 =$$

$$h_0^2 T^2 \int_0^1 (1 - m^2)(\alpha m + \tanh^{-1} m)\left(\alpha + \frac{1}{1 - m^2}\right) dm \qquad (3.104)$$

TAP argued that the physically relevant value of α was the one which minimized h_0 (the 'hole' dug in the distribution $p(\tilde{h})$ around $\tilde{h} = 0$ should be as narrow as possible); this gives $\alpha = 2(\ln 2)^{\frac{1}{2}} \approx 1.665$. Bray and

Moore (1979) used instead a marginal stability condition (discussed below), together with the equation corresponding to (3.101) for the fourth moment of m, to obtain an additional relation between α and h_0. This gives a different value $\alpha \approx 1.81$.

The solutions of the TAP equations are stationary points $\partial F/\partial m_i = 0$ of the free energy $F\{m_i\}$ (3.87). The stability of these solutions is governed by the Hessian

$$A_{ij} = \frac{\partial^2 \beta F}{\partial m_i \partial m_j} = -\beta J_{ij} - (\beta J_{ij})^2 m_i m_j$$

$$+\delta_{ij} \left[\sum_k (\beta J_{ik})^2 (1 - m_k^2) + \frac{1}{(1 - m_i^2)} \right] \tag{3.105}$$

The second term on the right hand side of (3.105) can be ignored, since from (3.2) J_{ij}^2 is of order N^{-1}. The third term is approximated as in (3.95) with (3.96).

The stability matrix (3.105) is connected with the susceptibility matrix $\tilde{\chi}_{ij}$ for a fixed set of bonds and for a single solution of the TAP equations in the following way. We have (for general field h_j)

$$\tilde{\chi}_{ij} = \frac{dm_i}{dh_j} = \frac{\partial m_i}{\partial h_j} + \sum_k \frac{\partial m_i}{\partial m_k} \frac{\partial m_k}{\partial h_j} \tag{3.106}$$

where the tilde indicates that we consider a single TAP solution with the order parameter $\tilde{q}$ (3.96). Differentiation of the TAP equations with respect to h_j then leads, with the definition (3.105) of the stability matrix, to

$$\beta \delta_{ij} = \sum_k A_{ik} \tilde{\chi}_{kj} \tag{3.107}$$

or, in matrix notation,

$$\tilde{\chi} = \beta \mathsf{A}^{-1} \tag{3.108}$$

The local susceptibility for a single TAP solution reads

$$\tilde{\chi}_{loc} = \frac{1}{N} \sum_i \tilde{\chi}_{ii} = \frac{\beta}{N} \sum_i \langle S_i^2 \rangle_c = \frac{\beta}{N} \sum_i (1 - m_i^2) = \beta(1 - \tilde{q}) \tag{3.109}$$

or

$$\tilde{\chi}_{loc} = \frac{\beta}{N} Tr(\mathsf{A}^{-1}) = \frac{\beta}{N} \sum_\lambda \frac{1}{r_\lambda} = \int_0^\infty dr \frac{\rho(r)}{r} \quad (N \to \infty) \tag{3.110}$$

where we have introduced the eigenvalues r_λ of the matrix A by means of the eigenvalue equation

$$(\mathsf{A} - r_\lambda \mathbf{1})\phi_\lambda = 0 \tag{3.111}$$

and the density of eigenvalues $\rho(r)$ for $N \to \infty$.

A stable solution must be a local minimum of the free energy, which requires that all eigenvalues r_λ be positive. For some solutions, there might be a minimum eigenvalue which is positive, while for other solutions the eigenvalue spectrum $\rho(r)$ could extend down to zero. In the latter case (which we shall argue occurs for the lowest-free-energy solutions on and below the AT line), there are modes which become 'soft' or 'massless'. Thus these solutions are marginally stable ('marginal') in the sense that one has a divergent (i.e. infinitely soft) response everywhere along and below this line. The physical quantity which exhibits this divergence is again the spin glass susceptibility (2.64)

$$\chi_{SG} = \frac{1}{N} \sum_{ij} \tilde{\chi}_{ij}\tilde{\chi}_{ji} = \frac{1}{N} \sum_i (\tilde{\chi}^2)_{ii} = \frac{\beta^2}{N} Tr(\mathsf{A}^{-2})$$

$$= \beta^2 \int_0^\infty dr \frac{\rho(r)}{r^2} \tag{3.112}$$

In order to show this we introduce the propagator

$$\mathsf{G}(r) = (r\mathbf{1} - \mathsf{A})^{-1} \tag{3.113}$$

with $\beta\mathsf{G}(0) = -\tilde{\chi}$ and with the eigenvalue density

$$\rho(r) = \frac{1}{N} \sum_\lambda \delta(r - r_\lambda) = \frac{1}{N\pi} Im \sum_\lambda \frac{1}{(r - r_\lambda - i\delta)}$$

$$= \frac{1}{N\pi} Im \, Tr\mathsf{G}(r - i\delta) \tag{3.114}$$

where $\delta = 0_+$ (Bray and Moore, 1979). The diagonal terms $G_{ii}(r)$ of $\mathsf{G}(r)$ are calculated by means of the so-called locator expansion. We define the 'locator' (essentially the diagonal terms of the matrix (3.1052))

$$f_i = \frac{1}{r - (\beta J)^2(1 - \tilde{q}) - (1 - m_i^2)^{-1}} \tag{3.115}$$

and expand $G_{ii}(r)$ in powers of βJ_{ij} (cf. Section 3.2)

$$G_{ii}(r) = [(f_i^{-1}\mathbf{1} + \beta J)^{-1}]_{ii}$$

$$= f_i - f_i \sum_j \beta J_{ij} f_j \beta J_{ji} f_i + f_i \sum_{jk} \beta J_{ij} f_j \beta J_{jk} f_k \beta J_{ki} f_i + \cdots \tag{3.116}$$

The dominant contributions to G_{ii} in the limit $N \to \infty$ are obtained by pairing the J_{ij}'s in all possible ways. We write (in generalization of the procedure we used to decouple the term $\sum_j J_{ij}^2 m_i$ in the linearized TAP equations (3.88))

$$\sum_j J_{ij}^2 f_j \approx J^2 \sum_j f_j \approx J^2 [f]_{av} \tag{3.117}$$

This self-averaging restricts our further calculation to fields and temperatures above the AT line or below if only a single valley is relevant (see Section 3.4). The calculation of $[f]_{av}$ indicated in Fig. 3.13 leads with $[G_{ij}]_{av} = \delta_{ij} [G]_{av}$ to

$$G_{ii}(r) = f_i - f_i^2 (\beta J)^2 [G]_{av} + f_i^3 (\beta J)^4 [G]_{av}^2 + \cdots$$

$$= \frac{1}{f_i^{-1} - (\beta J)^2 [G(r)]_{av}} \tag{3.118}$$

and to the self-consistency condition for $[G(r)]_{av}$

$$\frac{1}{N} \sum_i G_{ii}(r) = [G(r)]_{av} = \frac{1}{f_i^{-1} - (\beta J)^2 [G(r)]_{av}} \tag{3.119}$$

Equation (3.118) leads for $r = 0$ with (3.110), (3.114) and

$$[G(0)]_{av} = -\frac{1}{N} [Tr(\mathbf{A}^{-1})]_{av} = -(1 - \tilde{q}) \tag{3.120}$$

to the identity $G_{ii}(0) = -(1 - m_i^2)$ and to

$$f_i^{-1}(r) = r + (\beta J)^2 [G(0)]_{av} + G_{ii}^{-1}(0) \tag{3.121}$$

whence from (3.118)

$$G_{ii}(r) = \frac{G_{ii}(0)}{1 + G_{ii}(0)(r + (\beta J)^2 [G(0) - G(r)]_{av})} \tag{3.122}$$

We are interested in the density $\rho(r)$ for $r \to 0$. Expansion of (3.122) leads to

$$[\Delta G]_{av} \equiv \frac{1}{N} \sum_i (G_{ii}(0) - G_{ii}(r)) = [G^2(0)]_{av}(r + (\beta J)^2 [\Delta G]_{av})$$

$$- [G^3(0)]_{av}(r + (\beta J)^2 [\Delta G]_{av})^2 + \cdots \tag{3.123}$$

and in the limit $r \to 0$ to

$$[\Delta G]_{av} = \frac{r [G^2(0)]_{av}}{1 - (\beta J)^2 [G^2(0)]_{av}} \tag{3.124}$$

Figure 3.13: Graphs for the Green function G_{ii} in the thermodynamic limit. A dot connected to $2n$ lines carries a factor $(f_i)^{n+1}$. A shaded circle represents the average Green function $[G]_{av}$. Each loop then carries a factor $\beta^2 J^2 [G]_{av}$ (from Bray and Moore, 1979).

The density of eigenvalues $\rho(r)$ (3.114) extends for small r just to zero if the denominator of (3.124) vanishes:

$$1 = (\beta J)^2 [G^2(0)]_{av} = (\beta J)^2 [(1 - m_i^2)^2]_{av} \qquad (3.125)$$

This is exactly the condition for the AT line (3.27 or 3.65) if $\tilde{q}$ is identified with the solution q_{SK} of the SK equation (3.24). Thus the (single) TAP solution above the AT line indeed becomes marginally stable as the line is approached. We know already that the SK solution is unstable below the AT line. The solution of (3.125) then determines the *minimum* value of $\tilde{q}$ necessary for stability there. We argue that solutions with this value of $\tilde{q}$ will have the lowest free energies, so this $\tilde{q}$ can therefore be identified with the single-phase order parameter q_{EA}. This argument, together with the assumption of an internal field distribution (3.100) which is linear in the field, is just what Bray and Moore (1979) used to obtain the T-dependence of $\tilde{q}$ at low T that we described above.

Except in this low-T limit and right at the AT line, we do not yet have a way to evaluate the quantity $[m_i^4]_{av}$ which occurs in (3.125), so the utility of this equation for the determination of q_{EA} is limited. However, we will later be able to obtain an explicit expression for q_{EA} in zero field and slightly below T_f using the dynamical theory of Chapter 5.

The condition (3.125) leads with (3.123) to $\rho(r) \propto r^{\frac{1}{2}}$: There is no gap in the eigenvalue spectrum $\rho(r)$, and one has infinitely soft or 'massless' modes of the stability matrix. The investigation of the dynamics in Section 5.3 will show that these massless modes persist below the AT line and further support the identification of $\tilde{q}$ with q_{EA} (2.37). This is in agreement with numerical computation of $\rho(r)$ (Bray and Moore, 1979) and searches for TAP solutions to be discussed below.

A divergence similar to that in (3.124) appears in the spin glass suscep- tibility (3.112). In order to see this, we expand the definition (3.113) of the propagator G in powers of r

$$G(r) = -A^{-1} - rA^{-2} + \cdots \tag{3.126}$$

which leads with (3.112), (3.124) and

$$\frac{1}{N}Tr(G(0) - G(r)) \equiv [\Delta G]_{av} = r\beta^{-2}\chi_{SG} \tag{3.127}$$

to

$$\chi_{SG} = \frac{\beta^2[G^2(0)]_{av}}{1 - (\beta J)^2[G^2(0)]_{av}} \equiv \frac{\chi^{(2)}}{1 - J^2\chi^{(2)}} \tag{3.128}$$

Thus (3.125) also turns out to be the condition for the divergence of χ_{SG}. The numerator of (3.128) has a simple interpretation, as in (3.26):

$$\chi^{(2)} \equiv \beta^2[G^2(0)]_{av} = \beta^2[(1 - m_i^2)^2]_{av} = \frac{1}{N}\sum_i \chi_{ii}^2 \tag{3.129}$$

is the *local* spin glass susceptibility, which, in contrast to (3.128), does not diverge. The same is true for the local and total susceptibilities in an ideal ferromagnet. But while in an Ising ferromagnet the total susceptibility diverges only at $h \to 0$ and T_c; in a spin glass the divergence in χ_{SG} persists in finite field on and below the AT line. The dynamic theory discussed in section 5.3 will lead to results which reduce in the zero-frequency limit to those obtained here.

The marginal stability of TAP solutions can be checked numerically. Nemoto and Takayama (1985,1986) and Nemoto (1987) investigate numer- ically the stationary points of the TAP free energy $F\{m_i\}$ (3.87) as defined by minimizing $|\nabla F| \equiv [\sum_i (\partial F/\partial m_i)^2]^{\frac{1}{2}}$. For finite N they find solutions with $|\nabla F| = 0$ as well as solutions with $|\nabla F| \neq 0$. The latter turn out to be saddle points with $|\nabla F| \to 0$ as $N \to \infty$, which stick at the boundary of the region of stability as defined by the lowest eigenvalue $r_\lambda^{min} > 0$. In Fig. 3.14 this boundary is indicated by a dashed line. To the left of this stability limit r_λ^{min} would be negative, which is not permitted. The figure

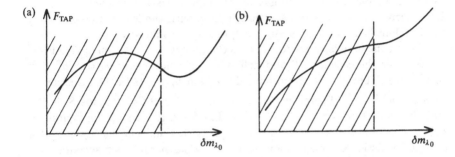

Figure 3.14: The free energy of Thouless, Anderson, and Palmer (TAP)
as a function of the mode $m_{\lambda min}$ when it has been minimized with respect
to magnetizations along all other eigenvector directions. Here $m_{\lambda min}$ is
the component of the magnetization of the spins along the eigenvector
corresponding to the smallest eigenvalue $r_{\lambda min}$ of the stability matrix
$\partial^2 F_{TAP}/\partial m_i \partial m_j$. The eigenvalue $r_{\lambda min}$ is negative on the left of the
stability limit, indicated by a broken line, and positive to the right.
(a) The case of a stable TAP solution, which is the minimum of F_{TAP}.
(A minimum within the shaded region would correspond to a solution for
which the TAP equations are invalid.)
(b) The situation when there is no TAP solution. As $N \to \infty$, the minimum in (a) approaches the stability limit and for (b) the value of ∇F
at the stability limit tends to zero (from Nemoto and Takayama, 1985;
Nemoto, 1987).

shows schematically the free energy (3.87) as a function of the mode $m_{\lambda_{min}}$
which belongs to the lowest eigenvalue r_λ^{min} of the stability matrix A with
$\partial F/\partial m_\lambda = 0$ for all other modes. In case (a) there is a stable TAP solution,
i.e a minimum of $F(m_{\lambda_0})$ with $r_\lambda^{min} > 0$, whereas in case (b) this minimum
with $|\nabla F| = 0$ tends to the point $r_\lambda^{min} = 0$ for $N \to \infty$, i.e. a marginally
stable saddle point.

Starting from the freezing temperature T_f, Nemoto and Takayama find
an increasing number of solutions with $|\nabla F| = 0$ or $|\nabla F| \to 0$ for $N \to \infty$,
80% of which are marginally stable. The eigenvalues of these marginally

stable solutions seem to vanish as a function of N like $N^{-2/3}$ (Nemoto and Takayama, 1986). We should stress that these results are based on a local stability analysis. It turns out that below T_f the TAP equations have a huge number N_s of such locally stable or marginally stable solutions (Tanaka and Edwards, 1980; De Dominicis et al, 1980; Bray and Moore, 1980). If the picture of a spin glass with many phases presented in Section 2.4 is correct, some of these solutions (those with the lowest free energies) can be identified with the 'states' or 'phases' introduced there.

We outline here how one formulates the analytic calculation of N_s. First we consider the case $T = 0$ and define a single-site energy $\epsilon_i = S_i h_i^{eff} \equiv S_i \sum_j J_{ij} S_j$. This energy is positive if the spin S_i points in the direction of its local field h_i^{eff}. Therefore the requirement $\epsilon_i > 0$ for all spins S_i is the stability condition against *single spin flips* for a given spin configuration $\{S_i\}$. It says nothing about stability against a simultaneous flipping of a cluster consisting of two or more spins, nor does it take into account that the local field itself changes during a spin flip.

The number of states subject to this condition is given by

$$N_s = Tr \int_0^\infty \prod_i d\epsilon_i \delta(\epsilon_i - \sum_j J_{ij} S_i S_j) \qquad (3.130)$$

The formal task now involves writing integral representations for the delta functions in (3.130) and evaluating the average over the distribution $P[J]$ of the bonds for large N by steepest descents. It turns out that N_s increases exponentially with the number N of spins. Hence $\ln N_s$ is proportional to N, i.e. an extensive quantity, so the average over the bond distribution should be performed on $\ln N_s$ rather than N_s itself. This would lead us back to introducing replicas, which the TAP approach is trying to avoid. However, it turns out that for most states (actually for all states with energy ϵ above a critical value ϵ_c)

$$[\ln N_s(\epsilon)]_{av} = \ln[N_s(\epsilon)]_{av} \qquad (3.131)$$

holds, and we can average N_s instead of $\ln N_s$.

The calculation of $[N_s]_{av}$ for the SK model is straightforward and leads to

$$[N_s]_{av} = \exp 0.1992N \qquad (3.132)$$

The average number of states $[N_s(\epsilon)]_{av}$ for a given energy ϵ can also be calculated analytically and leads to the function $g(\epsilon) = N^{-1} \ln[N_s(\epsilon)]_{av}$ indicated in Fig. 3.15. It turns out that only states with energy $\epsilon > \epsilon_c$ fulfil the condition (3.131). The broken curve indicates an extrapolation to states for which the 'direct' average is incorrect and where replicas should

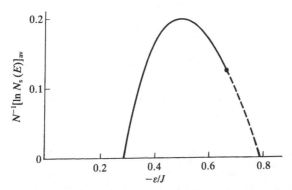

Figure 3.15: Average of the logarithm of the number of TAP solutions
for $h = T = 0$ as a function of energy. The total number of solutions is
dominated by the maximum value of the curve and is given by (3.132).
The broken line indicates an extrapolation to states for which the 'direct'
average $[N_s]_{av}$ is incorrect (from Bray and Moore, 1980).

be used. In fact the most important states (i.e. those which dominate
thermal averages) are these low-lying states. The corresponding part of the
function $g(\epsilon)$ can be calculated by means of the Parisi replica-symmetry-
breaking scheme that we used in Section 3.4.

At *finite temperatures* N_s can be defined as the number of solutions of
the TAP equations with $|\nabla F| = 0$ (or $|\nabla F| \to 0$ as $N \to \infty$). As mentioned
before, these solutions are *locally* stable or marginally stable but do not need
to be *global* minima of the free energy. The number of states $N_s(f)$ for a
given free energy per spin can again be calculated analytically (Bray and
Moore, 1980). One finds again a critical free energy f_c above which the
states are said to be 'uncorrelated', i.e. $[\ln N_s(f)]_{av} = \ln[N_s(f)]_{av}$, and
below which one has to introduce replicas. For $T \to 0$ the function $g(f)$
reduces to $g(\epsilon)$ (Fig. 3.15), and the total number of solutions approaches
(3.132).

The total number of solutions as a function of temperature is shown in
Fig. 3.16. For $T \geq T_f$ one has only the paramagnetic solution $m_i = 0$.
With decreasing temperature one has a sharp increase in the number of
solutions or a 'landscape' (Fig. 2.2) of the free energy with more and more

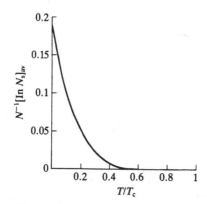

Figure 3.16: Logarithm of the total number of TAP solutions, divided by N, as a function of temperature (from Bray and Moore, 1983).

local minima. As we remarked in the discussion of this figure in Section 2.4, some of the barriers become infinitely high for $N \to \infty$, and the low-energy 'valleys' separated by such infinite barriers can be identified with the phases we discussed in the Parisi solution. The number N_0 of such solutions is considerably smaller than the number N_s. Indeed, Baldi and Baum (1986) have proved that ultrametricity does not allow N_0 to exceed N. TAP solutions which are not associated with the bottoms of these low-energy valleys could either correspond to high-energy valley bottoms (which would be stable forever but would not contribute to thermodynamics) or be metastable configurations within a valley (of either high or low energy).

There remains the thorny question of the relevance of these results to experiments. Do measurements on different timescales probe different TAP solutions? The answer is simple in principle for the infinite SK model: Any finite-time measurement probes a single phase or valley, since the barriers between valleys are infinite and the system can therefore never escape the valley it happens to find itself in. Within such a valley, there may be many metastable TAP solutions. Very short time measurements will only probe the neighbourhood of a single one of these, while measurements on successively longer timescales will probe more and more of the other TAP solutions in this valley, until an infinite-time measurement probes them all. For real systems with short range or RKKY forces, the current belief (see

Chapter 8) is that there are probably *not* many pure phases, just a pair of them as in a ferromagnet. However, it is certainly possible that qualitatively the same kind of relaxation that happens among TAP solutions within a valley of the SK model also happens among metastable configurations in the much bigger valleys in the short-range system. Very little theory has been done on these dynamics, in either the SK model or short-range ones, so we cannot be sure, but the general features of the experiments described in the introduction suggest strongly that we are seeing something of this sort. We will come back to this topic in greater detail in Chapter 9.

So far we have considered only Ising spins, but similar properties have been obtained for vector spins with infinte-range interactions (Bray and Moore, 1981). At first sight this might be surprising: In an Ising spin glass the necessary and sufficient condition for the existence of a metastable state was that the spin S_i at site i be parallel to its local field h_i^{eff} (see (3.130)). In vector systems this is a necessary but not a sufficient condition. A sufficient condition is the absence of negative eigenvalues of the Hessian or stability matrix (3.105), generalized for vector spins. One then finds a spectrum of states $[N_s(\epsilon)]_{av}$ which is similar to that shown in Fig. 3.15 for Ising spins. Again there is an upper limit for the energy ϵ, above which no such states exist, and a critical value ϵ_c below which $\ln[N_s(\epsilon)]_{av} \neq [\ln N_s(\epsilon)]_{av}$. Some of these states again must lie at the bottom of free energy valleys with infinite barriers between them, leading to nonergodicity, just as in the Ising case. This is consistent with the fact that the Parisi theory of Section 3.4 can be extended without essential change to vector spins (see (2.71-75) and Section 6.1).

Numerical computations are difficult to perform on the SK model because of the infinite-range interactions. We mentioned already the calculation of the internal field distribution $p(\tilde{h})$ (3.100), which agrees well with an exact solution of the Sompolinsky equations which will be discussed in Section 5.3 (Sommers and Dupont, 1984). A direct numerical solution of the mean field equations (3.86) *without* the Onsager term also yields interesting results. Fig. 3.17 shows the field-cooled and zero-field-cooled magnetizations $M_{fc}(T)$ and $M_{zfc}(T)$ obtained in this way. Both are rather similar to those measured in CuMn (Fig. 1.4). In addition, one observes hysteresis loops and thermoremanent and isothermal remanent magnetizations in these simulations which are similar to those in AuFe (Fig. 1.5) (Soukoulis et al, 1982, 1983a, 1983b). All these remanent effects are absent in the corresponding isotropic Heisenberg model, but can be restored by the addition of random anisotropy (Soukoulis et al, 1983b, 1984). An analytic treatment of the TAP equations without the Onsager term leads to a state with nontrivial broken ergodicity, but to a dynamical susceptibility $\chi''(\omega)$

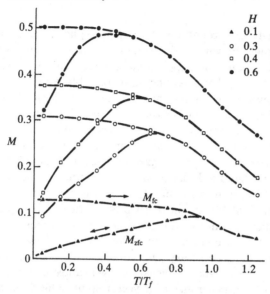

Figure 3.17:   Temperature dependence of field-cooled (fc) and zero-field-cooled (zfc) magnetization for various magnetic fields H (in units of J) where $T_f = 2J$. These are results of the numerical solution of the TAP equations (3.86) without the Onsager term (from Soukoulis and Grest, 1984).

which varies like $\omega^{\frac{1}{2}}$ for $\omega \to 0$ everywhere below T_f, in contrast to the behaviour we will find in the solution of the Sompolinsky equations (Bray, Sompolinsky and Yu, 1986).

3.6 Two soluble models

In this section we study briefly two models which are related to the SK model: the (infinite-range) spherical and random-energy models. They are both in a sense artificial: The spherical model is equivalent in most ways to taking an m-component SK model in the limit $m \to \infty$, while the random energy model is equivalent to an SK-like model with p-spin interactions, i.e.

$$H = - \sum_{\langle i_1 i_2 ... i_p \rangle} J_{i_1 i_2 ... i_p} S_{i_1} S_{i_2} \cdots S_{i_p} \tag{3.133}$$

instead of the usual two-spin interactions, in the limit $p \to \infty$.

The spherical model (Berlin and Kac, 1952) is of course familiar from the theory of phase transitions in nonrandom systems, where its importance lies in the fact that it is the simplest soluble model with nonclassical critical exponents (see, e.g. Ma, 1976). In the theory of spin glasses, the spherical model is of secondary importance, since it turns out not to display the novel phenomena of spin glass physics: broken ergodicity and replica symmetry breaking. The random-energy model, on the other hand, is simply soluble and does exhibit these properties. It therefore plays a role here somewhat analogous to that played by the spherical model in ordinary phase transition theory.

The spherical model

The original formulation of the spherical model for ferromagnetism was not as the $m \to \infty$ limit of the m-component spin model, but rather in terms of one-component spins which could take on any real values (not just ± 1), subject only to the so-called spherical constraint $\sum_i S_i^2 = N$. This is the formulation we follow here. Using the integral representation of the delta-function, the partition function can then be written (Kosterlitz et al, 1976)

$$
Z = \int_{-\infty}^{\infty} \prod_i dS_i \int_{c-i\infty}^{c+i\infty} \frac{dz}{2\pi i} e^{A[z,S]}
$$

$$
= \int_{-\infty}^{\infty} \prod_i dS_i \int_{c-i\infty}^{c+i\infty} \frac{dz}{2\pi i} \exp[z(N - \sum_i S_i^2)
$$

$$
+ \tfrac{1}{2}\beta \sum_{ij} J_{ij} S_i S_j] \tag{3.134}
$$

The Gaussian integrals can be carried out exactly; using the diagonalization (3.89) of the J_{ij} matrix we get

$$
Z = \frac{1}{2\pi i} \int_{c-i\infty}^{c+i\infty} dz \exp\left\{ N\left[z - \frac{1}{2N} \sum_\lambda \ln(z - \tfrac{1}{2}\beta J_\lambda) \right] \right\} \tag{3.135}
$$

In the large N limit this can be evaluated by steepest descents; the stationary point is the solution of

$$
1 = \frac{1}{N} \sum_\lambda \frac{1}{2z - \beta J_\lambda} = \int \frac{dJ_\lambda \rho(J_\lambda)}{2z - \beta J_\lambda} \tag{3.136}
$$

where $\rho(J_\lambda)$ is the density of eigenvalues J_λ. This equation just says that the spherical constraint is satisfied on the average: $1 = N^{-1} \sum_\lambda \langle S_\lambda^2 \rangle = N^{-1} \sum_i \langle S_i^2 \rangle$. So far this treatment is quite general. We now specialize to the infinite-range spin glass, using the eigenvalue density (3.93). Then (3.136) reduces to

$$z - [z^2 - (\beta J)^2]^{\frac{1}{2}} = (\beta J)^2 \tag{3.137}$$

In the high-temperature phase, the solution is very simple. (3.137) has the solution $z = \frac{1}{2}(1 + \beta^2 J^2)$ for $T > J = T_f$; the susceptibility is simply a Curie law since (3.136) is satisfied.

Below T_f there is no solution of (3.137); the exponent in (3.135) is maximized by keeping z at the rightmost branch point βJ. However, the lack of a solution of (3.136)–(3.137) means that the spherical constraint is violated: Evaluation of (3.136) for $T < T_f$ with $z = \beta J$ gives $\langle S_i^2 \rangle = T/T_f < 1$. To get a consistent theory we introduce a spontaneous magnetization $\langle S_{\lambda_0} \rangle = m_0$ in the mode with the largest eigenvalue $J_{\lambda_0} = 2J$ as follows. We express the general form (3.134) for the partition function in terms of the components S_λ, writing

$$S_{\lambda_0} = m_0 + \delta S_{\lambda_0} \tag{3.138}$$

in the contribution from the mode λ_0. The exponent is now

$$A[z, S] = zN - (z - \beta J)(m_0 + \delta S_{\lambda_0})^2 - \sum_{\lambda \neq \lambda_0} (z - \tfrac{1}{2}\beta J_\lambda) S_\lambda^2 \tag{3.139}$$

and if we evaluate it at the point $z = \beta J$ that maximizes A, the contribution from the magnetized mode vanishes, while the rest of A just gives the same result we obtained without the shift in S_{λ_0} This term then makes a contribution T/T_f to $\langle S_i^2 \rangle$, just as before, but we can now satisfy the constraint by choosing the value of the order parameter m_0 appropriately:

$$\langle S_{\lambda_0} \rangle = m_0 = N^{\frac{1}{2}} \left(1 - \frac{T}{T_f} \right) \tag{3.140}$$

Then we can simply evaluate the susceptibility

$$\chi_{loc} = \frac{\beta}{N} \sum_i \langle (S_i - \langle S_i \rangle)^2 \rangle = \frac{\beta}{N} \sum_{\lambda \neq \lambda_0} \langle S_\lambda^2 \rangle$$

$$= \frac{\beta}{N} \sum_{\lambda \neq \lambda_0} \frac{1}{2z - \beta J_\lambda} = \frac{1}{T_f} \tag{3.141}$$

Thus the susceptibility has exactly the same value as in the SK model, but the physics underlying this temperature dependence is entirely different. Here the reduction in χ from its high-temperature form below T_f is just analogous to that which occurs in an antiferromagnet below T_N. There is really just a single order parameter in this problem, $\langle S_{\lambda_0} \rangle$.

However, this problem is different from both the antiferromagnet and the SK model in that there is no transition in the presence of a uniform external field, i.e. no AT line. This can be seen as follows: When writing the exponent A in the partition function (3.134) with an extra term $\beta \sum_i h_i S_i = \beta \sum_\lambda h_\lambda S_\lambda$, one shifts the S_λ's to complete the square, giving an extra term

$$\delta A = \frac{1}{2} \sum_\lambda \frac{(\beta h_\lambda)^2}{2z - \beta J_\lambda} \tag{3.142}$$

in A, where $h_\lambda = h \sum_i \langle i | \lambda \rangle$ is the projection of the uniform external field onto mode λ. Then, after carrying out the integration over the S_i's and differentiating with respect to z to find the stationary point, we find that (3.136) is replaced by

$$1 = \frac{1}{N} \sum_\lambda \left[\frac{1}{2z - \beta J_\lambda} + \frac{(\beta h)^2}{(2z - \beta J_\lambda)^2} \right] \tag{3.143}$$

This extra term makes it always possible to find a solution with $z > \beta J$. Since the previous formal manifestation of the transition lay in the disappearance of the solution, it is therefore suppressed for any finite field.

One can also study the $m \to \infty$ limit of an m-component SK model, and in zero field the results turn out to be the same (de Almeida et al, 1978). In finite field, however, we expect that the m-vector model will have a transition: The components of the spin perpendicular to the field direction will not couple to the field to lowest order in their order parameters, so for small field they will have a T_f just like the zero-field case. (The same argument applies to finite m; we will discuss this transition in Chapter 5.)

To summarize, the spherical model does provide an example of a very simply soluble model, but does not give any insight into the properties which are special in spin glasses. By contrast, the model we now turn to turns out to be at least as simple to solve and to exhibit these properties in a nearly generic fashion.

The random energy model

We start by introducing the notion of the statistics of energy levels in a random statistical mechanical system. The simplest quantity one can consider is the probability distribution of the energy of a spin configuration. Denoting the spin configuration by $S_i^{(1)}$, or just $S^{(1)}$ for short, the probability density that the energy of this configuration is E can be written

$$P(E) = [\delta(E - H[S^{(1)}])]_{av} \tag{3.144}$$

For symmetrically distributed random interactions, moreover, this average is independent of the configuration for which it is evaluated. This is because for each set of J_{ij}'s and configuration $\{S_i\}$ we can make a gauge transformation like (2.82) to an arbitrary other configuration $\{S_i'\}$ with a different set of bonds $J_{ij}' = S_i S_i' J_{ij} S_j S_j'$ which occurs with equal probability in the bond distribution (provided only the distribution is symmetric). Therefore the average over the bond distribution for one spin configuration and that for another one are just rearrangements of the same sum of terms, so we can calculate (3.144), say, for the configuration with all spins up. For the SK model, for example, we get simply

$$P(E) = \frac{1}{(N\pi J^2)^{\frac{1}{2}}} \exp\left(-\frac{E^2}{NJ^2}\right) \tag{3.145}$$

For the random p-spin model (3.133) with a Gaussian distribution of couplings

$$P(J_{i_1\ldots i_p}) = \left(\frac{N^{p-1}}{\pi p!}\right) \exp\left[-\frac{(J_{i_1\ldots i_p})^2 N^{p-1}}{J^2 p!}\right] \tag{3.146}$$

we obtain exactly the same answer. Thus at this level all random p-spin models behave identically.

To see the difference between them, it is necessary to go to higher-order statistics, e.g. the joint probability $P(E_1, E_2)$ that a chosen pair of configurations $S_i^{(1)}$ and $S_i^{(2)}$ have energies E_1 and E_2, respectively:

$$P(E_1, E_2) = [\delta(E_1 - H[S^{(1)}])\delta(E_2 - H[S^{(2)}])]_{av} \tag{3.147}$$

Just as $P(E)$ was independent of the configuration for which it was defined, $P(E_1, E_2)$ depends only on how many spins in the two configurations are the same, i.e. on the overlap

$$q^{(1,2)} = \frac{1}{N}\sum_i S_i^{(1)} S_i^{(2)} \tag{3.148}$$

The result is

$$P(E_1, E_2, q) = \frac{1}{[N\pi J^2(1+q^p)(1-q^p)]^{\frac{1}{2}}}$$
$$\times \exp\left[-\frac{(E_1+E_2)^2}{2N(1+q^p)J^2} - \frac{(E_1-E_2)^2}{2N(1-q^p)J^2}\right] \tag{3.149}$$

Now the important thing which Derrida (1980, 1981) pointed out is that in the limit $p \to \infty$, unless $q = 1$ (i.e. unless the two configurations are the same except for a finite number of spins), $q^p \to 0$, so

$$P(E_1, E_2, q) \rightarrow P(E_1)P(E_2) \tag{3.150}$$

Similarly, all higher order joint distributions factorize in a corresponding way. In this sense, the large p limit of the random p-spin model (3.133) can be characterized as a 'random energy' model: The single energy-level distribution $P(E)$ contains all thermodynamic information about the system.

The thermodynamics is most easily found in the microcanonical ensemble. The average number of levels with energy E is just $P(E)$ multiplied by the total number 2^N of levels:

$$N(E) = \frac{1}{(\pi N J^2)^{\frac{1}{2}}} \exp N \left[\ln 2 - \left(\frac{E}{NJ} \right)^2 \right] \tag{3.151}$$

We notice a critical dependence on E (for large N): for $|E| > E_0 \equiv NJ(\ln 2)^{\frac{1}{2}}$, there is an exponentially large number of levels, and therefore a finite entropy

$$S(E) = N \left[\ln 2 - \left(\frac{E}{NJ} \right)^2 \right] \tag{3.152}$$

On the other hand, for $|E| > E_0$, there are *no* levels left in the thermodynamic limit, and therefore no entropy. The situation is illustrated in Fig. 3.18.

Using $T^{-1} = \partial S / \partial E$, one finds that the critical temperature is

$$T_f = \frac{J}{2(\ln 2)^{\frac{1}{2}}} \tag{3.153}$$

Temperatures below T_f correspond formally to being in the range of E where there are no states. Physically, the system must freeze into the last available state at $|E| = E_0$ at these temperatures. Thus the low-T phase is thermodynamically trivial and quite naturally has zero entropy.

These arguments are readily generalized to include an external field h; now the Gaussian distribution (3.145) is just shifted so as to have a nonzero mean Mh, where M is the magnetization. One finds a line of transitions as a function of h analogous to the AT line in the SK model, and, in the low-field limit, a temperature-independent susceptibility in the low-T phase.

The resemblance between the properties of this simple model and those found (with so much more effort) for the SK model is of course quite intriguing. The similarities and differences were illuminated very elegantly by Gross and Mézard (1984), by solving the $p \rightarrow \infty$ limit of the p-spin model (3.133) explicitly in the replica formalism. They find broken replica

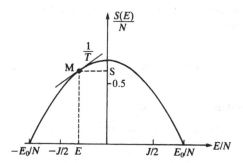

Figure 3.18: Entropy (logarithm of the level density (3.151)) for the random energy model (from Derrida, 1980).

symmetry, but only one level of the Parisi hierarchical replica-symmetry-breaking scheme is necessary, rather than the infinite number required in the SK ($p = 2$) case. In this sense the random-energy model is much simpler than the SK model; Gross and Mézard call it 'the simplest spin glass'. On the other hand, it also exhibits more generic behaviour: The transition is 'first order' in the sense that the plateau value $q(1) = q_{EA}$ of the Parisi order function jumps discontinuously at T_f. Such discontinuous behaviour occurs, in fact, for all $p > 2$; the SK case is the only one with a continuous q_{EA}.

We now outline their solution, beginning with the replicated partition function for general p:

$$[Z^n]_{av} = \int \prod dJ_{i_1\ldots i_p} P(J_{i_1\ldots i_p})$$

$$\times Tr_S \left[\exp \beta \sum_\alpha \left(\sum_{i_1 < \ldots < i_p} J_{i_1\ldots i_p} S_{i_1}^\alpha \cdots S_{i_p}^\alpha + h \sum_i S_i^\alpha \right) \right] \quad (3.154)$$

The averaging over the bonds is easily carried out, leading to

$$[Z^n]_{av} = Tr_S \exp \left[\tfrac{1}{4} \beta^2 N \left(n + \sum_{\alpha \neq \beta} (\frac{1}{N} \sum_{i=1}^{N} S_i^\alpha S_i^\beta)^p \right) \right]$$

$$+\beta h \sum_{i\alpha} S_i^\alpha \Bigg] \tag{3.155}$$

where we have set $J = 1$ for convenience. With the help of a set of Lagrange multipliers $\lambda^{\alpha\beta}$, this can be cast into the form

$$[Z^n]_{av} = e^{\frac{1}{4}nN\beta^2} \times$$

$$\int_{-\infty}^{\infty} \prod_{\alpha<\beta} dq^{\alpha\beta} \int_{-i\infty}^{i\infty} \prod_{\alpha<\beta} \frac{d\lambda^{\alpha\beta}}{2\pi} \exp[-NG(q^{\alpha\beta}, \lambda^{\alpha\beta})] \tag{3.156}$$

where

$$G(q^{\alpha\beta}, \lambda^{\alpha\beta}) = -\tfrac{1}{4}\beta^2 \sum_{\alpha\neq\beta} (q^{\alpha\beta})^p + \tfrac{1}{2} \sum_{\alpha\neq\beta} \lambda^{\alpha\beta} q^{\alpha\beta}$$

$$- \ln tr_S \exp[\tfrac{1}{2} \sum_{\alpha\neq\beta} \lambda^{\alpha\beta} S^\alpha S^\beta + \beta h \sum_\alpha S^\alpha] \tag{3.157}$$

In the high-temperature phase, we expect a replica-symmetric solution, so, setting $q^{\alpha\beta} = q$ and $\lambda^{\alpha\beta} = \lambda$ $(\alpha \neq \beta)$, we find, in the $n \to 0$ limit,

$$\frac{1}{n}G(q, \lambda) = \tfrac{1}{4}\beta^2 q^p - \tfrac{1}{2}\lambda q$$

$$- \int \frac{dz}{\sqrt{2\pi}} e^{-\frac{1}{2}z^2} \ln[2\cosh(\lambda^{\frac{1}{2}} z + \beta h)] \tag{3.158}$$

and the saddle point equations are

$$\tfrac{1}{2}\beta^2 pq^{p-1} = \lambda \tag{3.159}$$

and

$$q = \int \frac{dz}{\sqrt{2\pi}} e^{-\frac{1}{2}z^2} \tanh^2(\lambda^{\frac{1}{2}} z + \beta h) \tag{3.160}$$

So far this is for general p. We now look at the random-energy limit $p \to \infty$, where the only solution is $\lambda = 0$ and

$$q = \tanh^2 \beta h \tag{3.161}$$

This is the obvious paramagnetic solution, and the free energy is the same as that obtained from the Derrida solution in the simple random-energy formulation given above.

Actually, unlike the corresponding SK solution, this solution is locally stable against AT-like fluctuations at all temperatures. That is, the onset of replica symmetry breaking in this model is not signaled by an AT instability. This fact is connected with the discontinuous nature of the transition that we mentioned above.

Now we turn to the replica-symmetry-breaking solution, which will be relevant at low temperatures. Here we take without proof Gross and Mézard's result that in the $p \to \infty$ limit, only the first stage of Parisi replica symmetry breaking is necessary, so only five variational parameters need appear in our G: q_0 and q_1 (the values of $q^{\alpha\beta}$ in the off-diagonal and diagonal blocks, respectively), λ_0 and λ_1 (the corresponding values of the elements of the λ matrix). and m $(= m_1)$, the size of the blocks along the diagonal. We get

$$\frac{1}{n}G = \ln 2 - \tfrac{1}{4}\beta^2[mq_0^p + (1-m)q_1^p] + \tfrac{1}{2}[m\lambda_0 q_0 + (1-m)\lambda_1 q_1]$$

$$-\tfrac{1}{2}\lambda_1 + \frac{1}{m}\int \frac{dz_0}{\sqrt{2\pi}}e^{-\frac{1}{2}z_0^2}\ln\int\frac{dz_1}{\sqrt{2\pi}}e^{-\frac{1}{2}z_1^2}\times$$

$$\cosh^m[\lambda_0^{\frac{1}{2}}z_0 + (\lambda_1 - \lambda_0)^{\frac{1}{2}}z_1 + \beta h] \qquad (3.162)$$

Again, the saddle point equations are fairly simple at $p \to \infty$. From $\partial G/\partial q_i = 0$ we find

$$\lambda_i = \tfrac{1}{2}\beta^2 p q_i^{p-1} \qquad (3.163)$$

If these equations are to be satisfied with $q_0 < q_1 \leq 1$, we need to take $\lambda_0 = 0$, $q_1 = 1$ and $\lambda_1 = \infty$. Then differentiation of (3.162) with respect to λ_0 gives

$$q_0 = \tanh^2(\beta m h) \qquad (3.164)$$

and differentiation with respect to m yields

$$m^2\beta^2 = 4[\ln 2 + \ln\cosh(m\beta h) - m\beta h\tanh(m\beta h)] \qquad (3.165)$$

From this equation we can see that $m\beta$ is equal to a constant, independent of temperature, i.e. $m \propto T$. We can find the constant of proportionality by recognizing that this solution is acceptable only for $m \leq 1$. Thus $m\beta = \beta_c$, where β_c is found from (3.165) at the limit $m = 1$:

$$\beta_c^2 = 4[\ln(2\cosh(\beta_c h) - \beta_c h\tanh(\beta_c h)] \qquad (3.166)$$

In zero field, β_c will be recognized as the inverse of the critical temperature T_f (3.153) where the entropy vanished. In terms of T_f, we have $m = T/T_f$. In fact, the correspondence extends to finite field as well.

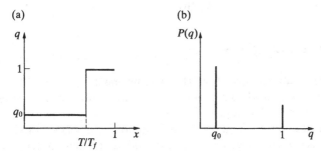

Figure 3.19: (a) $q(x)$ for the random energy model
(b) $P(q)$ for the random energy model.

The restricted nature of the replica symmetry breaking in this case means that the Parisi order function $q(x)$ consists of two flat portions at values $q_0 = \tanh^2(\beta_c h)$ and $q_1 = 1$, with a discontinuous jump between them at $x = m = T/T_f$ (see Fig. 3.19a). Thus the overlap distribution function $P(q)$ for the random energy model consists of a pair of delta-functions (Fig. 3.19b):

$$P(q) = \frac{T}{T_f}\delta(q - \tanh^2 \beta_c h) + \left[1 - \frac{T}{T_f}\right]\delta(q - 1) \qquad (3.167)$$

The peak at $q = 1$ tells us that the self-overlap of an individual state (q_{EA}) is as large as it possibly could be; this is in accord with our earlier picture of the system as completely frozen in a single configuration, with no thermal fluctuations, below T_f.

The other peak at q is equal to the overlap between different possible frozen configurations, and we see that this vanishes in zero field. That is, the different configurations are completely uncorrelated. In finite field, the peak moves to a value of q equal to the square of the net magnetization of the system; that is, the overlap is the minimum possible value, given the fact that the two configurations have the same magnetization. So this really is the simplest kind of broken ergodicity possible, with minimally correlated phases.

Finally, we comment on the fact that everywhere below T_f, the self-overlap or Edwards–Anderson order parameter is exactly 1, undergoing a discontinuous jump from 0 at T_f. In this sense, the transition could be called first order, as we mentioned above. However, one must remember that thermodynamic quantities involve integrals over x, and since the jump appears at T_f only at the single point $x = 1$, the transition is actually of second order in the formal thermodynamic sense. We will see in the next section that the generalization of the SK spin glass to Potts spins also exhibits this discontinuous behaviour.

3.7   The Potts glass

The foregoing example shows that the spin inversion symmetry possessed by Ising models may lead to rather special and exceptional properties not shared by more general spin-glass-like systems. Since there exist in nature systems such as quadrupole glasses which do not have this symmetry, it is interesting to see what one can learn about models of this more general sort. In this section we investigate this question, again at the mean field level, for a class of models called 'Potts glasses' (Elderfield and Sherrington, 1983a; Erzan and Lage, 1983; Gross et al, 1985). We will see that one can indeed find richer behaviour in these models than we found in the SK model. Their properties are in a sense intermediate between it and the random-energy model, in that both continuous and discontinuous behaviour in $q(x)$ can occur. At the end of the section we will return to the question of what these results tell us about systems we are more likely to encounter in the laboratory, such as quadrupole glasses.

A Potts model is a generalization of the two-state Ising model to an arbitrary number s of (discrete) states: $p_i = 1, \ldots s$, instead of $S_i = \pm 1$. Two Potts 'spins' p_i and p_j interact in the following way: Supposing the interaction strength is J_{ij}, the energy is equal to $-J(s-1)$ if the two spins are in the same state $(p_i = p_j)$ and equal to $+J_{ij}$ if they are in different states $(p_i \neq p_j)$:

$$H = -\tfrac{1}{2} \sum_{ij} J_{ij} \bigl(s \delta_{p_i p_j} - 1 \bigr) \tag{3.168}$$

(This particular form is convenient because it makes the average interaction energy zero as in the Ising model; thus the average energy vanishes in the high-T limit.) The case $s = 2$ is equivalent to an Ising model, of course, but the general case lacks inversion symmetry.

In a Potts ferromagnet, the ordered state is one in which a particular one of the s different states is more probable than the others on all sites. This order is measured by the order parameter $m_r = \langle \delta_{p_i r} \rangle - 1/s$, for one particular state r. In a Potts glass, we have random J_{ij}'s, and we look for a state with randomly frozen m_r's, characterized by a spin-glass-like order parameter

$$q_{rr'} = [(\langle \delta_{p_i r} \rangle - 1/s)(\langle \delta_{p_i r'} \rangle - 1/s)]_{av} \qquad (3.169)$$

This quantity has the symmetry $q_{rr'} = q(\delta_{rr'} - 1/s)$. It can thus be characterized completely by the single number q. In replica formalism, q becomes a matrix in replica indices:

$$q^{\alpha\beta} = \langle \delta_{p^\alpha p^\beta} \rangle - 1/s \qquad (3.170)$$

where the thermal average is with a replica effective Hamiltonian, as in (2.59–60). The term $1/s$ is just the correlation we would have if the configurations were completely random, so this has to be subtracted in (3.168) as in the expressions for m_r and $q_{rr'}$.

We can now write down a Landau expansion for the effective Hamiltonian in the same way we did for the SK model in (3.66). The important difference is that now we do in general get a term proportional to $\sum_{\alpha\beta}(q^{\alpha\beta})^3$, because the argument for its absence in the Ising case was based on inversion symmetry. Explicitly,

$$G[\mathbf{q}] = \lim_{n \to 0} \frac{s-1}{2n} \left[\theta tr\mathbf{q}^2 - \frac{1}{3}tr\mathbf{q}^3 - \frac{s-2}{6}\sum_{\alpha\beta}(q^{\alpha\beta})^3 \right.$$

$$\left. + \frac{y(s)}{6}\sum_{\alpha\beta}(q^{\alpha\beta})^4 \right] \qquad (3.171)$$

There are also other fourth-order terms (more of them, in fact, than there were in the SK case), but (as was the case there) they do not add any qualitatively new features to what we find from (3.171), so we again omit them. The values of the coefficients come from a systematic formal expansion of the exponent in the integral which gives the partition function when one goes over from the original p_i variables to the $q^{\alpha\beta}$'s, in analogous fashion to (3.34–37). We have from (3.66) that $y(s) = -1$ for $s = 2$, but $y(s)$ changes sign for $s = s^* = 2.8$ and is positive for larger s.

Having exhibited the effective Landau functional which is responsible for all the new features one finds in the Potts glass, we will now simply summarize the main features of the solutions. The reader should be able to verify many of these results by suitably generalizing the derivation given in Section 3.4 of the Parisi solution of the SK model.

We recall that it was the y term which was responsible for replica symmetry breaking in the SK model. Thus the presence of the analogous term at the cubic level in the Potts glass leads us to guess that replica symmetry

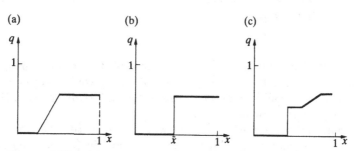

Figure 3.20: $q(x)$ in the Potts glass:
(a) for $s < s^*$
(b) for $s > s^*$, $T_2 < T < T_f$
(c) for $s > s^*$, $T < T_2 < T_f$.

breaking should be an even stronger effect here than there. And this is correct: One finds that the replica-symmetric state is more unstable than it was in the SK model, in that the negative eigenvalue of the stability matrix (3.59) is linear, rather than quadratic, in $T_f - T$.

Nevertheless, for $s < s^*$, the general features of the replica-symmetry-breaking solution are rather similar to those of the SK model. Fig. 3.20a shows the order parameter function $q(x)$ obtained in a Parisi parametrization of $q^{\alpha\beta}$ for s in this range. It rather resembles the SK solution in finite field, except that the first flat part of $q(x)$ is at $q = 0$ instead of a positive value. That is, in terms of the overlap distribution $P(q)$, (Fig. 3.21a) the corresponding delta-function spike is at zero overlap; many of the phases are uncorrelated with each other.

When y changes sign, however, the situation changes qualitatively. Now no Parisi–like solution with a continuous $q(x)$ is possible any more. What does work is a solution similar in form to that we presented for the random-energy model, with just one level of replica symmetry breaking and therefore a $q(x)$ with a discontinuous jump (Fig. 3.20b). The plateau value (i.e. the self-overlap q_{EA}) is proportional to $(T_f - T)/(4 - s)$, and the jump point $\bar{x} = (s - 2)/2$ (i.e. the ratio of the coefficients of the two cubic terms in (3.171)) for T near T_f. (Note however, that here, unlike in the random-energy model, the self-overlap is less than unity.) The corresponding $P(q)$

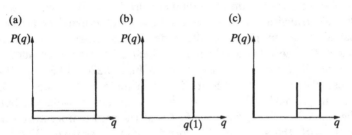

Figure 3.21: $P(q)$ in the Potts glass:
(a) for $s < s^*$
(b) for $s > s^*$, $T_2 < T < T_f$
(c) for $s > s^*$, $T < T_2 < T_f$.

is shown in Fig. 3.21b.

As $s \to 4$, the jump point $\bar{x}(T \to T_f^-) \to 1$. For larger s, the transition is *discontinuous*. (Note by comparison that the critical value of s above which the *ferromagnetic* Potts transition becomes discontinuous is $s = 2$.) Since a Landau expansion can only be used when the magnitude of the order parameter is small, the general case of a discontinuous transition cannot be analysed using (3.171). However, for s just a little above 4, the discontinuity is small, and one can use (3.171), finding a discontinuity $q(1, T_f) \propto (s - 4)$. The transition temperature T_f is higher than the temperature where the quadratic coefficient θ in (3.171) vanishes, by an amount $\propto (s - 4)^2$.

One can also solve the full problem in the limit $s \to \infty$; the result is just like that we found in the random-energy model; that is, the discontinuous jump in $q(1)$ at T_f assumes the maximum value of unity and the jump point $\bar{x} = T/T_f$.

As in the SK model, the free energy of the Potts glass states in all the above cases is *higher* than that of the analytic continuation of the paramagnetic phase. In the case where the transition is discontinuous, the two free energies cross at T_f. It seems paradoxical that the physical state should be the one of higher free energy when, as one can verify for $s > 4$, the eigenvalues of the stability matrix of both states are all positive. Nevertheless, we are forced to this conclusion if we want the physical state to change smoothly as s is varied through $s = 4$, since below $s = 4$ we have a continuous transition and must choose the nontrivial solution because the paramagnetic one has an AT instability. A plausible resolution lies in the possibility of a *nonperturbative* instability of the paramagnetic state below T_f, though this has not been demonstrated yet.

Gross et al (1985) also showed that for $s > s^*$, the Potts glass system also has a second transition at a temperature $T_2 < T_f$. They were able to study this transition systematically in a Landau-like framework only for s just above s^* by going to fifth order in the expansion in powers of $q^{\alpha\beta}$. Below T_2, one finds a solution in which $q(x)$ has a continuous part with positive slope for $x > \bar{x}$, as shown in Fig. 3.20c. This transition is continuous, like that of the SK model, even for $s > 4$. Away from the immediate neighbourhood of s^*, no explicit solution is available, but one infers the existence of a T_2 from the fact that the zero-temperature entropy of the state with the step–function form of $q(x)$ is negative. (For $s < s^*$, moreover, this 'T_2' transition is the only one, as we have seen above.)

The physical interpretation of this second transition for $s > s^*$ is that each of the many uncorrelated phases which become stable below T_f splits into an ultrametrically organized manifold of correlated states, in the same way that the single SK phase splits into many when one crosses the AT line. One sees in Fig. 3.21 that the delta-function spike in $P(q)$ at the self-overlap value which occurred between T_2 and T_f (Fig. 3.21b) now splits into two spikes at the minimum and maximum overlaps between pairs of the new phases, plus a continuous part in between (Fig. 3.21c).

Physical realizations of Potts glasses are unknown to us, but the Landau-like treatment of quadrupolar glasses, which do exist (albeit in short-range versions), is qualitatively the same as that of the Potts case. The difference lies in the particular values of the expansion coefficients and, therefore, of the critical values of the number of components where the qualitative character of the solution changes (analogous to s^* and 4 in the Potts case). Specifically, if the quadrupole variables are written (do not confuse the quadrupole Q with the spin glass order parameter q)

$$Q_{i,\mu\nu} = S_{i\mu}S_{i\nu} - \delta_{\mu\nu} \tag{3.172}$$

where $\mathbf{S}(x)$ is an m-component classical spin of unit length, and one looks for an isotropic quadrupole glass phase characterized by

$$q = \sum_{\mu\nu}[\langle (Q_{i,\mu\nu})^2 \rangle]_{av}, \tag{3.173}$$

the value of m above which $q(x)$ becomes discontinuous is $m^* \approx 2.67$, while the m above which the transition is discontinuous is $m_c \approx 3.4$.

One can of course only speculate on whether anything like these exotic properties of infinite-range models might be observable in physical quadrupole glasses. The lesson of this section is simply that glassy systems without inversion symmetry possess much richer possible properties than systems with this symmetry, a potentially useful warning to the wise experimentalist.

4

Introduction to dynamics

So far we have considered only the statics of spin glasses. However, the nonergodicity discussed in Sections 3.4 and 3.5 for the SK model suggests that the dynamics of spin glasses is rather unusual, at least below the freezing temperature. One expects transitions over energy barriers between the metastable states discussed in Section 3.5, leading to a new class of very slow relaxation times, some of which become infinite, at least in MFT. In this chapter we introduce the basic models and formal techniques necessary for describing the dynamics of spin glasses, corresponding to what we did for statics in Chapter 2. We focus on Ising systems here, both for the sake of simplicity and because, as we have mentioned, the anisotropic interactions found in most systems which one would expect to be Heisenberg spin glasses make them look rather Ising-like in many of their properties.

4.1 The Glauber model

The classical Ising model has no inherent dynamics, so to make a dynamical model one has to couple the spins to an additional 'heat bath' which induces spin flips (Glauber, 1963; Suzuki and Kubo, 1968). For metallic spin glasses this heat bath can be identified with the conduction electrons, which produce single-spin flip processes with the impurity spins via the exchange interaction J_{sd}. For a single magnetic impurity this leads to the so-called Korringa relaxation with relaxation time τ_0.

The dynamics of the spins are assumed to be Markoffian (each spins 'knows' only the state of all the other spins immediately before it flips; the spins have no 'memory') and are described by a master equation for the probability $P(S_1 \ldots S_N, t)$ of finding a spin configuration $S_1, \ldots S_N$ at time t. In addition one assumes that only a *single* spin is flipped at a given time. In real materials, one expects the that there may also be clusters of spins which flip simultaneously. The assumption of single-spin flips thus leads to slower processes than are observed in reality. This is especially true for RKKY spin glasses, in which nearest-neighbour spin pairs have a binding energy which is large compared to T_f. However, such effects lie outside the scope of MFT, since for infinite-range interactions the existence of well-separated clusters of this sort is not meaningful.

The master equation for single-spin flips reads

$$\frac{d}{dt} P(S_1 \ldots S_N, t) = -\sum_i W(S_i \rightarrow -S_i) P(S_1 \ldots S_i \ldots S_N, t)$$

$$+ \sum_i W(-S_i \rightarrow S_i) P(S_1 \ldots -S_i \ldots S_N, t) \tag{4.1}$$

The spin-flip transition probability $W(S_i \rightarrow -S_i)$ depends on the configuration of the surrounding spins and must obey the principle of detailed balance to ensure thermal equilibrium:

$$W(S_i \rightarrow -S_i) P_{eq}(S_1 \ldots S_N) =$$

$$W(-S_i \rightarrow S_i) P_{eq}(S_1 \ldots -S_i \ldots S_N) \tag{4.2}$$

with the canonical equilibrium probability

$$P_{eq}(S_1 \ldots S_N) = Z^{-1} \exp(-\tfrac{1}{2}\beta \sum_i \epsilon_i), \tag{4.3}$$

$$\left(Z = Tr \exp(-\tfrac{1}{2}\beta \sum_i \epsilon_i) \right)$$

Here $\epsilon_i = S_i h_i^{eff}$ is the local energy and $h_i^{eff} = h_i + \sum_j J_{ij} S_j$ is the effective local field defined in Section 3.5. Equations (4.2) and (4.3) lead to

$$\frac{W(S_i \rightarrow -S_i)}{W(-S_i \rightarrow S_i)} = \frac{\exp(-\beta\epsilon_i)}{\exp(\beta\epsilon_i)} = \frac{1 - S_i \tanh \beta\epsilon_i}{1 + S_i \tanh \beta\epsilon_i} \tag{4.4}$$

A possible solution of (4.4) introduced by Glauber is

$$W(S_i \rightarrow -S_i) = (2\tau_0)^{-1}(1 - S_i \tanh \beta\epsilon_i) \tag{4.5}$$

where $\tau_0(T)$ is the relaxation time of independent spins in zero field. We define the time-dependent expectation values for the spin S_i:

$$\langle S_i(t) \rangle = \sum_{\{S\}} S_i P(S_1 \dots S_N, t) \tag{4.6}$$

where the sum is taken over all spin configurations. One obtains from the master equation (4.1) and the Glauber transition rate (4.5) the equation of motion

$$\tau_0 \frac{d}{dt} \langle S_i \rangle = -(\langle S_i \rangle - \langle \tanh \beta h_i^{eff} \rangle) \tag{4.7}$$

For a one-dimensional ferromagnet with a periodic magnetic field $h_i(t)$ (4.7) can be solved exactly. In spin glasses and in higher dimensions no exact solution is possible.

A simple mean field approximation is obtained by the replacement

$$\langle \tanh \beta h_i^{eff} \rangle \rightarrow \tanh \langle \beta h_i^{eff} \rangle \tag{4.8}$$

Then the equation of motion reads

$$\left(1 + \tau_0 \frac{d}{dt} \right) \langle S_i \rangle = \tanh \beta \langle \bar{h}_i^{eff} \rangle \tag{4.9}$$

In the static limit, (4.9) just reduces to the standard Weiss mean field equations.

It is instructive to solve the mean field equation of motion for a ferromagnet above the Curie temperature T_c and in a small external field, where one can linearize. In k-space we then obtain

$$-\tau_0 \frac{d}{dt} m_k = [1 - \beta J(\mathbf{k})] m_k - \beta h_k \tag{4.10}$$

This leads with $h_k(t) = h_k e^{-i\omega t}$ to a susceptibility

$$\chi(\mathbf{k}, \omega) = \frac{dm_k}{dh_k} = \frac{\beta}{1 - i\omega\tau_0 - \beta J(\mathbf{k})} \tag{4.11}$$

For mode $\mathbf{k}$, the relaxation time is

$$\tau_k = \frac{\tau_0}{1 - \beta J(\mathbf{k})} \tag{4.12}$$

In a ferromagnet, $J(\mathbf{k})$ is largest at $\mathbf{k} = 0$, so the uniform mode has the largest χ and the longest relaxation time. As one approaches $T_c = J(0)$, this relaxation time diverges proportional to the uniform susceptibility, a simple example of 'critical slowing down' near a second-order phase transition. We will see that spin glasses also exhibit critical slowing down.

As in statics, this mean field calculation is only a qualitative guide to the singular behaviour of relaxation processes near a critical point. Fluctuation effects beyond MFT again give effects such as different values of critical exponents. We will discuss effects of this sort in spin glasses in Chapter 8.

We can now do the corresponding calculation for the SK Glauber model. Now one has in the static limit, in addition to the conventional mean field term h_i^{eff}, the Onsager term (the last term in the TAP equations (3.86)), so (4.9) would lead to the wrong transition temperature. Hence we define the local effective field

$$\bar{h}_i^{eff} = h_i + \sum_j J_{ij} s_j - \beta J^2 (1 - \tilde{q}) \qquad (4.13)$$

In the static limit, this leads back to the TAP equations. Compared to the statics, (4.13) involves an additional approximation: The Onsager term contains the local *static* zero-field susceptibility $\chi_{jj} = \beta(1-m_j^2)$ (see Section 3.5), and it is not obvious which susceptibility enters into the dynamics and for finite fields. We will use (4.9) with h_i^{eff} replaced by (4.13) only for vanishing external field and $T > T_f$, where $\tilde{q} = 0$ and our theory yields all essential features correctly. A more general treatment for this model can only be formulated by means of the path integral formulation of Sommers (1987), and will not be considered here.

To linearize the equation of motion we now use the transformation (3.89–91) to eigenstates of J_{ij} instead of Fourier transforming. The equation of motion for the 'staggered' magnetization m_λ then becomes

$$-\tau_0 \frac{d}{dt} m_\lambda = [1 - \beta J_\lambda + (\beta J)^2] m_\lambda - \beta h_\lambda \qquad (4.14)$$

which generalizes (3.92). This leads with $h_\lambda(t) = h_\lambda e^{-i\omega t}$ to the staggered susceptibility

$$\chi_\lambda(\omega) = \frac{dm_\lambda}{dh_\lambda} = \frac{\beta}{1 - i\omega\tau_0 - \beta J_\lambda + (\beta J)^2} \qquad (4.15)$$

The relaxation time τ_λ^{max} of the slowest mode (which has eigenvalue $J_\lambda^{max} = 2J$) diverges as one approaches $T_f = J$ like its susceptibility χ_λ^{max}:

$$\tau_\lambda^{max} \propto \frac{1}{(T - T_f)^2} \qquad (4.16)$$

Thus we have found critical slowing down in the SK spin glass analogous to that in the mean field ferromagnet (4.12).

The local susceptibility

$$\chi(\omega) = \frac{1}{N} \sum_{ii} \chi_{ii}(\omega, T) = \frac{1}{N} \sum_\lambda \chi_\lambda(\omega, T) \qquad (4.17)$$

contains the entire spectrum of relaxation times. Using again the distribution of eigenvalues $\rho(J_\lambda)$ (3.93) and the Hilbert transform

$$P \int_{-a}^{a} dx \frac{(a^2 - x^2)^{\frac{1}{2}}}{y - x} = \pi[y - (y^2 - a^2)^{\frac{1}{2}}] \tag{4.18}$$

leads (setting $T_f = 1$) to

$$\chi(\omega) = \tfrac{1}{2}\beta\{T^2(1 - i\omega\tau_0) + 1 - [(T^2(1 - i\omega\tau_0) + 1)^2 - 4T^2]^{\frac{1}{2}}\} \tag{4.19}$$

(Kinzel and Fischer, 1977; Fischer, 1983c).

The result (4.19) reduces at high temperature or frequency to the relaxation of independent spins

$$\chi(\omega, T) = \frac{\beta}{1 - i\omega\tau_0}, \qquad (T \gg T_f \text{ or } \omega\tau_0 \gg 1) \tag{4.20}$$

and at the freezing temperature with $\chi(\omega) = \chi'(\omega) + i\chi''(\omega)$ to $\chi'(0) = 1$ and to the dissipative contribution

$$\chi''(\omega) \propto \omega^{\frac{1}{2}} \tag{4.21}$$

This defines another 'dynamical critical exponent' $\nu = \tfrac{1}{2}$.

Thus MFT reveals a broad spectrum of relaxation times and critical slowing down at T_f. A qualitatively similar spectrum is actually found in metallic spin glasses, extending from the Korringa time $\tau_0 \approx 10^{-13}$s to macroscopic times, though it differs in detail from the MFT spectrum.

The Glauber model is the basis of nearly all Monte Carlo simulations, as discussed in the following section.

4.2 Monte Carlo calculations

Equilibrium quantities are expressed in statistical mechanics as averages over all the configurations of the system, with Boltzmann weights. These are obviously impractical to evaluate exactly for reasonably sized systems, since the number of states to be averaged over grows exponentially with the size of the system. If one tried to replace the exact average over all states by an average over a much smaller number of states chosen randomly with uniform *a priori* probability, the computation would still be impractical. The reason is that most of the states generated in this way would have very high energies and so would not make any important contribution to the thermal average. Most of the computation time would be wasted computing the value of the quantity in question in irrelevant states.

Metropolis et al (1953) showed that it is more useful to start with a certain state and generate all others recursively by stochastic transitions, defined by a transition probability $W(X \to X')$ between states $X = \{S_1, \ldots S_N\}$ and $X' = \{S_1', \ldots S_N'\}$ in which only a few spins are flipped. In most cases one assumes single spin-flips at randomly chosen lattice sites. The probabilities $W(X \to X')$ and $W(X' \to X)$ are connected by the condition of detailed balance. Except for the fact that the 'time' in this dynamical evolution is discrete rather than continuous, this is just the dynamics of the Glauber model of the preceding section. This means that the Glauber model can be solved to essentially arbitrary accuracy by such a computation. Such a tool is of course extremely valuable in testing approximate analytic theories, as well as in ascertaining how well the Glauber model describes particular experimental systems.

It is obvious that Monte Carlo simulations of spin glasses based on Glauber dynamics can be plagued by difficulties similar to those encountered in the interpretation of data on real materials: One has extremely long relaxation times near and below T_f, and it may be difficult to reach equilibrium or even to know whether one has reached it. Thus one may effectively have broken ergodicity within the available computer time, and time averages over this interval may not give equilibrium results even if there really is no true (infinite-time) broken ergodicity. An example of such a situation is shown in Fig. 4.1. There the spin autocorrelation function $C(t) = [\langle S_i(t) S_i(0) \rangle]_{av}$, which should go over to the spin glass order parameter q_{EA} at $t = \infty$, is plotted for the nearest-neighbour EA model with a symmetric Gaussian bond distribution for dimensionalities 2, 3 and 5 (Stauffer and Binder, 1981). At very long times one observes large fluctuations of $C(t)$, particularly at low temperatures. In the many-valley picture, the system gets stuck for a long time in a single valley if the adjacent free energy barriers are too high.

One possible way to get around this problem is to consider small systems of various sizes, all small enough that the system can reach all states many times within the 'measuring time'. The extrapolation to $N \to \infty$ should then yield the equilibrium average. (We have already seen an example of this in Fig. 3.11.) The difficulty with this approach is that even with the fastest computers available today one is restricted to fairly small systems below T_f.

Simulations of the SK model are particularly difficult and consumptive of computer time. This is because the coupling of each spin to every other one makes the calculation of the internal field h_i^{eff} which determines the transition probabilities (4.5) a lengthy one. Nevertheless, Monte Carlo calculations have yielded important results here, providing important checks on the static Parisi theory as well as information about its dynamics.

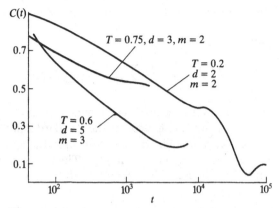

Figure 4.1: Semilogarithmic plot of the spin autocorrelation function $C(t)$ for the XY model ($m = 2$) and the Heisenberg model ($m = 3$) with a symmetric Gaussian distribution of nearest neighbour interactions for the dimensions $d = 2, 3$ and 5. At very late times, huge fluctuations occur from run to run, and an irregular behaviour results due to insufficient statistics.

For short range systems in 2 and 3 dimensions, Monte Carlo and other numerical methods are even more important, since so far we have neither a soluble theory nor even a systematic expansion in a small parameter to work with in such a system. We will examine the results of such calculations both in Chapters 5 and 6, where they will be compared with mean field theory to see to what extent short-range systems exhibit mean-field-like properties, and in Chapter 7, where we will be concerned with short-range systems in their own right.

4.3 Soft spins and Langevin models

The Glauber dynamics discussed in the preceding sections is extremely simple above the freezing temperature and well suited to Monte Carlo simulations. However, its extension to low temperatures is complicated since perturbation theory calculations are not straightforward to carry out. For this reason, it is often convenient to use instead a Langevin model, for which systematic perturbation theory formalisms are well developed (Martin et al, 1973; Ma and Mazenko, 1975; Ma, 1976; Bausch et al, 1976; De Dominicis and Peliti, 1978). These models also have the advantage in random systems that no replicas are needed (De Dominicis, 1978). They are formulated in terms of continuous rather than discrete variables, so one must relax the length constraints $S_i = \pm 1$ on the spins and introduce 'soft' spins σ_i with $-\infty < \sigma_i < \infty$ (Hertz and Klemm, 1979; Sompolinsky and Zippelius, 1981, 1982).

A convenient Hamiltonian for soft spins is

$$\beta H = \tfrac{1}{2} \sum_{ij} (r_0 \delta_{ij} - \beta J_{ij}) \sigma_i \sigma_j + \tfrac{1}{4} u \sum_i \sigma_i^4 - \beta \sum_i h_i \sigma_i \qquad (4.22)$$

where the interactions extend over z nearest neighbours with $z = N \to \infty$ in the SK model. For $u = -r_0 \to \infty$, one recovers the standard Ising model from (4.22). As for the Glauber model, we need to couple the spins to a thermal reservoir in order to get a dynamical model. This is most simply done by means of a Langevin equation (Ma, 1976; Hohenberg and Halperin, 1977)

$$\Gamma_0^{-1} \partial_t \sigma_i = -\frac{\partial \beta H}{\partial \sigma_i} + \xi_i(t)$$

$$= \beta \sum_j J_{ij} \sigma_j - r_0 \sigma_i - u \sigma_i^3 + \beta h_i + \xi_i(t) \qquad (4.23)$$

where $r_0 \Gamma_0 = \tau_0^{-1}$ is the relaxation rate of independent spins and the effect of the heat bath is represented by the time-varying Gaussian random noise field $\xi_i(t)$ driving the system. $\xi_i(t)$ is a Gaussian random variable with zero mean and variance

$$\langle \xi_i(t) \xi_j(t') \rangle = 2\Gamma_0^{-1} \delta_{ij} \delta(t - t') \qquad (4.24)$$

Here the brackets mean average over the noise. The physical picture of this model is of a particle with coordinates σ_i which feels a force pulling it downhill in a potential well $\beta H[\sigma]$ in the presence of friction (in the overdamped limit, where the Newtonian inertial term proportional to $\ddot{\sigma}_i$ can be ignored).

The fact that the same quantity Γ_0 apears both as the friction coefficient in the equation of motion (4.23) and in the expression (4.24) for the variance of the Langevin noise is called the 'Einstein relation' and is necessary in order that the model relax to equilibrium at temperature T.

The Ising model with a heat bath (with either soft or hard spins) contains only relaxational modes. Spin glasses with sufficiently small anisotropy should behave like vector spins with intrinsic dynamics (the precession of each spin around the direction of the local field acting on it). However, even the quantum-spin-dynamical Heisenberg model does not describe a spin glass completely, since it does not contain the exchange interaction J_{sd} between conduction electrons and localised spins. If the influence of this 'heat bath' is sufficiently strong, all spin waves become overdamped. In this case a vector model with Langevin dynamics like (4.23) leading to purely relaxational modes might be useful. (If the influence of the heat bath is weak, one must include the spin precessional dynamics which lead, e.g., to spin waves. In the long-wavelength, low-frequency limit, one can apply hydrodynamics or extensions thereof, as we do in Section 7.2.)

We now give a brief sketch of the functional integral formulation of the model (4.23)–(4.24). This gives a practical basis for carrying out systematic perturbation theory. Furthermore, it is a useful formal framework, within which one can make contact with statics in approximation schemes like MFT.

The quantities we are most interested in are the time-dependent two-spin correlation function

$$C_{ij}(t - t') = [\langle \sigma_i(t)\sigma_j(t')\rangle]_{av} \tag{4.25}$$

and the linear response function

$$T\chi_{ij}(t - t') \equiv G_{ij}(t - t') = \left[\frac{\partial \langle \sigma_i(t)\rangle}{\partial h_j(t')}\right]_{av} \tag{4.26}$$

where $\langle \cdots \rangle$ means averaging over the noise $\xi_i(t)$ introduced in (4.23). We will also need the Fourier transforms

$$G_{ij}(\omega) = \int_0^\infty dt e^{i\omega t}G_{ij}(t) , \qquad C_{ij}(\omega) = \int_{-\infty}^\infty dt e^{i\omega t}C_{ij}(t) \tag{4.27}$$

For the response function $G_{ij}(\omega)$ the Kramers–Kronig relations

$$Re\, G_{ij}(\omega) = -P\int_{-\infty}^\infty \frac{d\omega'}{\pi} \frac{Im G_{ij}(\omega')}{\omega - \omega'} ,$$

$$Im\, G_{ij}(\omega) = P\int_{-\infty}^\infty \frac{d\omega'}{\pi} \frac{Re G_{ij}(\omega')}{\omega - \omega'} \tag{4.28}$$

hold, since $G_{ij}(z)$ for complex z is an analytic function in the upper half z-plane. We now want to calculate these correlation and response functions by means of functional integrals.

To derive the formalism, we start with the fact that the Langevin noise $\xi_i(t)$ in (4.23) is Gaussian and uncorrelated in space and time. Making a discretization of time into little intervals of size $\Delta t \ll \tau_0$, its distribution is therefore

$$w[\xi] = \left(\frac{4\pi}{\Gamma_0 \Delta t}\right)^{-\frac{1}{2}(T/\Delta t)} \exp\left[-\frac{1}{4}\Gamma_0 \sum_{i,n} \xi_i^2(t_n)\Delta t\right] \tag{4.29}$$

where T is the length of the total time interval and n labels the discretized times. Taking the continuum limit gives us the distribution *functional*

$$w[\xi(t)] = W^{-1} \exp\left[-\tfrac{1}{4}\Gamma_0 \sum_i \int dt \xi_i^2(t)\right] \tag{4.30}$$

Here W is the obvious normalization factor obtained by integrating over all the $\xi_i(t)$:

$$W = \int D\xi_i(t) \exp\left[-\tfrac{1}{4}\Gamma_0 \sum_i \int dt \xi_i^2(t)\right] \tag{4.31}$$

Such a quantity, indicated by the notation $\int D\xi_i(t)$, is called a *functional integral* (or path integral). It is formally defined by

$$\int D\xi_i(t) \equiv \lim_{\Delta t \to 0} \int_{-\infty}^{\infty} \prod_{i,t_n} d\xi_i(t_n) \tag{4.32}$$

We now write $\xi_i(t)$ in terms of $\sigma_i(t)$, using the equation of motion (4.23), in equation (4.31) for W and change the integration variables from the $\xi_i(t)$ to the $\sigma_i(t)$:

$$W = \int D\sigma_i(t) J[\sigma] \times$$

$$\exp\left[-\tfrac{1}{4}\Gamma_0 \sum_i \int dt (\Gamma_0^{-1}\partial_t\sigma + \partial(\beta H)/\partial\sigma_i)^2\right] \tag{4.33}$$

where the Jacobian of the transformation of variables can be shown (De Dominicis and Peliti, 1978) to be expressible in the form

$$J[\sigma] = \exp\left[-\tfrac{1}{2}\sum_i \int dt \frac{\partial^2(\beta H)}{\partial\sigma_i^2(t)}\right]$$

$$= \exp\left[-\tfrac{1}{2}\sum_i \int dt (r_0 + 3u\sigma_i^2(t))\right] \tag{4.34}$$

The Jacobian plays a technical role in ensuring causality when one develops a perturbation theory for this formalism, but for us at this point the crucial observation is that it does not depend on the J_{ij}. This will be important in spin glasses when we come to averaging over the bond distribution. The quantity W will come to play a role similar to that played by the partition function in the equilibrium formalism, with the important difference that it does not depend on the value of the bonds J_{ij}, and so may be formally averaged over bonds without replicas.

The use of the generating functional W (4.33) is made technically easier if we exploit the Gaussian integral identity (3.10), introducing an auxiliary field $\hat{\sigma}_i(t)$:

$$W[l_i(t), \hat{l}_i(t)] = \int D\sigma_i(t)D\hat{\sigma}_i(t)J[\sigma]\times$$

$$\exp\{A[\sigma, \hat{\sigma}] + \sum_i \int dt(l_i(t)\sigma_i(t) + i\hat{l}_i(t)\hat{\sigma}_i(t))\} \qquad (4.35)$$

with the 'action'

$$A[\sigma, \hat{\sigma}] =$$

$$\sum_i \int dt[-\Gamma_0^{-1}\hat{\sigma}_i^2(t) - i\hat{\sigma}(t)(\Gamma_0^{-1}\partial_t\sigma_i(t) + \partial(\beta H)/\partial\sigma_i(t))] \qquad (4.36)$$

Here we have also introduced formal external ('source') fields $l_i(t)$ and $\hat{l}_i(t)$, coupled to σ_i and $\hat{\sigma}_i$, respectively, so that the (cumulant) correlation functions of the $\sigma_i(t)$ and $\hat{\sigma}_i(t)$ can be expressed as derivatives of $\ln W$:

$$\left[\frac{\delta^n\delta^m \ln W}{\delta\hat{l}_1(\hat{t}_1)\dots\delta l_m(t_m)}\right]_{l_i=\hat{l}_i=0} = \langle i\hat{\sigma}_1(\hat{t}_1)\dots\sigma_m(t_m)\rangle_c \qquad (4.37)$$

where the brackets mean averages weighted as in (4.35) for vanishing source fields.

The meaning of correlation functions obtained in this way for, say, a pair of σ's $\langle\sigma_i(t)\sigma_j(t')\rangle_c$ is evident. But what about averages containing $\hat{\sigma}$'s? The answer is that such quantities can be used as formal representations of response functions, e.g. from (4.23), (4.35) and (4.36), we obtain

$$\langle i\hat{\sigma}_i(t')\sigma_i(t)\rangle = \frac{\delta\langle\sigma_i(t)\rangle}{\delta\beta h_j(t')} \qquad (4.38)$$

That is, a factor $i\hat{\sigma}_i$ in a correlation function in this formalism acts like $\delta/\delta\beta h_i$. Causality requires that all response functions vanish for any of the times $\hat{t}_i$ larger than all t_j. In addition, one can show that

$$\langle i\hat{\sigma}_i(t)i\hat{\sigma}_j(t')\rangle = 0 \qquad (4.39)$$

for all values of the time arguments. One can see that this is true for the present model if $u = 0$ from the absence of a term proportional to σ_i^2 in

(4.36) and verify that it remains true term by term in perturbation theory in u.

We see that this formalism is highly analogous to what one has in many other problems in quantum field theory and statistical mechanics (such as the static Ising problem described in Section 3.2). We now sketch the derivation of the diagrammatic perturbation theory for calculating the response and correlation functions. For simplicity we omit the random part of $A[\sigma, \hat{\sigma}]$ containing the bonds J_{ij} and concentrate on dealing with the nonlinearity represented by the parameter u in the resulting symmetric single-site problem.

We begin by collecting the terms of quadratic order from the action (4.36):

$$A_0 = \sum_i \int dt[\Gamma_0^{-1}\hat{\sigma}^2(t) - i\hat{\sigma}(t)(\Gamma_0^{-1}\partial_t\sigma(t) + r_0\sigma(t))] \qquad (4.40)$$

(We have dropped site indices because we are now dealing with a single-site problem.) By itself, this part of the problem is exactly soluble because the exponent of the functional integral is a quadratic form; inverting the matrix of coefficients gives the response function $\langle i\hat{\sigma}(\omega)\sigma(-\omega)\rangle$

$$g_0(\omega) = \frac{1}{r_0 - i\omega\Gamma_0^{-1}} \qquad (4.41)$$

and the correlation function

$$c_0(\omega) = \langle \sigma(\omega)\sigma(-\omega)\rangle = \frac{2\Gamma_0^{-1}}{(\omega/\Gamma_0)^2 + r_0^2} \qquad (4.42)$$

To develop perturbation theory in the $ui\hat{\sigma}\sigma^3$ term in (4.36), we expand W (or its derivatives with respect to the auxiliary fields $l(t)$ and $\hat{l}(t)$) in powers of u. We obtain expressions involving powers of factors of σ and $i\hat{\sigma}$ at various times, integrated over the 'free' evolution probability $\exp A_0$. Like other forms of field theory, the present one lends itself very naturally to a diagrammatic formulation, which is sketched in Fig. 4.2. The response functions $\langle \sigma(t)i\hat{\sigma}(t')\rangle$ are represented by directed lines (with double arrows going from t' to t), while the correlation functions $\langle \sigma(t)\sigma(t')\rangle$ are indicated by lines with arrows coming in at both ends. At a u vertex (indicated by a dot), one $i\hat{\sigma}$ line (from a response function) comes in and three σ lines go out (to response or correlation functions). In the presence of a finite external field h, there is an extra diagrammatic building block, a short line with a single incoming arrow, which stands for $\langle \sigma \rangle$, the mean magnetization.

From these one can construct in the conventional fashion diagrams for the terms in the perturbation series for $G(\omega)$ and $C(\omega)$. By dividing the

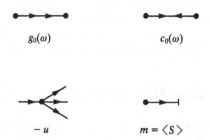

$g_0(\omega)$ $c_0(\omega)$

$-u$ $m = \langle S \rangle$

Figure 4.2: Diagrams for the (unaveraged) response and correlation functions, the bare vertex, and $\langle S \rangle = m$.

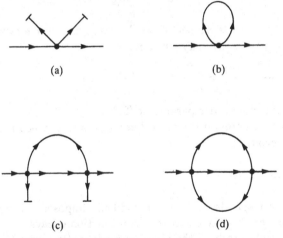

(a)   (b)

(c)   (d)

Figure 4.3: (a–d) Self-energy diagrams.

diagrams for G into reducible and irreducible ones (ones which can or cannot be disconnected by cutting a single g_0 line) one obtains in the standard way the Dyson equation

$$G^{-1}(\omega) = g_0^{-1}(\omega) + \Sigma(\omega) \tag{4.43}$$

A few of the low-order diagrams for Σ (irreducible G diagrams with the incoming and outgoing g_0 lines removed) are shown in Fig. 4.3.

It is straightforward to include the bond terms we have omitted from this discussion. The resulting diagrams can then be averaged over in the same way we did in the perturbation expansions we used for static problems in Sections 3.2 and 3.5. Alternatively, one can formally average W over the bond distribution and get an effective nonrandom problem, as we did via replicas for statics in Section 3.3. This is the approach we will take for the dynamics of the soft-spin SK model in the next chapter.

4.4   The fluctuation-dissipation theorem

There is a very useful generalization to dynamics of the fluctuation-response relation of equilibrium statistical mechanics (1.10). This is called the fluctuation-dissipation theorem (FDT) (Callen and Welton, 1951; for a general introduction and discussion see Kadanoff and Martin, 1962). For the classical models we consider in this book, it can be written

$$T\frac{\chi_{ij}(\omega) - \chi_{ij}(-\omega)}{i\omega} = \int_{-\infty}^{\infty} dt e^{i\omega t}(\langle S_i(0)S_j(t)\rangle - \langle S_i\rangle\langle S_j\rangle) \quad (4.44)$$

This form is valid for any given set of bonds; averaging over a symmetric bond distribution gives as the only surviving term

$$\frac{G''(\omega)}{\omega} = \frac{2T\chi_{ii}''(\omega)}{\omega} = \int_{-\infty}^{\infty} dt e^{i\omega t}(C_{ii}(t) - q) \quad (4.45)$$

with $q(T)$ the equilibrium order parameter (2.40).

The FDT can be written in the time domain in a way which makes its physical meaning clear:

$$\int_0^t G(t')dt' = C(0) - C(t) \quad (4.46)$$

That is, the system responds in time to a suddenly imposed external field in exactly the same as the way the correlation function decays.

This theorem applies to both the Glauber model of Section 4.1 and the soft-spin Langevin model of the preceding section. In the Glauber model, a necessary condition for its validity is the detailed balance condition (4.2). In the Langevin model, the corresponding condition is that the nonnoise part of $\partial_t \sigma$ be expressible as the derivative of some effective Hamiltonian, as in (4.23). The reader may easily check that the free response and correlation functions (4.41) and (4.42) of the version of the model discussed in the preceding section obey the FDT and that it is satisfied order by order in the perturbation theory described there.

If there is no broken ergodicity, the long time limit of $C(t)$ is just q, so the argument of the Fourier transform in (4.45) vanishes as $t \to \infty$. If there is broken ergodicity, however, the $t \to \infty$ limit of the expression in parentheses in (4.44) is $q_{EA} - q = \Delta$ (as defined in (2.43)). Thus $2G''/\omega$ acquires an extra piece proportional to $2\pi\Delta\delta(\omega)$. This means, in turn, using the first of the Kramers–Kronig relations (4.28), that the static susceptibility $\chi(0)$ acquires an extra contribution $\beta\Delta$. This is just the difference between the equilibrium $\chi = \beta(1-q)$ and the single-phase $\tilde{\chi} = \beta(1-q_{EA})$ we discussed in statics. Thus we see in this dynamical formulation that the parameter Δ is a direct measure of broken ergodicity, i.e. of the failure of the correlation function to decay to its statistical mechanical equilibrium value, even after infinite time.

This extra contribution to the susceptibility occurs only at *exactly* zero frequency, however. The easiest way to see this is to write the extra contribution to G'' as

$$\delta G''(\omega) = \Delta \lim_{\epsilon \to 0} \frac{\omega\epsilon}{\epsilon^2 + \omega^2} = \Delta \lim_{\epsilon \to 0} Im \frac{\epsilon}{\epsilon - i\omega} \tag{4.47}$$

or the extra contribution to G as

$$\Delta(\omega) \equiv \delta G(\omega) = \Delta \lim_{\epsilon \to 0} \frac{\epsilon}{\epsilon - i\omega} \equiv \Delta\delta_{\omega,0} \tag{4.48}$$

For ω exactly zero (more generally, for $\omega \ll \epsilon$), we then obtain a finite extra part $\delta G = \Delta$, while for finite frequency (i.e. $\omega \gg \epsilon$) $\delta G = 0$; hence the 'Kronecker delta' notation above.

It can sometimes be useful to think of (4.47)–(4.48) *with finite* ϵ as applying to a large but finite system in which $C(t)$ eventually relaxes to its equilibrium value q with a very long decay time ϵ^{-1}. This interpretation will be used in the Sompolinsky theory of SK spin glass dynamics that we present in Section 5.3.

5
Mean field theory II: Ising dynamics

Most of our knowledge of spin glass dynamics is within MFT. This chapter is therefore devoted to a fairly complete review of the dynamics of the soft-spin SK model introduced in Section 4.3, focusing especially on the description of the AT line as a dynamical instability and on the dynamical description of broken ergodicity below it. Now we should note that the SK model is not *a priori* expected to give a good description of experimental reality in two and three dimensions: We are now quite sure that the two-dimensional EA model does not order at all except at zero temperature, and in the three-dimensional system the observed T_f is much lower than mean field theory would predict, indicating relatively weak order. Nevertheless, effects can be observed, in both experiments and Monte Carlo simulations, which can be identified in a qualitative way as something like an AT line. Therefore we spend some time in this chapter reviewing the comparison between MFT and these measurements.

As we have noted, the dynamical theory as formulated here does not require replicas. There is, however, a similarity between the dynamical formalism and the replica approach: The bond averaging will now lead to quantities which play a role analogous to the matrix $q^{\alpha\beta}$ of the static theory but have two time arguments instead of the two replica indices. In both cases, one uses the method of steepest descent, which is exact for the SK model. Below T_f or below the AT line the solution of the static problem discussed in Section 3.4 required an order parameter function $q(x)$. In the corresponding dynamics problem Sompolinsky (1981) found a solution, stable in the static limit, with two order parameter functions $q(x)$ and $\Delta(x)$. Indeed, we will see that the two theories turn out to be physically equivalent.

5.1   The dynamical functional integral for the SK model

We now apply the general functional formulation that we studied in Section 4.3 to the SK model, following Schuster (1981) and Sompolinsky and Zippelius (1982). Our starting point is the generating functional $W[l, \hat{l}]$ given formally in (4.35) in terms of the action (4.36). The fact that from its definition (4.31) $W[l = \hat{l} = 0]$ is independent of the random variables J_{ij} now permits us to average it formally over the distribution $P[J]$ without any need to introduce replicas. Then the resulting functional derivatives (4.37) will yield directly *bond-averaged* correlation and response functions like (4.25) and (4.26).

The average of the functional W over the bond distribution (3.29) can easily be evaluated, with the result

$$
W[l, \hat{l}] = \int D\sigma D\hat{\sigma} \exp \left[L_0 + \frac{\beta J_0}{z} \sum_{\langle ij \rangle} \int dt i \hat{\sigma}_i(t) \sigma_j(t) \right.
$$

$$
+ \frac{\beta^2 J^2}{4z} \sum_{\langle ij \rangle} \int dt \int dt' i \hat{\sigma}_i(t) \sigma_j(t) (i \hat{\sigma}_i(t') \sigma_j(t')
$$

$$
\left. + i \hat{\sigma}_j(t') \sigma_i(t')) \right]
$$

(5.1)

where L_0 is the Lagrangian for noninteracting spins

$$
L_0 =
$$

$$
\sum_i \int dt [l_i \sigma_i + \hat{l}_i i \hat{\sigma}_i + i \hat{\sigma}_i (-\Gamma_0 \partial_t \sigma_i - r_0 \sigma_i - u \sigma_i^3 + \beta h_i + \Gamma_0^{-1} i \hat{\sigma}_i)] (5.2)
$$

We have replaced N in (3.29) by the number z of nearest neighbours. In this form (5.1) and (5.2) are quite general and hold for both nearest-neighbour and (by replacing z by N) infinite-range spin glasses or ferromagnets. As we remarked above, there is a strong similarity to the replica formalism: The bond averaging leads here to the coupling of spins with different time arguments, while in Section 3.3 it led to a coupling of spins in different replicas.

Now we go back to the SK model, letting $z = N \to \infty$, and ignore any tendency to ferromagnetic order, taking $J_0 = 0$. In the SK model all spins interact on the average in the same way and we can write (cf. 3.32)

$$
\sum_{i \neq j} i \hat{\sigma}_i(t) i \hat{\sigma}_i(t') \sigma_j(t) \sigma_j(t') = \tfrac{1}{4} [\sum_i (i \hat{\sigma}_i(t) i \hat{\sigma}_i(t')
$$

$$+\sigma_i(t)\sigma_i(t'))]^2 - \tfrac{1}{4}[\sum_i(i\hat{\sigma}_i(t)i\hat{\sigma}_i(t') - \sigma_i(t)\sigma_i(t'))]^2 + O(N) \quad (5.3)$$

and a similar expression for the second term in (5.1). A Gaussian transformation similar to that used in (3.34) then leads to

$$W = \int \prod_{r=1}^{4} DQ_r(t,t') \exp\left\{-\frac{N}{\beta^2 J^2}\int dt\,dt'[Q_1(t,t')Q_2(t,t')\right.$$

$$\left. +Q_3(t,t')Q_4(t,t') + \ln\int D\sigma D\hat{\sigma}\exp L(\sigma,\hat{\sigma},Q)]\right\} \quad (5.4)$$

where

$$L(\sigma,\hat{\sigma},Q) = L_0(\sigma,\hat{\sigma}) + \tfrac{1}{2}\int dt\,dt' \sum_i[Q_1(t,t')i\hat{\sigma}_i(t)i\hat{\sigma}_i(t')$$

$$+Q_2(t,t')\sigma_i(t)\sigma_i(t')+Q_3(t,t')i\hat{\sigma}_i(t)\sigma_i(t')+Q_4(t,t')i\hat{\sigma}_i(t')\sigma_i(t)] \quad (5.5)$$

As in the static problem, the exponent of the Q-integrals is proportional to N, so steepest descents can be used, replacing the $Q_r(t,t')$ by their stationary values (as in (3.43) and (3.50)). These stationary values are

$$Q_1^0(t,t') = \frac{\beta^2 J^2}{2N}\sum_i\langle\sigma_i(t)\sigma_i(t')\rangle = \tfrac{1}{2}(\beta J)^2 C(t-t') \quad (5.6)$$

$$Q_2^0(t,t') = \frac{\beta^2 J^2}{2N}\sum_i\langle i\hat{\sigma}_i(t)i\hat{\sigma}_i(t')\rangle = 0 \quad (5.7)$$

$$Q_3^0(t,t') = \frac{\beta^2 J^2}{2N}\sum_i\langle\sigma_i(t)i\hat{\sigma}_i(t')\rangle = \tfrac{1}{2}(\beta J)^2 G(t-t') \quad (5.8)$$

$$Q_4^0(t,t') = \frac{\beta^2 J^2}{2N}\sum_i\langle i\hat{\sigma}_i(t)\sigma_i(t')\rangle = \tfrac{1}{2}(\beta J)^2 G(t'-t) \quad (5.9)$$

where the brackets indicate averages which are calculated with the Lagrangian

$$L = L_0 + \tfrac{1}{2}(\beta J)^2 \times$$

$$\int dt\,dt' \sum_i[C(t-t')i\hat{\sigma}_i(t)i\hat{\sigma}_i(t') + 2G(t-t')i\hat{\sigma}_i(t)\sigma_i(t')] \quad (5.10)$$

The foregoing involved a good deal of technical algebra, but the result is simple to describe. Comparing (5.10) with the original Lagrangian (4.36), we see that we are now left with an effective single-site problem with a retarded self-interaction $(-\beta J)^2 G(t - t')$ and a noise $\phi(t)$ with statistics (given by the coefficient of $(i\hat{\sigma})^2$ in L (5.10))

$$\langle \phi(t)\phi(t') \rangle = 2\Gamma_0^{-1}\delta(t - t') + (\beta J)^2 C(t - t') \tag{5.11}$$

or, after Fourier transforming,

$$\langle \phi(\omega)\phi(-\omega) \rangle = 2\Gamma_0^{-1} + (\beta J)^2 C(\omega) \tag{5.12}$$

This noise is no longer white.

We can now handle this single-site problem in almost the same way we did that in Section 4.3; the nonwhite noise and retarded self-interaction do not introduce any complications. The equation of motion becomes, after Fourier transformation,

$$-\omega\Gamma_0^{-1}\sigma(\omega) = [-r_0 + \beta h_i(\omega) + (\beta J)^2 G(\omega)]\sigma(\omega)$$

$$-\frac{u}{(2\pi)^2}\int_{-\infty}^{\infty} d\omega_1 d\omega_2 \sigma(\omega_1)\sigma(\omega_2)\sigma(\omega - \omega_1 - \omega_2) + \phi(\omega) \tag{5.13}$$

We can then do the perturbation theory in u with a 'bare' propagator $G_0(\omega)$ defined by

$$G_0^{-1}(\omega) = r_0 - i\omega\Gamma_0^{-1} - (\beta J)^2 G(\omega) \tag{5.14}$$

instead of (4.41) and a bare correlation function which satisfies

$$C_0(\omega) = G_0(\omega)\langle \phi(\omega)\phi(-\omega)\rangle G_0(-\omega)$$

$$= |G_0(\omega)|^2[2\Gamma_0^{-1} + (\beta J)^2 C_0(\omega)] \tag{5.15}$$

so that

$$C_0(\omega) = \frac{2|G_0(\omega)|^2}{\Gamma_0[1 - (\beta J)^2|G_0(\omega)|^2]} \tag{5.16}$$

Except for these changes in the bare quantities, the diagrammatic perturbation theory, including in particular the self-energy diagrams of Fig. 4.3, is then the same as described in Section 4.3.

5.2   Dynamics of the phase transition

In this section we will investigate the dynamics of spin glasses for frequencies which are small compared to characteristic frequencies such as τ_0^{-1} but do not vanish as the size of the system approaches infinity. We will find the AT line as a dynamic instability line for $\omega \to 0$ and singular behaviour of the susceptibility $\chi(\omega, T, h)$ and the correlation funtion $C(\omega, T, h)$ along this line. We will also calculate the crossover from analytic behaviour of the dynamical susceptibility far above the AT line to nonanalytic behaviour on it.

We will also see that a naive approach to the static limit of this theory leads once again to the SK theory. We will then in the next section describe the Sompolinsky formulation of dynamics on timescales which become infinite in the thermodynamic limit, which reduces effectively to the Parisi result in the static limit.

All these results can also be obtained from the Glauber dynamics using the path integral formalism of Sommers (1987). Apparently the dynamics of the spin length, which are an artifact of the soft spins, is unimportant for the *critical* behaviour. Outside the critical region there might well be differences between the two approaches. At high temperatures the soft-spin dynamics becomes complicated and no explicit results are known. This is different for the Glauber dynamics, where the results of Section 4.1 might give a reasonably good approximation for all frequencies and for all temperatures above T_f.

To study the low-(but finite-)frequency dynamics, it is natural to define a generalised kinetic coefficient or damping function $\Gamma(\omega)$ by

$$\Gamma^{-1}(\omega) = i \frac{\partial G^{-1}(\omega)}{\partial \omega} \tag{5.17}$$

which reduces for independent spins ($\beta J \ll 1$) to $\Gamma^{-1} = \Gamma_0^{-1} = r_0 \tau_0$. If our experience with the Glauber Ising spin glass studied in the preceding section is a good guide, we should look for a divergence in $\Gamma^{-1}(0)$ at the spin glass transition.

From the Dyson equation (4.43) we obtain

$$\Gamma^{-1}(\omega) = \Gamma_0^{-1} + i \frac{\partial \Sigma}{\partial \omega} - i(\beta J)^2 \frac{\partial G}{\partial \omega}$$

$$= \frac{\Gamma_0^{-1} + i \partial \Sigma / \partial \omega}{1 - (\beta J)^2 G^2(\omega)} \tag{5.18}$$

The term $\partial \Sigma / \partial \omega$ turns out to be finite in the low-ω limit (as can be verified term by term in the series for Σ). Hence $\Gamma^{-1}(0)$ diverges when

$$(\beta J)^2 G^2(0) = 1 \tag{5.19}$$

In the Ising (hard spin) limit in zero field, we know that in the paramagnetic phase $\beta G((0) \equiv \chi(0) = \beta$, so this instability occurs at the temperature

$T_f = J$ we have identified as the spin glass transition temperature in the static theory, just as the dynamical instability of the Glauber spin glass of the preceding section did. Just above T_f, (5.18) then gives the 'critical slowing down'

$$\Gamma(0) \propto \frac{T^2 - T_f^2}{T^2} \qquad (5.20)$$

It is also simple to recover the $\omega^{\frac{1}{2}}$ behaviour in $G''(\omega)$ that we found for the Glauber model right at the transition. We will return to show this as a special case of the more general investigation we now make.

We now try to treat quite generally the dynamics of the soft-spin SK model in finite field, allowing a time-persistent part of the spin correlation function

$$C(t) = \tilde{C}(t) + q, \qquad \lim_{t \to \infty} \tilde{C}(t) = 0 \qquad (5.21)$$

Because the statistics (5.11) of the noise in the effective one-site problem obtained after bond averaging depend on $C(t)$, (5.21) means that when q is nonzero this noise contains a frozen part (a static random field). We separate out this part of ϕ, writing

$$\phi(\omega) = f(\omega) + z(\omega) \qquad (5.22)$$

where

$$\langle f(\omega)f(-\omega) \rangle = 2\Gamma_0^{-1} + (\beta J)^2 \tilde{C}(\omega) \qquad (5.23)$$

and

$$\langle z(\omega)z(-\omega) \rangle = 2\pi(\beta J)^2 q \delta(\omega) \qquad (5.24)$$

Consider first the static limit of this problem. Putting together the original single-site effective potential $\frac{1}{2}r_0\sigma^2 + \frac{1}{4}u\sigma^4$, the static limit of the effective retarded self-interaction which came from the bond averaging, and the static random field z, we have an effective Hamiltonian for $\omega \to 0$ of

$$\beta H_{eff} = \frac{1}{2}[r_0 - (\beta J)^2 G(\omega = 0)]\sigma^2 + \frac{1}{4}u\sigma^4 - (\beta h + z)\sigma \qquad (5.25)$$

This is for any r_0 and u. In the limit $u = -r_0 \to \infty$ it just becomes the Hamiltonian of an Ising spin in a field $h + Tz$. The average and mean square static magnetization for such a system are then simply

$$M(T, h) = [\tanh(\beta h + z)]_z$$

$$= (2\pi\beta^2 J^2 q)^{-\frac{1}{2}} \int_{-\infty}^{\infty} dz \exp[-\frac{1}{2}z^2/\beta^2 J^2 q] \tanh(\beta h + z)$$

$$= (2\pi)^{-\frac{1}{2}} \int_{-\infty}^{\infty} dz e^{-\frac{1}{2}z^2} \tanh \beta (Jq^{\frac{1}{2}}z + h) \qquad (5.26)$$

and

$$q(T, h) = [\langle\sigma\rangle^2]_z = (2\pi)^{-\frac{1}{2}} \int_{-\infty}^{\infty} e^{-\frac{1}{2}z^2} \tanh^2 \beta (Jq^{\frac{1}{2}}z + h) \qquad (5.27)$$

These are just the SK equations (3.51–53). Their region of validity is just what we found in Chapter 3: above the AT line. Thus in this region our dynamic theory also leads to the correct solution (and outside it to the same problems encountered in the replica-symmetric statics).

We will next see that the AT line manifests itself as a dynamic instability like that we found for the zero-field problem in (5.20). To show this we introduce a response function $g(\omega, z)$ which depends on the static random field z with $[g(\omega, z)]_z = G(\omega)$. A Dyson equation like (4.43) also holds for $g(\omega, z)$:

$$g^{-1}(\omega, z) = G_0^{-1}(\omega) + \Sigma(\omega, z) \qquad (5.28)$$

For the generalized kinetic coefficient $\Gamma(\omega)$ (5.17) we now obtain

$$\frac{\partial G}{\partial \omega} = \left(\frac{T}{J}\right)^2 \frac{i\Gamma_0^{-1}[g^2(\omega, z)]_z - \left[g^2(\omega, z)\frac{\partial\Sigma(\omega, z)}{\partial\omega}\right]_z}{\left(\frac{T}{J}\right)^2 - [g^2(\omega, z)]_z} \qquad (5.29)$$

The term $[g^2 \partial\Sigma/\partial\omega]_z$ turns out to be finite (see below). Hence $\Gamma^{-1}(\omega)$ would become infinite and would change its sign for $\omega \to 0$ if

$$(\beta J)^2 [g^2(0, z)]_z \geq 1 \qquad (5.30)$$

with

$$[g^2(0, z)]_z = [(\langle\sigma^2\rangle - \langle\sigma\rangle^2)^2]_z = 1 - 2q + [m^4]_z \qquad (5.31)$$

where the last equality is for Ising spins ($\sigma_i = \pm 1$). This is exactly the condition (3.27) or (3.65) for the AT line. The present analysis tells us that the damping rate $\Gamma(0)$ would become negative, indicating a dynamic instability, if the SK equations remained correct. We will show in the last section of this chapter that with the correct static limit as found in the Sompolinsky approach, the equality in (5.30) also holds for all temperatures and fields below the AT line. This says that the system relaxes more slowly than any exponential function of time everywhere in this region. This behaviour is the physical manifestation of the 'marginal' or 'massless' modes we discussed in Chapter 3.

We conclude this discussion of the static limit with the explicit results for the AT line near the freezing temperature. One has from (5.26), (5.27) and the corresponding equation for $[m^4]_z$, for $\theta \equiv (T_f - T)/T_f \ll 1$

$$q = \theta + \frac{1}{3}\theta^2 + \frac{h^2}{2\theta} \quad (\theta > 0) \tag{5.32}$$

$$q = \frac{h^2}{2\theta} \quad (\theta < 0) \tag{5.33}$$

and

$$[m^4]_z = 3q^2 - 8q^3 \tag{5.34}$$

Along the instability line, expansion of (5.30) yields

$$q = q_c = \theta_c + \theta_c^2 - \theta_c^3 + \cdots \tag{5.35}$$

Together with (5.33) this gives

$$\theta_c = \left(\frac{3}{4}\right)^{1/3} \left(\frac{h}{J}\right)^{\Theta}, \qquad \Theta = 2/3 \tag{5.36}$$

in agreement with the static result (3.28).

We turn now to dynamic properties, which requires calculation of the self energy $\Sigma(\omega, z)$ in (5.28). We begin by writing the Dyson equation for small ω in the form

$$g(\omega, z) = (g^{-1}(0, z) - \eta(\omega, z))^{-1}$$
$$= g(0, z) + g^2(0, z)\eta(\omega, z) + g^3(0, z)\eta^2(\omega, z) + \cdots \tag{5.37}$$

with

$$\eta(\omega, z) = (\beta J)^2 \Delta G(\omega) + \Delta\Sigma(\omega, z) + i\omega\Gamma_0^{-1} \tag{5.38}$$

and $\Delta G(\omega) = G(\omega) - G(0)$, $\Delta\Sigma(\omega, z) = \Sigma(\omega, z) - \Sigma(0, z)$. Averaging (5.37) over the static random field z leads to

$$\left(1 - (\beta J)^2 [g^2(0, z)]_z\right) \Delta G(\omega) =$$
$$(\beta J)^4 (\Delta G(\omega))^2 [g^3(0, z)]_z + [g^2(0, z)\Delta\Sigma(\omega, z)]_z + i\omega\Gamma_0^{-1} \tag{5.39}$$

The prefactor on the left hand side of (5.39) vanishes for $\omega = 0$ along the AT line, as we have just seen. If we could ignore the self energy $\Sigma(\omega, z)$, (5.39) would lead to $\Delta G(\omega) \propto (-i\omega)^{\frac{1}{2}}$ along this line, but this turns out to be correct only for zero field: The contribution to $\Gamma^{-1}(\omega)$ from $\partial\Sigma(\omega, z)/\partial\omega$ also diverges in general along the critical line. The self energy diagrams of Fig. 4.3 can be formally differentiated with respect to ω by differentiating the internal response and correlation function lines. The general form of the result is shown in Fig. 5.1, where the lines stand for full G's and C's,

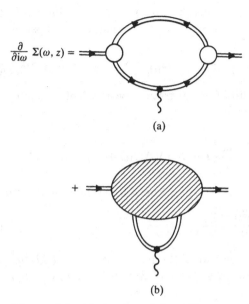

$$\frac{\partial}{\partial i\omega}\,\Sigma(\omega,z) =$$

(a)

$$+$$

(b)

Figure 5.1: Derivative of the self energy. The G-line with a dot is $\partial G(\omega,z)/\partial i\omega$, and the full (frequency dependent) irreducible vertices cannot be cut by a single line.

the circles indicate full vertices (including terms like those in Figs. 4.3a and c) and the line with the external wavy line means $\partial g(\omega,z)/\partial i\omega$. The diagram of Fig. 5.1b stands for all terms with more than a single C-line between the vertices. Evidently Fig. 5.1a is a generalisation of Fig. 4.3c. It turns out that Fig. 5.1b leads after integration to a term linear in ω for $\omega \to 0$ which renormalises the relaxation rate Γ_0 in (5.39). In the diagram of Fig. 5.1a we replace for small frequencies the vertices $\Gamma_3(\omega,z)$ by their static values $\Gamma_3(z)$. The most important part of Fig. 5.1 then reads

$$\frac{\partial \Sigma(\omega,z)}{\partial \omega} = \Gamma_3^2(z) \int_{-\infty}^{\infty} \frac{d\omega'}{\pi\omega'} Im\, g(\omega',z) \frac{\partial g(\omega+\omega',z)}{\partial \omega} \qquad (5.40)$$

The full static vertex is easily found by the same kind of argument we used to determine similar vertices in Section 3.2 (see Amit (1984) for a full discussion of formal properties of such quantities). We obtain

$$\Gamma_3(z) = \langle (\delta\sigma)^3 \rangle g^{-3}(0,z) \qquad (5.41)$$

with

$$\delta\sigma = \sigma - \langle\sigma\rangle \tag{5.42}$$

and with the static response function $g(0, z) = \langle(\delta\sigma)^2\rangle$. We write the response function $\Delta g(\omega, z) = g(\omega, z) - g(0, z)$ by means of the Dyson equation (5.28) with (5.14) for $\omega \to 0$

$$\Delta g(\omega, z) = (G_0^{-1}(0) - G_0^{-1}(\omega))g^2(0, z) = (\beta J)^2\Delta G(\omega)g^2(0, z) \tag{5.43}$$

and insert (5.43) into (5.40) after integration over ω. This leads to

$$g^2(0, z)\Delta\Sigma(\omega, z) =$$

$$(\beta J)^4\langle(\delta\sigma)^3\rangle^3 \int_{-\infty}^{\infty} \frac{d\omega'}{\pi\omega'} Im\, G(\omega')(\Delta G(\omega + \omega') - \Delta G(\omega')) \tag{5.44}$$

We define as a measure of the distance from the AT line the quantity

$$2\epsilon \equiv \frac{1 - (\beta J)^2[g^2(0, z)]_z}{(\beta J)^4[g^3(0, z)]_z} \tag{5.45}$$

and include in the relaxation rate Γ the factor $[g^2(0, z)]_z/(\beta J)^4[g^3(0, z)]_z$ and the contribution of the self-energy of Fig. 5.1. Equation (5.39) then leads to

$$2\epsilon\Delta G(\omega) = (\Delta G(\omega))^2$$

$$+2A \int_{-\infty}^{\infty} \frac{d\omega'}{\pi\omega'} Im\Delta G(\omega')(\Delta G(\omega + \omega') - \Delta G(\omega')) + \frac{i\omega}{\Gamma} \tag{5.46}$$

with the constant

$$2A = \frac{[\langle(\delta\sigma)^3\rangle^2]_z}{[g^3(0, z)]_z} = \frac{[2m^2(m^2 - 1)^2]_z}{[(1 - m^2)^3]_z} \tag{5.47}$$

where the second equation holds for the Ising limit $\sigma_i = \pm 1$. (The preceding equations hold for *all* values of the parameters r_0 and u in the model (4.22), including the limit $u = -r_0 \to \infty$.)

In order to solve (5.46) for $\omega \to 0$ along the AT line $\epsilon = 0$ we make the ansatz for the response function

$$\Delta G(\omega) = -R(-i\omega/\Gamma)^\nu$$

$$= R\sin(\tfrac{1}{2}\pi\nu)\left(-\cot(\tfrac{1}{2}\pi\nu)\left|\frac{\omega}{\Gamma}\right|^\nu + i\,\mathrm{sgn}\,\omega\left|\frac{\omega}{\Gamma}\right|^\nu\right) \tag{5.48}$$

with some constant R and where the exponent $\nu > 0$ has to be determined. This ansatz obeys the Kramers–Kronig relations (4.28) since $\Delta G(\omega)$ is an analytic function in the upper half plane of complex frequencies ω and has a cut along the imaginary axis. We have to consider two cases: For $h = 0$ the prefactor $A \approx [m^2]_z = q$ vanishes, one has from (5.46) $R = 1$ and $\Delta G(\omega) = -(-i\omega/\Gamma)^{\frac{1}{2}}$ or $\nu = \frac{1}{2}$, in agreement with our result (4.21) for the Glauber model. For $h \neq 0$ a self-consistent solution is obtained for $\nu < \frac{1}{2}$: The term $i\omega/\Gamma$ in (5.46) can then be ignored and there remains the equation

$$(-i\omega)^{2\nu} + 2A \int_{-\infty}^{\infty} \frac{d\omega'}{\pi\omega'} Im(-i\omega')^{\nu}[(-i\omega - i\omega')^{\nu} - (-i\omega)^{\nu}] = 0 \quad (5.49)$$

which determines the exponent $\nu(T, h)$. Both the real and imaginary parts of (5.49) yield

$$A = \frac{2\pi \cot \pi\nu}{B(\nu, \nu)} \quad (5.50)$$

with the Beta function (Abramowitz and Stegun, 1965, p.258)

$$B(z, \omega) = \int_0^1 dx x^{z-1} (1 - x)^{\omega - 1} = \frac{\Gamma(z)\Gamma(\omega)}{\Gamma(z + \omega)} \quad (5.51)$$

where $\Gamma(z)$ is the Gamma function. This leads for small fields, with $2A \approx 2[m^2]_z = 2q \approx 2\epsilon$ from (5.32) to

$$\nu = \frac{1}{2} - \frac{A}{2\pi} = \frac{1}{2} - \frac{\theta}{\pi} \quad (5.52)$$

which justifies our assumption $\nu < \frac{1}{2}$. The result (5.52) is rather surprising: one has a critical exponent which varies as a function of temperature and field along the AT line. In the large-field limit, $\nu \to 0.395$.

Sompolinsky and Zippelius (1982) have also calculated ν below the AT line, assuming the marginality which we will prove in the next section. They obtained the same result (5.52); i.e. to this order, at least, ν depends only on temperature and not on field. They also made an argument based on marginal stability, similar to that used by Bray and Moore (1979) that we referred to in Section 3.5 in the calculation of the internal field distribution from TAP equations, which gives $\nu \to \frac{1}{4}$ as $T \to 0$ in zero field. If both these results are correct, the field-independence of ν seen near T_f is not exact for general $T < T_f$.

With this singular behaviour of the response function $G(\omega)$, the fluctuation–dissipation theorem (4.45) implies that $\tilde{C}(\omega) \propto \omega^{\nu-1}$ or, in the time domain, that the correlation function $\tilde{C}(t)$ decays like $t^{-\nu}$ at long times.

The crossover in the behaviour of the response and correlation functions from this singular form to one which is analytic as $\omega \to 0$ can also be calculated. We consider first the case $h = 0$, for which one has from (5.46) with $A = 0$

$$\Delta G(\omega) = \epsilon - (\epsilon^2 - i\omega/\Gamma)^{\frac{1}{2}} \qquad (h=0, \quad T \geq T_f, \quad \omega \ll \Gamma) \quad (5.53)$$

Inserting $[g^2(0,z)]_z = [g^3(0,z)]_z = 1$ and $(\beta J) = 1$ into (5.45) for ϵ leads (again setting $T_f = 1$) to

$$\chi(\omega) - \chi(0) = \tfrac{1}{2}\beta\{T^2 - 1 - [(T^2 - 1)^2 - 4i\omega/\Gamma]^{\frac{1}{2}}\} \quad (5.54)$$

which agrees with the result (4.19) we obtained for the Glauber model in the limit $\omega\tau_0 \to 0$, $T \to 1$ if the single-spin relaxation time τ_0 is identified with $2/\Gamma$. Apparently the dynamics of the fluctuations in the spin length $|\sigma|$, which is an artefact of the soft-spin model, does not modify the critical and crossover behaviour of the system. This is a familiar feature near a critical point (see Section 8.1), where one often also works with spins of variable length.

According to (5.53) the dynamical susceptibility scales like

$$\Delta G(\omega) = \epsilon f\left(\frac{\omega}{\Gamma\epsilon^{1/\nu}}\right), \qquad h=0, \quad \nu = \tfrac{1}{2} \quad (5.55)$$

as a function of frequency and temperature as defined in (5.45). This result can be generalized for $h \neq 0$ and $\nu \neq \tfrac{1}{2}$ (Sommers and Fischer, 1985) to

$$\Delta G(\omega) = \epsilon f\left(\frac{\omega}{\Gamma\epsilon^{1/\nu}}\right) + \epsilon^2 g\left(\frac{\omega}{\Gamma\epsilon^{1/\nu}}\right) + \cdots \quad (5.56)$$

with $f(x) \propto -(ix)^\nu$ and $g(x) \propto (-ix)^{2\nu}$.

The crossover from $\Delta G(\omega) \propto \omega^{\frac{1}{2}}$ at $T = T_f$ to $\Delta G(\omega) \propto \omega$ for $T \gg T_f$ in zero field is best seen in the imaginary part $G''(\omega) = T\chi''(\omega)$ of the dynamical susceptibility (Fig. 5.2). For conventional susceptibility measurements the condition $\omega \ll \Gamma$ always holds (in contrast to neutron scattering, see Section 9.2), and one has a very abrupt increase near T_f.

Despite the fact that mean field theory is not really expected to describe the transition or AT line in real three-dimensional spin glasses, a number of the qualitative aspects of MFT, especially at finite frequencies, appear to characterize real systems as well. Fig. 5.2 also shows, as inset, data on $Eu_{0.2}Sr_{0.8}S$ (Hüser et al, 1983). The overall behaviours of the theoretical and experimental data look fairly similar. However, on closer examination, there are important differences: The critical exponent $\nu = \tfrac{1}{2}$ predicted by MFT is not observed. The data of Hüser et al can be fitted with $\nu = 0.05$, and more precise data on $Eu_{0.4}Sr_{0.6}S$ shown in Fig. 5.3 suggest $\nu = 0.1$ (Paulsen et al, 1987).

Another discrepancy between MFT and experiment can be seen in Figs. 5.2 and 5.3. For small ω the theory predicts a *frequency-independent* maximum in the real and imaginary parts of the susceptibility, whereas real

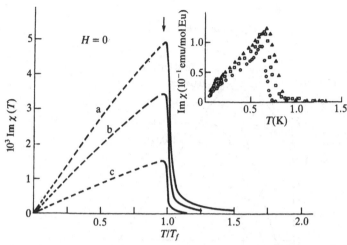

Figure 5.2: Temperature dependence of the imaginary part of the dynamic susceptibility in MFT for zero field and frequency $\tilde{\omega} = \omega/\Gamma_0$ where Γ_0 is the single-spin relaxation rate: $\tilde{\omega} = 10^{-4}$ (a), 5×10^{-5} (b), and 10^{-5} (c). The arrow indicates the AT line. The inset shows the experimental data on $Eu_{0.2}Sr_{0.8}S$ (from Hüser et al, 1983):

o: 10.9 Hz

□: 261 Hz

△: 1962 Hz

(From Fischer and Kinzel, 1984).

spin glasses exhibit noticeable frequency dependence (strongest in systems with short-range interactions, weaker in RKKY systems). The frequency dependence of the 'cusp' in the ac susceptibility was considered for a long time to be evidence against a sharp phase transition. However, the data of Fig. 5.3 seem to indicate that the 'cusp' persists even in the extrapolation to $\omega = 0$.

In MFT we defined the AT line by the onset of irreversibility as defined by the broken ergodicity parameter Δ in (2.43) and (4.48). This onset should be seen in the appearance of remanence, which suggests ways to measure the AT line experimentally. For example, as we noted in the Introduction, the thermoremanent magnetization (measured after field cooling and switching the field off) can frequently be fit by

$$M_{TRM} = M_0 - S(T, H) \ln t \qquad (5.57)$$

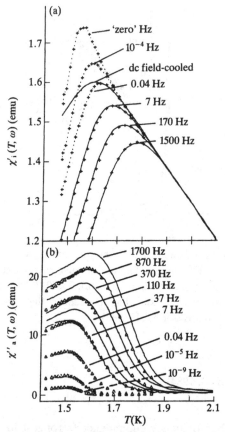

Figure 5.3: Real and imaginary parts of the ac susceptibility of $Eu_{0.40}Sr_{0.60}S$ for various frequencies and for temperatures near T_f (solid lines). The crosses in (a) and triangles in (b) are the predictions if one fits the data by the power law (5.48) with temperature dependent parameters $R(T)$ and $\nu(T)$. The data for the dc susceptibility M_{FC}/H for $H = 0.5$ Oe are also indicated (from Paulsen et al, 1987).

with the 'magnetic viscosity' $S(T, H)$ as prefactor of the logarithmic term, or a power law

$$M_{TRM} = M_0 \left(\frac{t_0}{t}\right)^a \tag{5.58}$$

One can then look for a line in the $h-T$ plane where S or $a \to 0$. An effective

AT line can also be defined by the onset of a difference between the field-cooled and zero-field-cooled magnetizations. Another (and perhaps better) criterion is given by the 'differential end point', below which the equilibrium and nonequilibrium responses to an additional small field imposed after field cooling differ. If there were true broken ergodicity and one could really measure both equilibrium and nonequilibrium (i.e. single-valley) responses, the difference between the two responses would be a direct measure of the parameter Δ.

Examples of 'AT lines' obtained either from $S(T, H) \to 0$ (within the times $30s < t < 1200s$) or from $\chi''(\omega, T, H) \to 0$ for $\nu = \omega/2\pi = 11.3$ Hz and 11.3 kHz are shown in Fig. 5.4. One observes a strong time or frequency dependence of the coefficient of h^Θ in (5.36), though the exponent $\Theta = 0.64...0.77$ is fairly near the theoretical value $\Theta = 2/3$. The vanishing of the exponent $a(T, H)$ in (5.58) leads to a similar line in $Fe_{0.16}Cr_{0.84}$ (Palumbo, Parks, and Yeshurun, 1982). The differential branch point method has been applied to amorphous $(Fe_{0.64}Mn_{0.36})_{75}P_{16}B_6Al_3$ and yields $\Theta = 0.75 \pm 0.15$. However, the coefficient $3/4$ in (5.36) has to be replaced by a factor of about 15. This prefactor is found empirically to be larger in systems with a tendency to ferromagnetic interactions and for larger spin values, but this effect does not seem sufficient to explain the discrepancy. We think one should not take this kind of problem very seriously, since three-dimensional systems are so far from mean field behaviour anyway. The agreement between the MFT and measured effective values of the exponent Θ is probably just fortuitous, as we discuss at greater length in chapter 8.

A crossover from the field-temperature dependence of the frequency-dependent effective AT line $\theta_c \propto h^{2/3}$ for $\omega \to 0$ to analytic behaviour $\theta_c \propto h^2$ at large ω has been calculated in MFT (Fischer, 1983b). This crossover has also been measured in Monte Carlo simulations on a model with short-range interactions (Young, 1983a) and in experiments on EuSrS (Paulsen et al, 1984, Rajchenbach and Bontemps, 1983). In the theory one calculates the lines of constant average relaxation times

$$\tau_{av}(T, h) = (1 - q) \int_0^\infty dt\tilde{C}(t) = G^{-1}(0) \left(\frac{\partial G(\omega)}{\partial i\omega} \right)_{\omega=0} \qquad (5.59)$$

with $G(0) = 1 - q$ and $\tilde{C}(t)$ from (5.21), which for $\tau_{av} \to \infty$ reduces to the AT line. In the data on $Eu_{0.4}Sr_{0.6}S$ (Paulsen et al, 1984), critical lines are determined from the inflection point of $\chi''(T, H)$ for a given frequency ω. The crossover is clearly seen for higher frequencies. In addition, one has a frequency-dependent freezing temperature as defined by the maximum of $\chi'(T, H)$ which is not predicted in MFT.

In this discussion we have compared theory for Ising models with experiments on systems which are at least superficially much more Heisenberg-

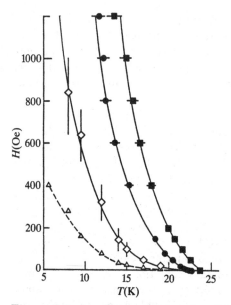

Figure 5.4: Lines along which irreversibility vanishes on different time scales in $Fe_{10}Ni_{70}P_{20}$: $S(H)$ is the magnetic viscosity defined in (5.57).
■: $\chi'' \to 0$ (11.3 kHz)
●: $\chi'' \to 0$ (11.3 Hz)
◇$S \to 0$ for $30s < t < 1200s$
△: maximum of $S(H)$
(from Salamon and Tholence, 1983).

like. We will show in subsequent chapters that Heisenberg spin glasses with sufficient random anisotropy behave like Ising systems. Whether or not the anisotropy is large enough to allow us to invoke this result in all these experimental systems is not yet clear.

5.3   Sompolinsky theory

The solution we obtained above for the dynamic response function $G(\omega) = T\chi(\omega)$ cannot be correct below the AT line since it leads to a negative kinetic coefficient $\Gamma(\omega)$ (5.29), i.e. to exponential increase of the local magnetization in response to an initial perturbation. As a first step to remedy this insufficiency Sompolinsky introduced an anomalous response term $\Delta(\omega)$ (4.48) which according to (4.45) describes the nonergodicity. However, he also noted that this leads to an ambiguity: In the self energy diagram of Fig. 4.3b there enters the product $G(\omega + \omega')C(\omega')$ (see (5.44)). Below the AT line, this contains the product $q\delta(\omega')\Delta\delta_{\omega+\omega',0}$, which is ill defined for $\omega \to 0$. The limit in which the width of $\Delta(\omega)$ is small compared to that of $q\delta(\omega)$ leads back to the solution $\Delta = 0$, which we have seen to be acceptable only above the AT line. The opposite limit leads to a finite Δ but again to a negative relaxation rate for $\omega \to 0$. This dilemma is not unexpected since we know from Section 3.4 that a stable static solution can only be obtained by the introduction of an order parameter *function* $q(x)$ instead of just two parameters. Since any dynamic theory has to reduce in the static limit to the correct equilibrium solution of the system, one is faced with the problem of constructing a theory for spin glass dynamics which includes Parisi's theory discussed in Section 3.4, i.e. physically, the infinitely many phases.

Sompolinsky considered a finite system with energy barriers which are hierarchically ordered, leading to a hierarchy of relaxation times, all of which go to infinity in the thermodynamic limit. To make the argument simpler, we will initially treat the hierarchy as discrete, taking a continuum limit later. Specifically, we assume for a system with N spins that for two different relaxation times in the hierarchy t_i, t_j, with $i < j$, the ratio $t_i/t_j \to \infty$ as $N \to \infty$, as well as that each time should separately go to infinity. (The convention is that the relaxation times with the *lowest* indices are the longest.) Then one will observe a sequence of time intervals $t_{i+1} \ll t \ll t_i$ over which the processes with relaxation times less than t will have relaxed completely, while those with longer times will essentially not be relaxed at all. The configurations of the system observed for early times are confined to relatively small regions of configuration space, while those for the longest times approach true (all-valley) equilibrium. In the thermodynamic limit, each of these partially relaxed states becomes a stable broken-ergodicity state, characterized by an order parameter

$$q(t_i) = [\langle S_j(t_i)S_j(0)\rangle]_{av} \tag{5.60}$$

The continuum limit of this set of order parameters, with i replaced a variable x between 0 and 1, will become the Sompolinsky order function corresponding to Parisi's $q(x)$.

We can represent the time-persistent piece of the correlation function $C(\omega)$ as an infinite sum:

$$C(\omega) = \tilde{C}(\omega) + q(\omega) \tag{5.61}$$

with

$$q(\omega) = \sum_{i=1}^{k} 2\pi q_i' \delta^i(\omega) \qquad (5.62)$$

where $\delta^i(\omega)$ is a delta–like function of width $\omega_i \equiv t_i^{-1}$. (The number k of levels in the dynamical hierarchy will eventually be taken infinite.) Thus

$$q(t_j) = \sum_{i=1}^{j} q_j' \qquad (5.63)$$

Similarly, we write the time-persistent noise (5.24) as $\sum_{i=1}^{k} z_i$ with

$$\langle z_i(\omega) z_i(-\omega) \rangle = 2\pi (\beta J)^2 q_i' \delta^i(\omega) \qquad (5.64)$$

As with the $q(t_i)$, we label the z_i in the continuum limit by the variable x: $z_i \to z(x)$.

The shortest of the t_i (labelled in the continuum limit by $x = 1$) is the time the system remains in a single valley, confined by barriers which become infinite in the thermodynamic limit. This is the infinite-time ($\omega \to 0$) limit of the dynamics of the infinite system. Thus we identify

$$q(x = 1) = q_{EA} \qquad (5.65)$$

For $x = 0$, at the 'longest infinite time' the system is assumed to be in full thermal equilibrium. One might then guess $q(x = 0) = q$, the equilibrium statistical mechanics order parameter. However, Sompolinsky assumed instead that

$$q(x = 0) = 0 \qquad (5.66)$$

(in zero field) or, according to (5.60), a state in which all correlations have decayed.

This raises some questions: A simple interpretation would be that the longest of the times t_i is equal to the ergodic time τ_e within which all spins flip and the net magnetization can undergo reversals. Such a time can certainly be reached if one assumes first $t \to \infty$ and then $N \to \infty$. In an ideal ferromagnet the ergodic time will be well separated from all other characteristic times of the system. In a spin glass one cannnot exclude *a priori* the possibility that the broad spectrum of relaxation times extends to the ergodic time without a gap. However, this seems to contradict evidence from Monte Carlo simulations on small samples $N = 16$ to 192) at $T = 0.4T_f$ and $0.6T_f$ (MacKenzie and Young, 1982,1983). As we already mentioned in Chapter 3, these simulations indicate a rather well-defined maximum relaxation time τ_{max} for processes which do not change the sign of the total magnetization, with $\ln \tau_{max} \propto N^{1/4}$, while $\ln \tau_e \propto N^{1/2}$. (Compare this with the ideal ferromagnet, where $\ln \tau_e \propto N$ in MFT.) An additional argument for $\tau_{max} \ll \tau_e$ comes from the Parisi theory (Section 3.4). There, for a small external field, all states are described by a *positive* order function $q(x)$, and it is only for $h = 0$ that the full interval $-1 \leq q(x) \leq 1$ is needed. The Sompolinsky theory contains only positive $q(x)$ and turns out to be equivalent to the Parisi one (see below). Hence the reversed-magnetization states have been excluded, so it does not seem to make sense to assume that the decay of correlations at the longest times involves changes in the sign of the total magnetization.

A possible consistent interpretation of $q(t_x)$ is in terms of a system which relaxes in response to a very small change of field (or temperature) (which needs only to be $\gg T/N^{1/2}$) (Hertz, 1983a,b; Fischer and Hertz, 1983). As we argued in Section 3.4, because of the extreme sensitivity of the free energy landscape to h and T, the average overlap between initial and final phases at different fields or temperatures is $q(0)$, not $q(1)$. Therefore, the average correlation with the initial configuration does indeed go to zero at infinite time for $h = 0$ (or, at general h, to $q(0)$).

We make an ansatz for the anomalous response $\Delta(\omega)$ of (4.48) which is analogous to what we wrote for $q(\omega)$ in (5.62):

$$\Delta(\omega) = -\sum_{i=1}^{k} \Delta'_i \delta^i_{\omega,0} \qquad (\Delta'_i < 0) \tag{5.67}$$

where $\delta^i_{\omega,0}$ is defined as in (4.48) with ϵ replaced by ω_i. In the continuum limit the anomalous response also goes over to a function $\Delta(x)$. Sompolinsky defines this function with the boundary condition

$$\Delta(x = 1) = 0 \tag{5.68}$$

That is, in the shortest 'infinite' time the system is unable to overcome any infinite free energy barrier. The nonergodicity function $\Delta(x)$ is then defined as the deviation from the state $x = 1$ and increases with increasing timescales. This is the opposite of the way which would be most natural in the picture we have given of ergodicity breaking (we would naturally have $\Delta \neq 0$ when the system is in a state with spin correlation function $q(1) = q_{EA} > q$; see the definition (2.43) of Δ), but in this section we adopt the convention Sompolinsky established, which has been followed in subsequent literature. Thus the anomalous response or broken ergodicity parameter observed for times between t_{j+1} and t_j is

$$\Delta(t_j) = -\sum_{i=j}^{k} \Delta'_i \tag{5.69}$$

Equivalently, in terms of the function $\Delta(x)$, the local susceptibility measured at frequency $\omega_x = t_x^{-1}$ of

$$\chi(x) \equiv \int_0^{t_x} \chi_{ii}(t) = \beta(1 - q_{EA} + \Delta(x)) \qquad (h \to 0) \qquad (5.70)$$

and $\chi(0) = \beta(1 - q_{EA} + \Delta(0)) = \beta(1 - q)$, the true equilibrium susceptibility. However, the relation $q_{EA} - q = \Delta$ cannot be generalized similarly: $q(1) - q(x) \neq \Delta(x)$, since $q(0) \neq q$.

Like Parisi's, Sompolinsky's derivation of the self-consistency equations for $q(x)$ and $\Delta(x)$ are heuristic. It also has an ultrametric structure implicitly built into it in the assumption of the strong hierarchy of timescales, i.e. that $t_x \gg t_{x'}$ in the large N limit if $x < x'$ ($q(x) < q(x')$) (Mézard et al, 1987): In general the transition time from state α to state β via state γ must be at least that for the direct transition from α to β:

$$t_{\alpha\gamma} + t_{\gamma\beta} > t_{\alpha\beta} \qquad (5.71)$$

For the strong hierarchy, where one of any two different relaxation times is always much larger than the other, this becomes

$$\max(t_{\alpha\gamma}, t_{\gamma\beta}) > t_{\alpha\beta} \qquad (5.72)$$

or, assuming that the times depend reciprocally on the overlap,

$$q_{\alpha\beta} > \max(q_{\alpha\gamma}, q_{\gamma\beta}) \qquad (5.73)$$

which is just the inequality that defines an ultrametric space.

The introduction of the expressions (5.62) and (5.67) resolves the ambiguity in the product $\delta(\omega)\Delta\delta_{\omega,0}$. One has $\delta^i(\omega)\delta^j_{\omega,0} = \delta^i(\omega)$ if $i < j$ and zero if $i > j$ since $\omega_i/\omega_j \ll 1$ for $i < j$. Any ambiguity about the term with $i = j$ becomes irrelevant when the number k of levels in the hierarchy becomes infinite, with each q_i' and Δ_i' infinitesimal.

We now try to generalize our previous treatment of the static limit to the 'states' i which would be observed for $t_{i+1} \ll t \ll t_i$. The time-persistent effective field which appears in the effective Hamiltonian (5.25) is now given by

$$\beta H[z] = \beta h + z(\omega) + (\beta J)^2 \sigma(\omega) \Delta(\omega) \qquad (5.74)$$

on a timescale ω^{-1}. (The second term comes from the anomalous response part of the term $-\frac{1}{2}(\beta J)^2 G(\omega = 0)\sigma^2$ in (5.25).) Taking the continuum limits of (5.63) and (5.67), we have

$$q(x) = q(0) + \int_0^x dy\, q'(y) \qquad (5.75)$$

and

$$\Delta(x) = -\int_x^1 dy \Delta'(y) \tag{5.76}$$

The effective field (5.74) at a timescale t_x is then given by

$$\beta H_x[z] = \beta h + \int_0^1 dx z(x) + z_0(t_x^{-1}) - (\beta J)^2 \int_0^1 dx \Delta'(x) m_x[z] \tag{5.77}$$

with $z_0 \equiv z(0)$, where $m_x[z]$ is the local magnetization produced by the part of $H_x[z]$ which remains frozen on the timescale t_x.

It is convenient to rescale the $z(x)$ by a factor $\beta J(2\pi q'(x))^{1/2}$ so they become univariate Gaussian variables:

$$\langle z(x)z(x')\rangle = \delta(x - x'), \qquad \langle z_0^2 \rangle = 1, \qquad \langle z_0 z(x)\rangle = 0 \tag{5.78}$$

Then the effective field becomes

$$H_x[z] = h + J \left(\int_0^1 dx z(x)(2\pi q'(x))^{\frac{1}{2}} + z_0(t_x^{-1})(2\pi q(0))^{\frac{1}{2}} \right)$$

$$- \beta J^2 \int_0^1 dx \Delta'(x) m_x[z] \tag{5.79}$$

and $m_x[z]$ is given by

$$m_x[z] = \int_{-\infty}^{\infty} \prod_{y>x} \frac{dz(y)}{2\pi} \exp\left[-\frac{1}{2}\int_x^1 dy z^2(y)\right] \tanh(\beta H[z]) \tag{5.80}$$

where we have specialized to the hard-spin Ising limit by writing $\tanh \beta H$ for the magnetization produced by a field H acting on a single spin.

The self-consistent equation for $q(x)$ is then

$$q(x) = \langle m_x^2[z]\rangle_z \tag{5.81}$$

where the average $\langle\cdots\rangle_z$ means over the remaining $z(y)$ for $y < z$. The magnetization is given by a similar average with only one power of m_x; a look at (5.80) shows that it is independent of x. Differentiating it with respect to an auxiliary field $h(x)$ leads for $h = 0$ to

$$1 - q_{EA} + \Delta(x) = (2\pi\beta^2 J^2 q'(x))^{-\frac{1}{2}} \left\langle \frac{\delta m}{\delta z(x)} \right\rangle_z \tag{5.82}$$

The integrals (5.26) and (5.27) of the conventional theory are now replaced by functional integrals.

These 'Sompolinsky equations' can also be derived from a formal static free energy functional

$$-\beta f = \tfrac{1}{4}(\beta J)^2 \left[(1 - q(1))^2 + 2 \int_0^1 dx \Delta'(x) q(x) \right]$$

$$+ \langle \ln(2 \cosh \beta H[z]) \rangle_z + \tfrac{1}{2}(\beta J)^2 \langle \int_0^1 dx \Delta'(x) m_x^2[z] \rangle_z \qquad (5.83)$$

by differentiation with respect to $q'(x)$ and $\Delta'(x)$. That (5.83) at its extremum describes a free energy becomes clear if we compare it with Parisi's theory. Parisi did not derive an explicit expression for his order parameter function $q(x)$. However, one can verify that (5.83) leads to Parisi's result if the functions $\Delta(x)$ and $q(x)$ are connected by

$$\Delta'(x) = -xq'(x) \qquad (5.84)$$

One can postulate such a relation since without it the scale of x is not fixed: The equations (5.81)–(5.83) still hold if x is replaced by some monotonic function $u(x)$ with $u(0) = 0$ and $u(1) = 1$ if $z(x)$ is replaced by $z(u)(du/dx)^{1/2}$. One says that the functional (5.83) is 'gauge invariant'. Parisi's solution is obtained by choosing a special 'gauge'.

The free energy (3.50) of the SK solution is recovered from (5.83) if we set $\Delta'(x) = q'(x) = 0$ or $\Delta = 0$, $q = \text{const.}$

Sompolinsky's approach has the advantage that no replicas are needed. Furthermore, there is at least a heuristic derivation by Sommers, De Dominicis, and Gabay (1983) of the Sompolinsky equations from TAP equations.

These equations are in general extremely complicated functional integral equations, and no general solution is known. However, one can expand the right hand side of the equation (5.81) for $q(x)$ for $x \to 0$, which leads for $h = 0$ and all temperatures below T_f to

$$q(x) = (J\chi(0))^2 q(x) + O(q^2(x)) \qquad (5.85)$$

or to a *constant* static susceptibility: $\chi(0) = J^{-1}$, as we mentioned in connection with the Parisi result (see below (3.78)). Expansion near T_f yields (Sompolinsky 1981; Sommers, 1985)

$$q(1) = \theta + \theta^2 - \theta^3 \qquad (5.86)$$

which agrees with (5.35) along the AT line.

The condition $\Delta(x) \to 0$ for $x \to 1$ in the expansion of (5.82) leads to

$$\Delta(x) = (\beta J)^2 (1 - 2q(1) + \langle m^4[z] \rangle_z)\Delta(x) + O(\Delta^2(x)) \qquad (5.87)$$

and agrees with the condition (5.30) for the AT instability. But here we see that this condition also holds everywhere below the AT line, as we promised in Section 3.4. Furthermore, this marginality holds at each timescale (Sommers, 1983; Hertz, 1983b). The local frozen-in magnetization $m_x[z]$ and the local frozen-in field distribution have been calculated by Sommers and Dupont (1984). The latter agrees with that obtained numerically from the TAP equations by Palmer and Pond (1979).

Below the AT line one has irreversibility which is most easily seen in the remanent magnetization, including its time dependence. So far only the thermo-remanent magnetization $M_{TRM} = -df/dh$ has been calculated (Elderfield, 1983). Here the total derivative indicates that one has to include contributions from the field dependence of $q(x)$ and $\Delta(x)$. The corresponding uniform susceptibility after field cooling

$$\chi_{fc} = \frac{\beta}{N} \sum_{ij} [\langle S_i S_j \rangle - \langle S_i \rangle \langle S_j \rangle]_{av} \qquad (5.88)$$

is difficult to calculate directly: In a finite field one has nondiagonal terms $\chi_{ij} \neq 0$ whereas (5.70) describes only the *local* susceptibility.

Both M_{TRM} and the isothermal remanence M_{IRM} of the SK model have been calculated by Monte Carlo simulations (Kinzel, 1986a). Below T_f one observes a power law of the form (5.58) with an exponent $a \approx 0.4$, which is consistent with the picture of a marginally stable state extending down to $T = 0$. We postpone discussion of experimental results for the time-dependent magnetization to Section 9.2.

6

Mean field theory III: vector spins

So far we considered only Ising spins $S_i = \pm 1$ or the soft-spin variant thereof, which have no intrinsic dynamics. More realistic is a model with vector spins $S_{i\mu}, \mu = 1, ..., m$ with the Hamiltonian

$$\mathcal{H} = -\tfrac{1}{2} \sum_{ij,\mu} J_{ij} S_{i\mu} S_{j\mu} - \sum_{i\mu} h_\mu S_{i\mu} \tag{6.1}$$

where the Heisenberg model with $m = 3$ is a special case. The dynamics of the Heisenberg model are defined by the quantum mechanical equation of motion for symmetric exchange $J_{ij} = J_{ji}$

$$\partial_t \mathbf{S}_i = (i/\hbar)[\mathcal{H}, \mathbf{S}_i] = \sum_j J_{ij} \mathbf{S}_i \times \mathbf{S}_j + \mathbf{S}_i \times \mathbf{h} \tag{6.2}$$

with the spin commutators ($\epsilon_{\mu\nu\rho}$ is the totally antisymmetric tensor)

$$[S_{i\mu}, S_{j\nu}] = i\hbar\epsilon_{\mu\nu\rho} S_{i\rho} \delta_{ij} \tag{6.3}$$

One easily proves that the exchange interaction $\mathcal{H}_{ex}$ (the first term on the left hand side of (6.1) leaves the total spin $\mathbf{S}_{tot} \equiv \sum_i \mathbf{S}_i$ unchanged:

$$\partial_t \mathbf{S}_{tot} = (i/\hbar)[\mathcal{H}_{ex}, \mathbf{S}_{tot}] = 0 \tag{6.4}$$

For a field in direction $\mu = z$ the component $S_{tot,z}$ remains constant and the length of each spin is also conserved:

$$[\mathcal{H}, S_{tot,z}] = [\mathcal{H}, S_i^z] = 0 \tag{6.5}$$

Henceforth we consider only *classical* spins and use (6.2) in this sense.

Even for classical spins the dynamics (6.2) are extremely complicated, apart from certain limits such as hydrodynamics (see Section 7.2). It can be somewhat simplified if one considers 'soft' spins. This is physically sensible when the distance over which significant magnetization variations occur is very large, such as in the Landau–Ginzburg theory of second-order phase transitions. This also occurs in hydrodynamics.

The model (6.1) is not a complete description of a spin glass. Most spin glasses contain additional random or nonrandom anisotropies in the form of single-ion (crystal field) anisotropy or anisotropic exchange interactions (Section 6.2). Another type of anisotropy has been predicted by Dzyaloshinskii (1958) and Moriya (1960) for ferromagnets and is particularly important in metallic spin glasses (Section 6.3). Furthermore, in metallic spin glasses the interaction between the localized spins and the conduction electrons is also important for the dynamics: The exchange mechanism is of the RKKY type described in Section 2.1, where a localized spin polarizes the surrounding conduction electrons via an exchange interaction, and they in turn polarize a second localized spin. The exchange interaction between conduction electrons and the impurity spins is important for the transport properties of spin glasses, leading, for instance to a characteristic temperature dependence of the electrical resistivity (Section 10.3). But most important for us here is the fact that it also destroys the conservation (6.4) of the total (localized) spin S_{tot}, since a conduction electron can flip a single spin S_i without any change of the states of all other spins S_j. One has therefore to add to the equation of motion (6.2) a stochastic term or 'heat bath' which describes the influence of the conduction electrons on the localized spins in the same way as we did in the case of the Ising model (Section 4.1). A model with Heisenberg spins in a heat bath will be considered in Section 7.2.

If the interaction with the heat bath is sufficiently strong, one can neglect the processional dynamics (6.2) completely. The resulting vector spin glass model is very similar to the Glauber model discussed in Section 4.1. It ignores completely spin waves and has only relaxational dynamics. In Sections 6.1 and 6.4 we consider this model with and without additional anisotropy.

A vector spin glass is described at sufficiently high temperatures and fields by an order parameter $\mathbf{q}$ with the components q_μ, $\mu = 1, \ldots m$. This parameter $\mathbf{q}$ is a straightforward generalization of the parameter q (3.46) in Ising spin glasses. In a magnetic field the longitudinal and transverse components q_L and q_T of $\mathbf{q} = (q_L, q_T, \ldots q_T)$ become different. For a field $\mathbf{h}$ in direction $\mu = 1$ one has

$$q_\mu = N^{-1} \sum_i [\langle S_{i\mu} \rangle^2]_{av} = q_T + (q_L - q_T)\delta_{\mu,1} \qquad (6.6)$$

and, in addition, a third nontrivial spin glass parameter (Gabay and Toulouse, 1981)

$$p_\mu = N^{-1} \sum_i [\langle S_{i\mu}^2 \rangle]_{av} = 1 + (m\delta_{\mu,1} - 1)X \qquad (6.7)$$

since $\langle S_{i\mu}^2 \rangle \neq 1$. The quadrupolar deformation parameter X vanishes for $\mathbf{h} = 0$; (6.7) holds for the spin normalization

$$\sum_{\mu=1}^m S_{i\mu}^2 = m \qquad (6.8)$$

i.e. a spin $\mathbf{S}_i$ has length $m^{\frac{1}{2}}$.

6.1 The Gabay–Toulouse line

Our first task will be the calculation of the static properties of vector spins in an external field $h_\mu = h\delta_{\mu,1}$ and for a field-temperature region in which a solution with the order parameters (6.6) and (6.7) is correct. We know already from Ising spin glasses that such a solution possibly holds only at sufficiently high fields and temperatures and is limited by a critical line. The free energy is again best calculated by introducing replicas. The same steps as described in Section 3.3 for Ising spins lead to the free energy per spin (Cragg, Sherrington and Gabay, 1982; Gabay, Garel and De Dominicis, 1982)

$$\beta f(T, h) = \lim_{n \to 0} \frac{1}{n} [-(\beta J/2)^2 \sum_{\alpha\beta\mu\nu} (q_{\mu\nu}^{\alpha\beta})^2 + \ln Tr \exp \mathcal{H}_{eff}] \qquad (6.9)$$

with the effective Hamiltonian

$$\mathcal{H}_{eff} = \frac{1}{2}(\beta J)^2 \sum_{\alpha\beta\mu\nu} q_{\mu\nu}^{\alpha\beta} S_\mu^\alpha S_\nu^\beta + \beta h \sum_\alpha S_1^\alpha \qquad (6.10)$$

Here $\mu, \nu = 1, \ldots m$ are components and $\alpha, \beta = 1, \ldots n$ replica indices. In deriving (6.9) and (6.10) we assumed a symmetric distribution of bonds (3.17 with $J_0 = 0$). In contrast to (3.43) the replica sums still include the terms $\alpha = \beta$. With the field direction as the only distinguished direction we can chose a coordinate system in spin space for which the tensor $q_{\mu\nu}^{\alpha\beta}$ is diagonal: $q_{\mu\nu}^{\alpha\beta} = \delta_{\mu\nu} q_\mu^{\alpha\beta}$. Making a replica-symmetric ansatz in analogy with (3.48) we can identify the order parameters (6.6) and (6.7):

$$q_\mu^{\alpha\beta} = q_\mu = \langle S_\mu^\alpha S_\mu^\beta \rangle \tag{6.11}$$

for $\alpha \neq \beta$, while

$$q_\mu^{\alpha\alpha} = p_\mu = \langle (S_\mu^\alpha)^2 \rangle \tag{6.12}$$

where the average $\langle \cdots \rangle$ is defined as in (3.39) but with (3.42) replaced by (6.10).

This replica-symmetric solution is correct above and along an instability line first detected by Gabay and Toulouse (1981) which will be calculated below. Performing the limit $n \to 0$ in (6.9) and (6.10) leads to

$$\lim_{n\to 0} \frac{1}{n} \sum_{\alpha\beta\mu\nu} (q_{\mu\nu}^{\alpha\beta})^2 = -(q_L^2 - q_T^2 + m(q_T^2 - 1) - m(m-1)X^2) \tag{6.13}$$

and

$$\sum_{\alpha\beta\mu\nu} q_{\mu\nu}^{\alpha\beta} S_\mu^\alpha S_\nu^\beta = [-q_L + (1 + (m-1)X)] \sum_\alpha (S_1^\alpha)^2$$

$$+[-q_T + (1-X)] \sum_\alpha \sum_{\mu\neq 1} (S_\mu^\alpha)^2 + q_L (\sum_\alpha S_1^\alpha)^2 + q_T \sum_{\mu\neq 1} (\sum_\alpha S_\mu^\alpha)^2 \tag{6.14}$$

and with

$$\exp\left[\frac{r^2}{2}\left(\sum_\alpha S_\mu^\alpha\right)^2\right] = (2\pi)^{-\frac{1}{2}} \int dt_\mu \exp[-\tfrac{1}{2}t_\mu^2 + rt_\mu \sum_\alpha S_\mu^\alpha] \tag{6.15}$$

to the free energy

$$-\beta f(T, h) = (\beta J/2)^2 [q_L^2 - q_T^2 + m(1 - q_T)^2$$

$$+m(m-1)X^2 - 2mX] + \int \prod_\mu \frac{dt_\mu}{\sqrt{2\pi}} \exp(-\tfrac{1}{2}t_\mu^2) \ln \tilde{Z} \tag{6.16}$$

Here,

$$\tilde{Z} = Tr \exp(\sum_\mu a_\mu S_\mu + bS_1^2) \tag{6.17}$$

$$a_\mu = \beta J q_\mu^{\frac{1}{2}} + \beta h \delta_{\mu,1} \tag{6.18}$$

$$b = \tfrac{1}{2}(\beta J)^2 (q_T - q_L + mX) \tag{6.19}$$

The parameters q_μ and p_μ can easily be expressed in terms of $\tilde{Z}$. One has

$$q_\mu = \left\langle \left(\tilde{Z}^{-1} \partial \tilde{Z}/\partial a_\mu \right)^2 \right\rangle_t \tag{6.20}$$

and

$$p_\mu = \langle \tilde{Z}^{-1} \partial^2 \tilde{Z}/\partial a_\mu^2 \rangle_t \tag{6.21}$$

where $\langle \cdots \rangle_t$ means the Gaussian average which enters into (6.16). Of special interest are the local spin glass susceptibility (see (3.129))

$$\chi_{\mu\nu}^{(2)} = \beta^2 [(\langle S_\mu S_\nu \rangle - \langle S_\mu \rangle \langle S_\nu \rangle)^2]_{av}$$

$$= \left\langle \left(\tilde{Z}^{-1} \partial^2 \tilde{Z}/\partial a_\mu \partial a_\nu - \tilde{Z}^{-2} (\partial \tilde{Z}/\partial a_\mu)(\partial \tilde{Z}/\partial a_\nu) \right)^2 \right\rangle_t \tag{6.22}$$

and the local linear susceptibility

$$\chi_\mu = \beta(p_\mu - q_\mu) \tag{6.23}$$

for vector spins.

The trace in (6.17) is an integration over an m-dimensional hypersphere. One has

$$\tilde{Z} = \int d^m S \delta(S_\perp^2 - (m - S_1^2)) \exp(\sum_\mu a_\mu S_\mu + b S_1^2) =$$

$$\frac{1}{2} \int_{-\sqrt{m}}^{\sqrt{m}} dS_1 (m - S_1^2)^{(m-3)/2} \exp(a_1 S_1 + b S_1^2)$$

$$\times \int d^{m-2}\Omega \exp(r S_\perp \cos \theta) \tag{6.24}$$

with

$$r = (a_2^2 + \cdots + a_m^2)^{\frac{1}{2}} = \beta J q_T^{\frac{1}{2}} \left(\sum_{\mu \neq 1} t_\mu^2 \right)^{\frac{1}{2}} \tag{6.25}$$

The integral in (6.24) over the $(m-2)$-dimensional surface of a sphere in $m - 1$ dimensions is tabulated (Abramovitz and Stegun, 1965, p. 376, eq. 9.6.18)

$$\int d^{m-2}\Omega \exp(r S_\perp \cos \theta) =$$

$$(2\pi)^{(m-1)/2} I_{(m-3)/2}(r S_\perp)/(r S_\perp)^{(m-3)/2} \tag{6.26}$$

where $I_n(z)$ are Bessel functions of the first kind. This leads to the partition function (Cragg et al, 1982)

$$\tilde{Z} = \tfrac{1}{2}(2\pi)^{(m-1)/2} r^{(3-m)/2} \times$$

$$\int_{-\sqrt{m}}^{\sqrt{m}} dS_1 e^{aS_1 + bS_1^2} (m - S_1^2)^{(m-3)/4} I_{(m-3)/2}(r(m - S_1^2)^{\frac{1}{2}}) \quad (6.27)$$

Gabay and Toulouse (1981) made the important observation that there is a transition line in the $h - T$ plane for the onset of the transverse order parameter (6.20) for $\mu = 2, 3, \cdots m$. For small fields this follows easily from (6.20) and (6.21) by expansion of the Bessel function in (6.26)

$$I_{(m-3)/2}(z)/z^{(m-3)/2} = \left(\frac{2}{\pi}\right)^{\frac{1}{2}} \frac{1}{1 \cdot 3 \cdots (m-3)} \times$$

$$\left(1 + \frac{z^2}{2(m-1)} + \frac{z^4}{8(m-1)(m+1)} + \cdots\right) \quad (6.28)$$

All quantities can now be expressed by the integrals

$$P_{mr} = \int_{-\sqrt{m}}^{\sqrt{m}} dS_1 \exp(a_1 S_1 + b S_1^2)(m - S_1^2)^{(m-3+r)/2} S_1^r \quad (6.29)$$

Examples are the order parameters

$$q_L = \langle (P_{01}/P_{00})^2 \rangle_{t_1} \quad (6.30)$$

and

$$p_1 = \langle (P_{02}/P_{00}) \rangle_{t_1} \quad (6.31)$$

The notation $\langle \cdots \rangle_{t_1}$ means a one-dimensional Gaussian average (like those which occurred in the Ising problem (3.52–53)) over the remaining integration variable t_1 in (6.16). The condition $q_T \to 0$ leads to

$$(m-1)^2 = (\beta J)^2 \langle (P_{20}/P_{00})^2 \rangle_{t_1} \quad (6.32)$$

or with $k_B T_f = J$ to the 'Gabay–Toulouse' (GT) line in the field-temperature plane (Cragg et al, 1982).

$$T_{c1}/T_f = 1 - \left(\frac{h}{k_B T_f}\right)^2 \left(\frac{m^2 + 4m + 2}{4(m+2)^2}\right) \quad (6.33)$$

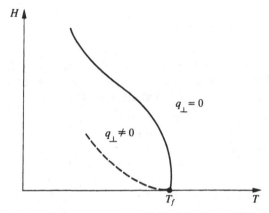

Figure 6.1: Sketch of the Gabay–Toulouse (GT) line for an infinite range vector spin glass. The low-temperature phase has nonzero transverse spin glass ordering. The Almeida–Thouless (AT) line, strictly speaking, no longer occurs, but there is a well-defined crossover region, indicated by the dashed line, which follows a similar curve.

$(\beta h \ll 1)$. This line varies as $\theta_c \equiv (T_f - T)/T_f \propto h^2$, in contrast to the AT line for Ising spins with $\theta_c \propto h^{2/3}$ from (3.28) or (5.36). It describes the freezing-in of the *transverse* spin components: $q_T = 0$ above $T_{c1}(h)$ and $q_T \neq 0$ below. The freezing temperature $T_{c1} \to T_f$ for $h = 0$, in agreement with that of Ising spins.

In this replica-symmetric approximation the GT line is a line of conventional second-order phase transitions: The order parameter q_T vanishes continuously at $T_{c1}(h)$, and the transverse spin glass susceptibility ((2.64), generalized to vector spins) diverges along this line, as shown below. The longitudinal order parameter q_L remains finite for all temperatures for $h \neq 0$. However, as we shall now show, the GT line is also an instability line with respect to replica symmetry breaking. In this sense it is similar to the AT line.

Consider small deviations of the order parameters from their replica-symmetric values q_μ and p_μ

$$q_\mu^{\alpha\beta} = q_\mu + \eta_\mu^{\alpha\beta} \tag{6.34}$$

$$q_\mu^{\alpha\alpha} = p_\mu + \epsilon_\mu^{\alpha} \tag{6.35}$$

and calculate the resulting change in the free energy. We do not go through the calculation (see Cragg et al, 1982) since the result is derived more easily from the dynamics discussed below. One finds eigenvalue equations with the eigenvectors c_μ^λ:

$$\sum_\nu (\delta_{\mu\nu} - J^2 \chi_{\mu\nu}^{(2)}) c_\nu^\lambda = \epsilon_\lambda c_\mu^\lambda \tag{6.36}$$

where the local spin glass susceptibility $\chi_{\mu\nu}^{(2)}$ (6.22) is calculated from the replica-symmetric solution. The latter becomes unstable if one of the eigenvalues ϵ_λ becomes negative. For $T > T_{c1}(h)$ all eigenvalues are positive and the replica-symmetric solution is stable. Near T_{c1} one has $\chi_{\mu\nu}^{(2)} = \chi_{\nu\mu}^{(2)}$ and $\chi_{21}^{(2)} = \chi_{31}^{(2)} = \cdots = \chi_{m1}^{(2)}$ and eigenvectors $c_2^\lambda = c_3^\lambda = \cdots c_m^\lambda$ which are symmetric in the plane perpendicular to the field. This leads from (6.36) to the eigenvalue equation for the *transverse* mode $\mu = 2$

$$1 - \epsilon_2 - \chi_{22}^{(2)} - \chi_{23}^{(2)}(m - 2) - \chi_{21}^{(2)} c_1/c_2 = 0 \tag{6.37}$$

with

$$\frac{c_1}{c_2} = \frac{2q_T(m - 1)}{(m + 2)^2} \tag{6.38}$$

from (6.22) with (6.26) to (6.30) and (6.31). One finds the eigenvalue

$$\epsilon_2 = -\frac{4q_T^2}{(m + 2)^2}\left(m + 1 + \sqrt{2}\frac{m - 1}{(m + 2)^2}\frac{|h|}{J} + O(h^2)\right) \tag{6.39}$$

which vanishes on the GT line (6.33) where $q_T \to 0$. Below this line the replica-symmetric solution becomes unstable and has to be replaced by a Parisi-type solution which describes broken ergodicity. One now has for vector spins below $T_{c1}(h)$ the same many-valley structure as discussed in Section 3.4 for Ising spins. This replica symmetry breaking is still weak just below the GT line (Elderfield and Sherrington 1982b), i.e. the Parisi functions $q_L(x)$ and $q_T(x)$ are nearly constant for all values of $x \in [0, 1]$. If one ignores it and calculates the eigenvalue ϵ_1 for the longitudinal mode within the replica-symmetric scheme, one finds a second instability line where $\epsilon_1 = 0$:

$$T_{c2}/T_f = 1 - \left(\frac{h}{J}\right)^{2/3}[(m + 1)(m + 2)/8]^{1/3} \tag{6.40}$$

which has exactly the field-temperature dependence of the AT line (3.28) and for $m = 1$ even the same coefficient. One has below this line much stronger replica symmetry breaking of the longitudinal component q_L than in the field-temperature region between $T_{c1}(h)$, and $T_{c2}(h)$. But the transition at $T_{c2}(h)$ is not sharp because replica symmetry has already been broken at $T_{c1}(h)$, so $T_{c2}(h)$ is better thought of as a crossover line from weak to strong irreversibility (Elderfield and Sherrington, 1982a,b; Gabay et al, 1982; Elderfield and Sherrington, 1984; Goltsev, 1984). The calculations in these references are based on Sompolinsky equations which are extensions of (5.74–82) to vector spins. The solutions of these equations have the property that $\epsilon_2 = 0$ for all T and h below the GT line, in analogy to (5.87) for Ising spins. One has 'massless' modes as discussed in Sections 3.4–5 and first predicted by Bray and Moore (1979). The susceptibility (6.23) has to be replaced (for $h \to 0$) by

$$\chi_\mu = \beta \left(p_\mu - \int_0^1 dx q_\mu(x) \right) \tag{6.41}$$

Both the GT line and the crossover line $T_{c2}(h)$ are indicated in Fig. 6.1. A similar line will also appear in a system with finite spontaneous magnetization (see Fig. 3.7 for Ising spins) and will be discussed in connection with reentrant transitions in Section 11.3.

The relaxational dynamics of a vector spin glass with infinite range interactions can be calculated exactly in the same way as for Ising spins (see Section 5.2) (Fischer, 1984). Above the GT line the limit $\omega \to 0$ leads back to the partition function (6.27) and to the order parameters (6.30) and (6.31). A possible dynamic instability can be found by the investigation of the kinetic coefficients $\Gamma_\mu(\omega)$ defined by

$$\Gamma_\mu^{-1}(\omega) = i \frac{\partial G_\mu^{-1}(\omega)}{\partial \omega} \tag{6.42}$$

which generalizes (5.14). In analogy with the Ising case, the propagator $G_\mu(\omega) = T\chi_\mu(\omega)$ is defined as an average of a propagator in a static external vector random field $\mathbf{z}$:

$$G_\mu(\omega)\delta_{\mu\nu} = [g_{\mu\nu}(\omega, \mathbf{z})]_z \tag{6.43}$$

where $g_{\mu\nu}(\omega, \mathbf{z})$ satisfies a (matrix) Dyson equation like (5.28):

$$\mathbf{g}^{-1}(\omega, \mathbf{z}) = \mathbf{G}_0^{-1}(\omega) + \mathbf{\Sigma}(\omega, \mathbf{z}) \tag{6.44}$$

with the bare propagator (see (5.14))

$$\mathbf{G}_0^{-1}(\omega) = (r_0 - i\omega\Gamma_0)\mathbf{1} - (\beta J)^2 \mathbf{G}(\omega) \tag{6.45}$$

and the matrix self energy $\mathbf{\Sigma}(\omega, \mathbf{z})$. Equation (5.29) now reads (in differentiated form) with the local spin glass susceptibility $\chi_{\mu\nu}^{(2)}(\omega) = \beta^2 [g_{\mu\nu}^2(\omega, \mathbf{z})]_z$ (Fischer, 1984).

$$\sum_\nu (\delta_{\mu\nu} - J^2 \chi_{\mu\nu}^{(2)}(\omega)) \frac{\partial G_\nu}{\partial \omega} =$$

$$i(\Gamma_0\beta^2)^{-1} \sum_\nu \chi^{(2)}_{\mu\nu}(\omega) - ([\mathbf{g}(\omega,\mathbf{z})(\partial\Sigma(\omega)/\partial\omega)\mathbf{g}(\omega,\mathbf{z})]_z)_\mu$$

$$\equiv f_\mu(\omega) \tag{6.46}$$

The diagonalization of the left hand side of (6.46) for $\omega = 0$ leads back to (6.36) or equivalently

$$\left(\frac{\partial G_\lambda(\omega)}{\partial\omega}\right)_{\omega=0} = \epsilon_\lambda^{-1} f_\lambda(0) \tag{6.47}$$

with

$$G_\mu(\omega) = \sum_\lambda G_\lambda(\omega)c_\mu^\lambda \tag{6.48}$$

and

$$f_\mu(\omega) = \sum_\lambda f_\lambda(\omega)c_\mu^\lambda \tag{6.49}$$

Above the GT line $\epsilon_2 = 0$ all nondiagonal susceptibilities $\chi^{(2)}_{\mu\nu}(0)$ with $\mu \neq \nu$ vanish and $\lambda = 2$ can be identified with the transverse mode. One has for the inverse relaxation rate from (6.42) and (6.47) $\Gamma_{\lambda=2}(0) = \Gamma_T(0) \to 0$, i.e. critical slowing down, indicating a dynamic instability. This instability is connected with the divergence of the transverse spin glass susceptibility

$$\chi_{SG}(0) = \frac{\chi^{(2)}_\perp(0)}{1 - J^2\chi^{(2)}_\perp(0)} \tag{6.50}$$

which replaces (3.128).

The *linear* susceptibility $\chi_\mu(\omega) = \beta G_\mu(\omega)$ behaves differently along the GT line from that of Ising spins along the AT line. For Ising spins we had $\chi(\omega) \propto \omega^\nu$ for $\omega \to 0$, with a field- and temperature-dependent exponent ν. The transverse susceptibility $\chi_T(\omega)$ in the same limit is proportional to $\omega^{1/2}$ everywhere along the GT line, including the point $h = 0$, $T = T_f$, as one would expect for a line of conventional second-order phase transitions (Fischer, 1984). This is because the field does not couple to the transverse components. The longitudinal susceptibility $\chi_L(\omega)$ is analytic everywhere along and above the GT line, except at the point $h = 0$, $T = T_f$. Here one has a crossover from $\chi_L(\omega) \propto \omega$ to $\chi_L(\omega) \propto \omega^{1/2}$ for $\omega \to 0$.

Experimentally, the GT line is less apparent than the AT line. The anomalies discussed below refer either to local susceptibilities or to the spin glass susceptibility. The latter cannot be measured directly in a finite magnetic field. For $h \neq 0$ the susceptibility $\chi \equiv \sum_{ij} \chi_{ij}$ differs from its local counterpart $\chi_{loc} = N^{-1} \sum_i \chi_{ii}$ since the nondiagonal terms χ_{ij} are nonzero. Hence the GT line can best be seen (if it exists) in Mössbauer data, which are sensitive to the local field on a Mössbauer atom, and in the onset of irreversibility, as discussed in connection with the AT line.

Unfortunately the Mössbauer experiments which have been performed were on 'reentrant' systems in which one has, with decreasing temperature, first a ferromagnetic state with nonzero spontaneous magnetization and only at lower temperatures a spin-glass-like state. Even these measurements, with a few exceptions, have only been made in zero field and therefore do not yield information about the GT line (see Section 11.3). In spin glasses without double transitions (say in AuFe with Fe concentrations less than 8at%) fewer Mössbauer data are known, and the field dependence has not yet been investigated systematically. For a more complete discussion of Mössbauer data on spin glasses see Section 9.1.

For this reason the best evidence for a possible GT line comes from torque and transverse ac susceptibility experiments, which both measure the onset of irreversibility. In both experiments anisotropy is crucial and leads to a crossover from the field/temperature relation (6.33) $\theta_{GT} \equiv (T_{c1} + T_f)/T_f = (h/h_1)^2$ with $h_1 = 10k_BT_f/\sqrt{23}\mu_B$ to that of Ising spins $\theta_{AT} = (h/h_2)^{2/3}$ with $h_2 = 2k_BT_f/\sqrt{3}\mu_B$. This anisotropy couples the macroscopic response in a uniform field to the microscopic degrees of freedom (Kotliar and Sompolinsky, 1984).

We concentrated in this section mostly on the GT line, which is specific to vector spin glasses. Many other properties are similar to those of Ising spins. The extension of the SK theory without replica symmetry breaking leads to very similar static properties such as the cusp in the linear susceptibility and in the specific heat. Furthermore, as we mentioned at the end of Section 3.5, one can derive TAP equations for vector spins, finding a number of metastable states or local minima of the free energy at $T = 0$, $h = 0$ which grows exponentially with the number of spins: $[N_s]_{av} \propto \exp 0.0233$ for $m = 2$ and $\exp 0.0084$ for $m = 3$ (Bray and Moore, 1981). This large number of states should lead to the same nonergodic effects as in Ising spin glasses (see Section 3.5). The Sompolinsky equations can also be generalized to vector spins (Goltsev, 1984) and lead to the Parisi functions $q_L(x)$, $q_T(x)$ mentioned above.

6.2 Anisotropy: nonrandom

Anisotropy has a profound influence on many properties of spin glasses. In the simplest case it is due to the crystal field and has its origin in the electrostatic interaction between the 3d or 4f electrons of the magnetic atom and its neighbouring ions. This electrostatic interaction lifts the orbital degeneracy of the 3d or 4f electrons. The perturbation is described by a 'crystal field' Hamiltonian which depends on products of the total angular momenta $\mathbf{J}$ and is invariant with respect to the symmetry of the crystal. In systems with spin and orbital moments one also has spin-orbit coupling. The spins follow the motion of the orbital momenta and also align themselves with respect to the crystal axes in such a way that the energy becomes a minimum. One describes this by a spin Hamiltonian with an additional crystal field term where the 'spins' in reality mean total angular momenta (see e.g. White, 1983). An example is a hexagonal crystal with the lowest-order anisotropy term $-D\sum_i S_{zi}^2$ with the z-direction parallel to the easy axis. The total spin–glass Hamiltonian reads

$$\mathcal{H} = -\tfrac{1}{2}\sum_{ij} J_{ij}\mathbf{S}_i \cdot \mathbf{S}_j - D\sum_i S_{zi}^2 - \mathbf{h}\cdot\sum_i \mathbf{S}_i \qquad (6.51)$$

The constant D is allowed to have any magnitude or sign and the exchange J_{ij} is assumed to be random with the distribution $P(J_{ij})$ (3.29). This model should be suitable for dilute alloys of Mn, Zn, Mg, and Cd with hexagonal lattice structure. In the limit $D \to \infty$ all spins are parallel or antiparallel to the z-direction and (6.51) describes an Ising model. For $D \to -\infty$ one has a planar (XY) model with all spins in the $x - y$ plane, and for $D = 0$ the Heisenberg model. One expects a phase diagram in the $D - T$ plane with a paramagnetic region (P) for high temperatures, a 'longitudinal' spin glass phase (L) with $q_L \neq 0$, $q_T = 0$, and a transverse spin glass phase (T) with $q_T \neq 0$, $q_L = 0$.

The calculation of this phase diagram and of the longitudinal and transverse susceptibilities is a simple extension of that for the Heisenberg model (Section 6.1) (Cragg and Sherrington, 1982; Roberts and Bray, 1982). We normalize the spin length as in (6.8) and assume $h = 0$ and the c–axis in direction $\mu = 1$. As before, in the replica-symmetric regime one has the order parameters $\mathbf{q} = (q_L, q_T, q_T)$ (6.6), the quadrupolar deformation tensor X (6.7), and the static local susceptibilities (6.23). The free energy is again calculated by means of the replica trick, leading to (6.9)–(6.21), where the field term is replaced by $\beta D \sum_\alpha (S_1^\alpha)^2$. This is because neither the field nor the anisotropy terms are involved in the various manipulations described in Section 3.3. The coefficients (6.18) and (6.19) now read

$$a_\mu = \beta J q_\mu^{\frac{1}{2}} t_\mu \qquad (6.52)$$

$$b = \tfrac{1}{2}(\beta J)^2 (q_T - q_L + mX) + \beta D \qquad (6.53)$$

The phase boundaries are calculated from $q_T = 0$ and $q_L = 0$, respectively, and are represented in Fig. 6.2.

In addition to the phases P, L and T one has a phase LT with $q_L \neq 0$ and $q_T \neq 0$. The static susceptibilities $\chi_L(T)$ and $\chi_T(T)$ have cusps at the temperatures where $q_L \to 0$, $q_T = 0$ and $q_T \to 0$, $q_L = 0$. However, the replica-symmetric solution becomes unstable below these lines and q_L and q_T have to be replaced by Parisi functions $q_L(x)$ and $q_T(x)$ as in Section 3.4 (Elderfield and Sherrington, 1983b; Elderfield, 1984). The susceptibilities (for $h \to 0$) have to be calculated from (6.41) and turn out to be constant (Elderfield and Sherrington, 1983b) as in the case of Ising spins (see (5.85)).

The inclusion of a magnetic field in direction $\mu = 1$ leads in the replica-symmetric theory to a rounding of the cusp in $\chi_L(T)$ with q_L nonzero everywhere and reduces the q_T-ordering temperature. For small fields h this reduction is proportional to h^2 and therefore similar to a GT line (Section 6.1). The longitudinal replica-symmetry breaking transition follows an $h^{2/3}$ law as one expects for an AT line (Cragg and Sherrington, 1982).

The phase boundaries L–LT and T–LT in Fig. 6.2 lie entirely in a region in which only solutions of the Parisi type are stable. These lines have been extrapolated from properties near the freezing temperature $k_B T / J = 1$ and the points $d_{\pm}$ for $T = 0$. Below the boundaries P–L and P–T one again has massless modes and a number of metastable states which (at least for $T = 0$) grows exponentially with the number of spins, as calculated by the method described in Section 3.5 (Roberts, 1983). We conclude that in spin glasses with additional uniaxial anisotropy all major spin glass properties predicted by the SK model are preserved. Monte Carlo simulations on systems with random nearest neighbour exchange interactions differ in detail from the phase diagram Fig. 6.2 but retain the feature of a double transition for a fixed parameter D (Morris and Bray, 1984).

These predictions have been tested experimentally for a variety of hexagonal systems such as dilute ZnMn, CdMn, and MgMn (Albrecht et al, 1982) and Er, Dy, Tb, or Gd in Y or Sc (Fert et al, 1982; Baberschke et al ,1984: Wendler et al, 1984). ZnMn is 'Ising–like', i.e. the Mn moments tend to lie parallel to the c–axis, CdMn is 'XY–like' with the easy axis in the basal plane $D < 0$ and MgMn is nearly isotropic or 'Heisenberg-like'. In all these systems one observes a more or less pronounced cusp in the zero-field cooled susceptibility, either for $\chi_{\perp}$ or $\chi_{\parallel}$ i.e. for a field either perpendicular or parallel to the c-axis. If a cusp is observed in $\chi_{\perp}$, $\chi_{\parallel}$ usually behaves smoothly and vice versa. Fig. 6.3 shows the phase diagram obtained for Sc- and Y-based alloys. In these systems the host has been kept constant, but different types of impurities have been added. Here, instead of the cusp temperature the temperature at which the field-cooled susceptibility differs from the zero-field-cooled susceptibility has been chosen. This difference describes the onset of remanence and irreversibility effects as described in Section 5.2). The field dependence of the thermo-remanent and isothermal

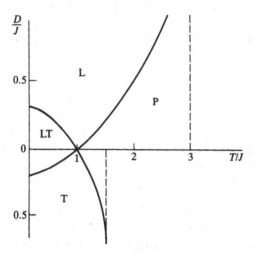

Figure 6.2: Sketch of the phase diagram of an an infinite–range vector spin glass model with uniaxial anisotropy D (P paramagnetic phase, L longitudinal phase, T transverse phase, LT a phase in which all components order) (following Roberts and Bray, 1982, and Cragg and Sherrington, 1982).

remanent magnetizations have been investigated for $\underline{Y}$Er (Fert et al, 1982), which is Ising-like and behaves similarly to $\underline{Au}$Fe and $\underline{Y}$Dy and $\underline{Y}$Tb, which are XY-like. In the latter case the field dependence is somewhat different. The remanence is pronounced only for the field direction in which the cusp appears. In some of the systems one also observes AT or GT lines as predicted by the theory. At higher concentrations these systems often show a rather complicated magnetic long-range order.

Other types of anisotropies also have been investigated. In the case of cubic anisotropy (Roberts, 1982) one has to lowest order in the spin variables a term proportional to $\sum_{i,\mu} S_{i,\mu}^4$ (see Section 12.1). A model in which the anisotropy constant D in (6.51) is assumed to vary randomly from site to site with a distribution $P(D) = (1-y)\delta(D-D_1) + y\delta(D-D_2)$ with parameters y, D_1 and D_2 leads to a rich variety of phase diagrams (Viana and Bray, 1983, 1985). A model with different longitudinal and transverse parts of the exchange interaction

$$-\sum_{ij}(J_{ij}^L S_{i1}S_{j1} + J_{ij}^T \sum_{\mu \neq 1} S_{i\mu}S_{j\mu}) \tag{6.54}$$

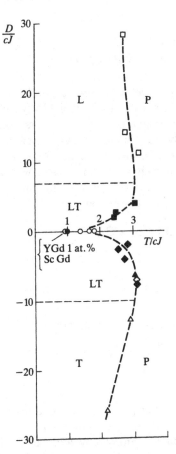

Figure 6.3: Experimental phase diagram for Sc- and Y-based alloys. To compare with theory (broken lines as defined in Fig. 6.2) the spins $\mathbf{S}_i$ in (6.51) have to be replaced by the total angular momenta $\mathbf{J}_i$, the exchange parameters J_{ij} by $J_{ij}(g_J - 1)^2$ with the de Gennes factor $(g_J - 1)^2$, and the anisotropy constant D by the crystal-field coefficient B_{20}. Dilution of Er, Dy, or Gd is taken into account by replacing the width J by cJ. Since higher crystal-field coefficients are not completely negligible, the ratio D/cJ has been replaced in Fig. 6.3 by $\chi_a(T_f) = 3[(\chi_\parallel - \chi_\perp)/c(\chi_\parallel + 2\chi_\perp)]_{T=T_f}$. The experimental data correspond to YEr ($\square$), ScEr($\blacksquare$), YDy ($\diamond$), ScDy($\blacklozenge$), YTb ($\triangle$), ScTb ($\blacktriangle$), YGd ($\circ$), and ScGd ($\bullet$) alloys (from Baberschke et al, 1984).

both with a Gaussian distribution, also leads to a phase diagram with phases P, L, T and LT similar to that shown in Fig. 6.2 (Bray and Viana, 1983).

6.3   the Dzyaloshinskii–Moriya interaction

So far we considered mostly single-site uniaxial anisotropy and its effect on static spin glass properties. Dzyaloshinskii (1958) and Moriya (1960) observed that the exchange interaction might also contain an anisotropic term. Dzyaloshinskii predicted, purely on grounds of symmetry, a term $\mathcal{H} \propto \mathbf{D} \cdot (\mathbf{S}_1 \times \mathbf{S}_2)$ in magnetic crystals with low symmetry. Moriya found a microscopic mechanism which leads to such a term in systems with spin–orbit coupling. Owing to this spin–orbit interaction one has an indirect coupling between the spins. The resulting Hamiltonian is

$$\mathcal{H} = -\tfrac{1}{2} \sum_{ij} J_{ij} \mathbf{S}_i \cdot \mathbf{S}_j - \tfrac{1}{2} \sum_{ij} \mathbf{D}_{ij} \cdot (\mathbf{S}_i \times \mathbf{S}_j) \tag{6.55}$$

with

$$J_{ij} = J_{ji}, \quad \mathbf{D}_{ij} = -\mathbf{D}_{ji}, \tag{6.56}$$

where $\mathbf{D}_{ij}$ is a pseudovector (in spin space) whose components $D_{ij,\rho}$ are elements of the antisymmetric tensor $D_{ij\mu\nu}$:

$$D_{ij\mu\nu} = \epsilon_{\mu\nu\rho} D_{ij,\rho}, \qquad \mu, \nu, \rho = 1, 2, 3 \tag{6.57}$$

This Hamiltonian is no longer is rotation invariant. It turns out that this anisotropy is *unidirectional* (see Section 7.3), in contrast to the uniaxial anisotropy discussed in the preceding section: It depends on the direction of the spins (i.e. $\mathbf{S}_i \times \mathbf{S}_j$) and not on the anisotropy axis as in (6.51). The second term on the right hand side of (6.55) is commonly called the Dzyaloshinskii–Moriya (DM) interaction.

An important mechanism which leads to the DM interaction has been found by Fert and Levy (1980) (Levy and Fert, 1981). Experiments by Prejean, Joliclerc and Monod (1980) on C̲uMn indicated that additional impurities with orbital moments strongly modified the remanent magnetization. This modification can be explained by additional anisotropy (see below). Fert and Levy therefore considered a model in which one has local exchange between a conduction electron at $\mathbf{r}$ with spin $\mathbf{s}$ and two localized spins $\mathbf{S}_A$ and $\mathbf{S}_B$ at sites $\mathbf{R}_A$ and $\mathbf{R}_B$, and in addition spin–orbit scattering with a nonmagnetic impurity at site $\mathbf{R} = 0$. This leads to the interaction terms

$$V = -J_{sd}\delta(\mathbf{r} - \mathbf{R}_A)\mathbf{s} \cdot \mathbf{S}_A - J_{sd}\delta(\mathbf{r} - \mathbf{R}_B)\mathbf{s} \cdot \mathbf{S}_B + \lambda(\mathbf{r})\mathbf{s} \cdot \mathbf{l} \quad (6.58)$$

with the spin–orbit coupling $\lambda(\mathbf{r})$ and the orbital moment $\mathbf{l}$. The interaction (6.58) leads in second-order perturbation theory to processes in which an electron is polarized by spin $\mathbf{S}_A$, scattered by the orbital moment $\mathbf{l}$, and finally polarizes spin $\mathbf{S}_B$. The interaction energy clearly depends on the positions $\mathbf{R}_A$ and $\mathbf{R}_B$ and is no longer rotation invariant. This process is one order higher than the RKKY interaction mentioned at the beginning of this chapter. For many electrons the spin $\mathbf{S}$ is replaced by the spin density operator $\mathbf{S}(\mathbf{r})$, which is expressed in terms of the electron field operators $\Psi(\mathbf{r})$ as $\Psi^\dagger(\mathbf{r})\mathbf{s}\Psi(\mathbf{r})$ (the resulting interaction is integrated over all $\mathbf{r}$; see Kittel (1964), p. 360). The interaction (6.58) is obtained by calculating the change of the electron ground state energy due to the interaction V in second-order perturbation theory. This leads to the RKKY interaction for spins $\mathbf{S}_A$, $\mathbf{S}_B$ (the first term in (6.58) and (with $\mathbf{s} = \frac{1}{2}\hbar\vec{\sigma}$, where $\vec{\sigma}$ are the Pauli matrices), using

$$(\mathbf{S}_A \cdot \vec{\sigma})(\mathbf{S}_B \cdot \vec{\sigma}) = \mathbf{S}_A \cdot \mathbf{S}_B + i(\mathbf{S}_A \times \mathbf{S}_B) \cdot \vec{\sigma} \quad (6.59)$$

and

$$Tr_S(\mathbf{S}_A \cdot \mathbf{s})\mathbf{s}(\mathbf{S}_B \cdot \mathbf{s}) = \frac{3}{8}\hbar^3 i(\mathbf{S}_A \times \mathbf{S}_B) \quad (6.60)$$

to (Levy and Fert, 1981)

$$\mathbf{D}_{ij} = V_1 \sin[(1 + \gamma)\phi] \times$$

$$\sum_n \frac{\sin[k_F(R_{ij} + R_{in} + R_{jn}) + \phi](\hat{\mathbf{R}}_{in} \cdot \hat{\mathbf{R}}_{jn})\hat{\mathbf{R}}_{in} \times \hat{\mathbf{R}}_{jn}}{(1 + \gamma)R_{ij}R_{in}R_{jn}} \quad (6.61)$$

Equation (6.61) holds for impurities on the sites n which have Z_d d-electrons with spin–orbit interaction λ_d and which lead to resonance scattering with the phase shift $\phi = \pi Z_d/10$ by the conduction electrons. The other quantities in (6.61) are

$$R_{in} = |\mathbf{R}_i - \mathbf{R}_n|, \quad \hat{\mathbf{R}}_{in} = \mathbf{R}_{in}/R_{in} \quad (6.62)$$

$$V_1 = \left(\frac{135\pi}{32}\right)\left(\frac{\lambda_d J_{sd}^2}{\epsilon_F k_F^3}\right) \quad (6.63)$$

$$\gamma = J_{sd}(R_{ij} + R_{in} + R_{jn}) \quad (6.64)$$

where ϵ_F is the Fermi energy and k_F the Fermi momentum. A full derivation of (6.61) is very tedious. The DM interaction in (6.55) is random since the nonmagnetic impurities are assumed to be randomly distributed. Levy and Fert (1981) estimated the ratio V_1/V_0 of the strength of the DM interaction to that of the RKKY interaction, obtaining 0.2 for Pt and 0.11 for Co in $\underline{C}$uMn (see also Goldberg, Levy and Fert, 1985). While the RKKY interaction is invariant under rotation of the spin system, the DM interaction depends on the orientation of $\mathbf{S}_i \times \mathbf{S}_j$ with respect to the local axis $\hat{\mathbf{R}}_i \times \hat{\mathbf{R}}_j$. The macroscopic measure of the anisotropy is $K_1 = \sum_{ij} \mathbf{D}_{ij}^2$; it varies linearly with the impurity concentration. We will discuss macroscopic anisotropy in Section 7.3.

6.4   Random DM anisotropy

The DM interaction discussed in Section 6.3 leads to random anisotropy and seems to be responsible for many low-temperature properties of spin glasses. We saw already in Section 6.2 that nonrandom uniaxial anisotropy can reduce the number of effective spin components. An extreme example is the limit $D \to \infty$, for which (6.51) describes an Ising spin glass, and for which all transverse spin degrees of freedom are suppressed. We will show that random DM anisotropy has a similar effect. In addition, it strongly modifies the critical behaviour near the freezing temperature. In three-dimensional systems with short-range interactions it is probably even responsible for the existence of a sharp phase transition, which almost surely is absent in isotropic Heisenberg spin glasses (see Chapter 8). At low temperature ($T \ll T_f$) the DM interaction leads to a breaking of the global rotation invariance and to macroscopic anisotropy as observed in ESR, NMR, remanence, and torque experiments (see Section 7.3).

The inclusion of random DM anisotropy in the MFT developed in Chapters 3 and 5 and the beginning of this chapter is straightforward (Kotliar and Sompolinsky, 1984; Fischer, 1985). We will consider only the dynamic approach, which in the limit of $t \to \infty$ or zero frequencies yields all static properties of interest. It is convenient to introduce soft spins $\vec{\sigma}$ with m components and write the DM interaction (6.55) in the form

$$\mathbf{D}_{ij} \cdot \vec{\sigma}_i \times \vec{\sigma}_j = \sum_{\mu\nu} \sigma_{i\mu} D_{ij\mu\nu} \sigma_{j\nu} \tag{6.65}$$

Each component $D_{ij\mu\nu}$ is assumed to be random with the distribution

$$P(D_{ij\mu\nu}) = \left(\frac{N}{2\pi D^2}\right)^{\frac{1}{2}} \exp\left[\frac{-N(D_{ij\mu\nu})^2}{2D^2}\right] \tag{6.66}$$

in analogy to (3.29). The total soft-spin Hamiltonian for a magnetic field in direction $\mu = 1$ reads (see Section 5.1)

$$\beta\mathcal{H} = -\tfrac{1}{2}\beta\sum_{ij} J_{ij}\vec{\sigma}_i\cdot\vec{\sigma}_j + \sum_i \left(\tfrac{1}{2}r_0\vec{\sigma}_i^2 + \frac{u}{4m}(\vec{\sigma}_i^2)^2 - \beta h\sigma_{i,1}\right)$$

$$-\tfrac{1}{2}\beta\sum_{ij\mu\nu}\sigma_{i\mu}D_{ij\mu\nu}\sigma_{j\nu} \tag{6.67}$$

We are mainly interested in the local susceptibility tensor

$$T\chi_{\mu\nu}(t-t') \equiv G_{\mu\nu}(t-t') = \left[\frac{\partial\langle\sigma_{i\mu}(t)\rangle}{\partial\beta h_{i\nu}(t')}\right]_{av} \tag{6.68}$$

and the autocorrelation function

$$C_{\mu\nu}(t-t') = [\langle\sigma_{i\mu}(t)\sigma_{i\nu}(t')\rangle]_{av} \tag{6.69}$$

where $[\cdots]_{av}$ means averaging over the distributions (3.29) with $J_0 = 0$ and (6.66). We proceed as in Section 5.2 and find the effective noise (see (5.12)) (Fischer, 1985)

$$\langle\phi_{i\mu}(\omega)\phi_{i\nu}(-\omega)\rangle = 2\Gamma_0^{-1}\delta_{\mu\nu} + (\beta J)^2[(1-d^2\delta_{\mu\nu})C_{\mu\nu}(\omega)$$

$$+d^2\sum_\rho C_{\rho\rho}(\omega)] \tag{6.70}$$

The equation of motion (5.13) is replaced by (omitting the index i)

$$\sigma_\mu(\omega) = G_\mu^0(\omega)[\phi_\mu(\omega) + \beta h(\omega)]$$

$$-\frac{u}{m}\sum_\nu\int\frac{d\omega_1 d\omega_2}{(2\pi)^2}\sigma_\mu(\omega_1)\sigma_\nu(\omega_2)\sigma_\nu(\omega-\omega_1-\omega_2) \tag{6.71}$$

with $d = D/J$ and with the inverse bare propagator

$$\begin{aligned}G_\mu^0(\omega) &= \{r_0 - i\omega\Gamma_0^{-1} - (\beta J)^2[(1-d)^2 G_\mu(\omega)\\ &\quad +d^2\sum_\nu G_\nu(\omega)]\}^{-1}\end{aligned} \tag{6.72}$$

In (6.71) and (6.72) we have chosen a coordinate system in which the response and correlation functions are diagonal

$$G_{\mu\nu}(\omega) = G_\mu(\omega)\delta_{\mu\nu}, \quad C_{\mu\nu}(\omega) = C_\mu(\omega)\delta_{\mu\nu} \tag{6.73}$$

From (6.70) to (6.72) it is obvious that the anisotropy mixes the longitudinal mode $\mu = 1$ with the transverse modes $\mu = 2,\ldots m$. The full propagator $G_\mu(\omega)$ is again defined by the Dyson equation

$$G_\mu^{-1}(\omega) = (G_\mu^0(\omega))^{-1} + \Sigma_\mu(\omega) \tag{6.74}$$

where the self energy $\Sigma_\mu(\omega)$ is calculated by summing perturbation theory in u (see Section 4.3).

Above the critical line (to be determined below) the static properties are determined by the order parameters q_μ (6.6) for soft spins, together with the quadrupolar deformation tensor X (6.7), with

$$C_\mu(t) = \tilde{C}_\mu(t) + q_\mu, \quad \tilde{C}_\mu(\infty) = 0 \tag{6.75}$$

We found in Section 6.1 that the GT line was defined by the divergence of one component of the relaxation time $\Gamma_\mu^{-1}(\omega)$ (6.42) for $\omega \to 0$. The same criterion leads for $h = \omega = 0$ with (6.72), (6.74) and $G(\omega = h = 0, T_f) = 1$ (from $\chi_\mu = T_f^{-1}$ at T_f) to a freezing temperature

$$T_f(d) = J[1 + (m-1)d^2]^{\frac{1}{2}} \tag{6.76}$$

We see that the DM anisotropy shifts T_f to higher temperature. The critical line $\Gamma_\mu(0) = 0$ is calculated as in Section 6.1. One has

$$\sum_\nu \{\delta_{\mu\nu} - J^2[(1-d^2)\chi_{\mu\nu}^{(2)}(\omega) - d^2 \sum_\rho \chi_{\mu\rho}^{(2)}(\omega)]\}\frac{\partial G_\nu(\omega)}{\partial \omega} = f_\mu(6.77)$$

with f_μ from (6.45). A linear combination of $\Gamma_\mu^{-1}(0)$ diverges if one of the eigenvalues ϵ_λ of the eigenvalue equation

$$\det\{(1-\epsilon_\lambda)\delta_{\mu\nu} - J^2[(1-d^2)\chi_{\mu\nu}^{(2)}(0) - d^2 \sum_\rho \chi_{\mu\rho}^{(2)}(0)]\} = 0 \tag{6.78}$$

vanishes. The local spin glass susceptibility $\chi_{\mu\nu}^{(2)}$ (6.22) is found from (6.24) where (6.18) and (6.19) are replaced by

$$a_L = \beta J Q_L^{\frac{1}{2}} t_1 + \beta h, \quad a_T = \beta J Q_T^{\frac{1}{2}} \tag{6.79}$$

$$Q_L = q_L + d^2(m-1)q_T, \quad Q_T = q_T + d^2[q_L + (m-2)q_T] \tag{6.80}$$

and

$$b = \frac{1}{2}(\beta J)^2(1-d^2)(q_T - q_L + mX) \tag{6.81}$$

The anisotropy couples the longitudinal and transverse spin glass order parameters q_L and q_T. In isotropic vector spin glasses (Section 6.1) the critical line (the GT line) could be determined from $q_T \to 0$ or the 'freezing' of the transverse spin components. In the presence of random DM anisotropy and a field h one has $q_L \neq 0$ and $q_T \neq 0$ for all temperatures. The field induces

not only a longitudinal magnetization but also a transverse local component. The critical line $T_f(h)$ remains defined by (a) a dynamic instability
and (b) the onset of broken ergodicity, as manifested, e.g. by irreversibility
effects. At high temperatures and fields all eigenvalues ϵ_λ of (6.78) are positive and the solution with order parameters q_L and q_T is stable. The line
$T_f(h)$ where the first eigenvalue vanishes is indicated in Fig. 6.4. Below
$T_f(h)$ ergodicity is broken and one has to introduce Parisi order parameter functions $q_L(x)$ and $q_T(x)$ and the nonergodicity parameters $\Delta_L(x)$
and $\Delta_T(x)$. These parameters are determined by Sompolinsky's equations
(5.75) to (5.77), where the anisotropy enters into the time persistent part
of the noise (6.70) and into the susceptibility (5.70) (Kotliar and Sompolinsky 1984). Eq. (6.78) with $\epsilon_\lambda = 0$ can then be derived from the condition
$\Delta(x) \rightarrow 0$ in the some way as in (5.87). One arrives at the interesting
result that the soft mode eigenvalue remains zero everywhere below $T_f(h)$:
One has a marginally stable phase everywhere in this region, with algebraic
decay of the spin autocorrelation function $\tilde{C}_\lambda(t)$ (6.75) and a power law for
the response function (6.68) $G_\lambda(\omega) - G_\lambda(0) \propto \omega^{\nu_\lambda}$ where $\nu_\lambda = \nu_\lambda(T, h, d)$.

One can distinguish three regions in Fig. 6.4. If the anisotropy energy
is small compared to the magnetic energy $(d \ll (h/J)^{\frac{5}{2}})$ one has essentially
a GT line as for isotropic systems. The transverse components of the order parameter are small $(q_L \gg q_T)$, but nonergodicity is strongest in the
transverse directions: $\Delta_T \gg \Delta_L$. The line $T_f(h)$ behaves as $h^2 \propto \tau_c^1$ where
$\tau_c^1 = (\tilde{T}_f - T_f)/T_f(0)$ does not extrapolate to $T_f(0)$ but to a lower temperature $\tilde{T}_f$. In the intermediate region $(h/J)^{\frac{5}{2}} \ll d \ll (h/J)^{\frac{2}{3}}$ the transition
is still mostly triggered by q_T but one has $\tau_c \propto (d^2 h/J)^{\frac{1}{3}}$. In the strong
anisotropy region $d \gg (h/J)^{\frac{2}{3}}$,

$$\left(\frac{h}{J}\right) = \frac{4q_L^3}{m+2} \approx \frac{4\tau^3}{m+2}, \quad \tau = 1 - \frac{T}{T_f} \qquad (6.82)$$

which reduces for $m = 1$ to the AT line for Ising spins. In this region one
has $q_L = q_T = q_{Ising}$ and all local spin glass susceptibilities $\chi_{\mu\mu}^{(2)} = \chi^{(2)}$
are equal along this part of the critical line. The spins behave like Ising
spins (as first predicted by Bray and Moore (1982)). This also holds for
the dynamic critical exponent ν_λ for $\lambda = 2$ which varies in this region with
the field h as

$$\nu = \frac{1}{2} - \frac{\sqrt{m}}{\pi}\left[\frac{1}{4}(m+2)\left(\frac{h}{J}\right)^2\right]^{\frac{1}{3}} \qquad (6.83)$$

and reduces for $m = 1$ with (6.82) to the result (5.52) for Ising spins.

In Fig. 6.4 a second line is also indicated which is obtained by putting
the second eigenvalue of (6.78) equal to zero, assuming that the nonergod-

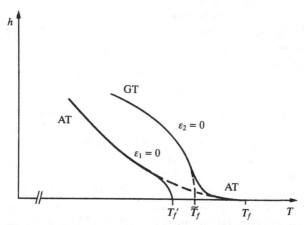

Figure 6.4: Schematic plot of the critical lines of an infinite-range vector spin glass model with random anisotropy. The part 'GT' is similar to the Gabay–Toulouse line in vector spin glasses without anisotropy, but extrapolates to a temperature $\tilde{T}_f \neq T_f$. The parts 'AT' have the field-temperature dependence of the Almeida–Thouless line in Ising spin glasses; ϵ_1 and ϵ_2 are the eigenvalues of the eigenvalue equation (6.78). The lower line is presumably a crossover line from weak to strong nonergodicity (from Fischer, 1985).

icity parameter $\Delta(x)$ can be ignored. This line has in an isotropic spin glass the character of an AT line and is a crossover line to strong irreversibility, as discussed in Section 6.1. The same presumably holds true for anisotropic spin glasses where the line has the field-temperature dependence indicated.

One can imagine that the crossover from Heisenberg to Ising properties also exists for other types of random anisotropy. In amorphous magnets the crystal-field anisotropy discussed in Section 6.2 becomes more or less random, and one can show (Fischer and Zippelius, 1986) that if it is large enough this random local anisotropy also leads to an instability line of the AT type (see also Section 12.2).

Even for symmetric distributions of the exchange and anisotropy $P(J_{ij})$ (3.29) and $P(D_{ij\mu\nu})$ (6.66) the DM interaction leads to different local and uniform susceptibilities below T_f. The reason is the appearance of macroscopic anisotropy, which breaks the global rotation invariance. The *local* susceptibility χ_μ is given by (6.41) and depends only implicitly via p_μ and $q_\mu(x)$ on the anisotropy parameter D. The uniform susceptibility

$\chi_\mu^{tot} = \sum_{ij} \chi_{ij\mu\mu}$ depends explicitly on D. For $D = 0$ one has (owing to rotation invariance) $\chi^{tot} = M/h$ and for $D \neq 0$ a transverse susceptibility (Kotliar and Sompolinsky, 1984)

$$\chi_T^{tot}(x = 1) = \frac{M}{h} - \Delta_T(0) \left(1 + \frac{hM_r}{K_1}\right)^{-1} \qquad (6.84)$$

with the remanent magnetization $M_r \propto h\Delta_L$ and the macroscopic aniso-tropy constant

$$K_1 = -D^2 \int_0^1 dx \left\{ [2q_T(x) + q_L(x)] \frac{d\Delta_T}{dx} + q_T(x) \frac{d\Delta_L}{dx} \right\} \qquad (6.85)$$

In thermal equilibrium $(x = 0)$ the system reaches all states and the sus-ceptibility remains $\chi^{tot}(x = 0) = M/h$ with $M = [\langle S_L \rangle]_{av}$ (see Section 7.2).

6.5 Dynamics with conserved magnetization

As noted at the beginning of this chapter, the standard dynamics (6.2) of Heisenberg spins conserve the total magnetization of the system. Al-though this conservation is broken by s–d interactions in RKKY glasses and, more generally, by dipolar forces and crystal field or random (e.g. Dzyaloshinskii–Moriya) anisotropies, it is of interest to examine the ideal case to see whether the conserving dynamics introduce new features into the physics with possible experimental consequences. In this section we study this question. The conclusion will turn out, at least for the aspects of the problem that we examine, to be that nothing dramatically new hap-pens. But this in itself is interesting because it is in strong contrast to the situation in the coresponding problems in ferro- and antiferromagnets.

Our starting point is a Langevin model with soft 3-component spins which is the obvious generalization of our soft-spin Ising model (4.22–23) except that the kinetic coefficient Γ_0 becomes wavenumber dependent, pro-portional to k^2 as $\mathbf{k} \to 0$. Thus we write the equation of motion as

$$\frac{\partial \sigma_i}{\partial t} = \Gamma_0 \nabla^2 \frac{\partial H}{\partial \sigma_i} + \eta_i(t) \qquad (6.86)$$

In our model, the expression '∇^2' means a lattice second derivative opera-tor; its Fourier transform is

$$-K^2(\mathbf{k}) \equiv -2 \sum_{\mu=1}^{3} (1 - \cos k_\mu) \xrightarrow{k \to 0} -k^2 \qquad (6.87)$$

The presence of the ∇^2 factor in the equation of motion means that this model will exhibit diffusive rather than simple relaxational behaviour at long wavelengths. Again, an Einstein relation between the kinetic coefficient in the dissipative term in the equation of motion and the mean square amplitude of the noise $\eta(t)$ holds:

$$\sum_j e^{-i\mathbf{k}\cdot\mathbf{r}_{ij}} \langle \eta_{i\mu}(t)\eta_{j\nu}(t')\rangle = 2T\Gamma_0 K^2(\mathbf{k})\delta(t-t')\delta_{\mu\nu} \qquad (6.88)$$

Thus both the noise and the decay term in the equation of motion (6.86) vanish at $k = 0$, guaranteeing the conservation of the total magnetization. Actually, in our analysis of this model the vector nature of the spins will play no essential role, and the following discussion applies equally well to a hypothetical Ising model (or one with any number of components) in which the total spin is conserved. (Of course, it is hard to imagine a real Ising-like spin system with a conserved magnetization, but if spin-glass-like behaviour ever occurred in a system where the effective 'spin' was a particle density operator, the present model would be appropriate.)

We now study the onset of critical slowing down on approaching the spin glass transition from above T_f in this model in the same way that we did in Sections 5.1 and 5.2 for its nonconserving Ising counterpart, working in the mean field limit of large lattice coordination number or dimensionality. The theory has nearly the same structure as we found there; the basic difference is that the 'bare' propagator defined in (5.14) after the random bonds were integrated out of the problem is no longer site-diagonal. It is replaced by

$$G_0(\mathbf{k},\omega) = \left[\frac{-i\omega}{\Gamma_0 K^2(\mathbf{k})} + r_0 - (\beta J)^2 G(\omega) \right]^{-1} \qquad (6.89)$$

where

$$G(\omega) \equiv \frac{1}{N}\sum_k G(\mathbf{k},\omega) \qquad (6.90)$$

is the average single-site response function (which, as in the nonconserved theory, has to be self-consistently determined). The diffusive form of $G_0(\mathbf{k},\omega)$ reflects the spin conservation.

As in the previous discussion, in the mean field limit above T_f the self-energy corrections do not have any singular frequency dependence at low frequencies. Furthermore, it is simple to verify that they do not have any singular dependence on $\mathbf{k}$, either (in particular, no terms proportional to k^{-2} at small k). Therefore in this limit we may write

$$G^{-1}(\mathbf{k},\omega) = \frac{-i\omega}{\Gamma_0 K^2(\mathbf{k})} + r + \frac{(\beta J)^2}{N}\sum_k [G(\mathbf{k},0) - G(\mathbf{k},\omega)] \qquad (6.91)$$

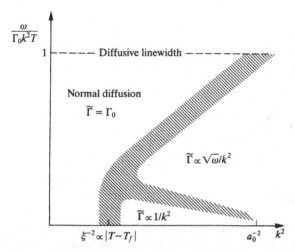

Figure 6.5: Regions of (k, ω) space with different effective dynamics for the purely dissipative model (6.86) with (6.88) (from Hertz and Klemm, 1983).

where $G^{-1}(\mathbf{k}, 0) = r$ (independent of $\mathbf{k}$) is the static limit of the response function.

The absence of self-energy terms proportional to k^{-2} means that the transport coefficient Γ_0 is not changed from its bare value. That is, if we write the exact G^{-1} formally in terms of an effective $\mathbf{k}$- and ω-dependent generalized transport coefficient $\Gamma(\mathbf{k}, \omega)$,

$$G^{-1}(\mathbf{k}, \omega) = \frac{-i\omega}{\Gamma(\mathbf{k}, \omega) K^2(\mathbf{k})} + r(\mathbf{k}, \omega) \qquad (6.92)$$

then $\Gamma(\mathbf{k}, \omega)$ is exactly Γ_0 at $k = 0$. However, the essential point is that the region of k-space in which this is even approximately true shrinks drastically as one approaches the spin glass transition. This is because, as in the corresponding situation described in Section 5.2, the coefficient of $-i\omega$ in the second term of (6.91) blows up proportional to χ_{SG}; this was the origin of the critical slowing down (5.20) as $T \to T_f$. So here in the conserving model we find that this critical slowing down occurs just as it does in the nonconserving one, for k of the order of or greater than a characteristic inverse length $\kappa \propto \chi_{SG}^{-1/2}$. The k^{-2} behaviour of the first term in (6.91) means that the critical slowing down does not occur at extremely long wavelengths ($k \ll \kappa$), but the important point is that the closer one gets to

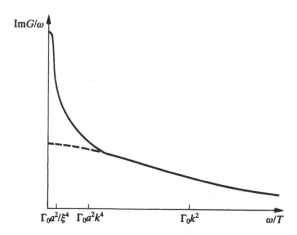

Figure 6.6: Schematic picture of the neutron line shape expected in a purely dissipative model. The broken line is the continuation of a normal diffusive Lorentzian (from Hertz and Klemm, 1983).

the transition, the smaller this hydrodynamic region is. In most of k-space, our diffusive model behaves the same in the long-time limit of its dynamics near T_f as the old relaxational one.

The discussion thus far concerns only the zero-frequency limit of the effective wavenumber-dependent kinetic coefficient $\Gamma(\mathbf{k}, \omega) K^2(\mathbf{k})$, i.e. terms in $G^{-1}(\mathbf{k}, \omega)$ which are linear in ω. We know, again from our examination of the nonconserving model, that for higher frequencies (and all the way down to $\omega = 0$ right at T_f), the dominant behaviour in G^{-1} is proportional to $(-i\omega)^{1/2}$. Again, the important qualitative comparison is between this term and the bare one $-i\omega/\Gamma_0 K^2(\mathbf{k})$. The bare term dominates, indicating qualitatively normal hydrodynamic behaviour, for $\omega \ll \omega_k \propto k^4$. In the opposite limit $\omega \ll \omega_k$, one finds the critical dynamics of the relaxational model (provided, as argued above, that $k \ll \kappa$). The situation in different regions of $(\mathbf{k}, \omega)$-space is summarized in Fig. 6.5. One sees, for example, that for $k \ll \kappa$ the correlation function $C(\mathbf{k}, \omega) = (T/\omega) ImG(\mathbf{k}, \omega)$ observed in a hypothetical neutron scattering experiment can have, in principle, three distinct characteristic frequency regions (Fig. 6.6): a high-frequency one $\omega \ll \omega_k$ where one observes the conventional hydrodynamic Lorentzian line of width $\propto \Gamma_0 K^2(\mathbf{k})$, an intermediate one in which one sees the $\omega^{-1/2}$-lineshape characteristic of the critical dynamics of the relaxational model at T_f, and, finally, a rounding-off of the $\omega^{-1/2}$-singularity below a characteristic frequency $\propto (T - T_f)^2$, again just as in the relaxational model.

In this discussion we have neglected the characteristic dynamics of vector spins, i.e. the precession of the spin vector around the direction of the local magnetic field as expressed in the equation of motion (6.2). What happens if we then add such a term to the equation of motion (6.86)? We know that, in ferro- and antiferromagnets, the inclusion of such precessional terms makes a big difference in the nature of the singularities in the transport coefficients. The calculations for the spin glass are somewhat messy and we do not reproduce them here, because they turn out not to be nearly so dramatic as in the ordered magnets. To summarize the results briefly, formal perturbation theory in the strength of the precessional term leads to a correction to $\Gamma(0,0)$ proportional to $d^{-1}\ln(T - T_f)$ (where d is the dimensionality of the lattice). Similarly, right at T_f, $\Gamma(0,\omega) \propto d^{-1}\ln|\omega|$. A self-consistent treatment of higher-order contributions weakens these singularities to $[d^{-1}\ln(T - T_f)]^{1/2}$ and $[d^{-1}\ln|\omega|]^{1/2}$, respectively. It does not seem as if such weak singularities could be observed very easily, since they would too easily be obscured by any of the nonconserving forces present in all real systems.

Here we have only discussed these models above T_f. Below T_f spin conservation leads to spin waves. We will discuss these excitations together with other low-temperature properties in the next chapter.

7

Short-range interactions: low-temperature properties

The mean field theory discussed in the preceding chapters revealed a surprisingly rich structure and in particular a very complex ordered phase with many 'pure' states. However, we know from periodic systems that the mean field theory does not always give the right answer. In this and the next two chapters, we will discuss alternative approaches to the study of spin glasses with short-range interactions. The most important one will be the renormalization group which led to a deep understanding of phase transitions (and in particular critical behaviour) of nonrandom systems and which has also turned out to be extremely useful for the study of spin glasses. Together with Monte Carlo simulations and experimental data, this will give us a fairly complete picture, at least of static spin glass properties below T_f. Critical behaviour near T_f and scaling arguments for $T < T_f$ have also been considered for the spin glass dynamics but here the situation seems to be less clear. Depending on the system, the remanent magnetization decays with quite different decay laws (see Chapter 1), which indicates that not all dynamic processes are universal. However, all spin glasses have a huge range of characteristic relaxation times, ranging from 10^{-13}s (the Korringa relaxation in a metallic spin glass) to $10^{-6} - 10^{-8}$s. This long-time limit is not inherent to the system but is simply determined by the patience of an experimentalist (or the average time of a student's thesis). Fortunately, a large number of different types of experiments are available which probe a system in rather different time windows (see Chapter 9).

The physical origin of these many time scales is presumably frustration, which is inherent to all spin glasses and which will be discussed in the next section. In mean field theory, we could distinguish between intra- and inter-valley transitions, though a large part of the latter should be separated by infinitely high energy barriers. For hydrodynamics and its extensions (as discussed in Sections 7.2 and 7.3), this concept still seems to be useful. There is also some evidence for it from Monte Carlo simulations of RKKY spin glasses (Walker and Walstedt 1977) and from experimental data discussed in Chapter 10. However, in real spin glasses, one expects energy barriers of all sizes and in connection with dynamical scaling we will make no distinction between 'fast' and 'slow' modes.

7.1   Frustration and local gauge theories

The concept of frustration, which we already introduced in Section 2.5 and which we will now discuss in more detail, enables us to distinguish between 'trivial' and 'nontrivial' disorder and helps us in characterizing the ground state of a spin glass. For Ising spins, it is based on the local 'gauge' transformation

$$S_i' = \tau_i S_i, \quad J_{ij}' = \tau_i J_{ij} \tau_j, \quad \tau_i, \tau_j = \pm 1 \tag{7.1}$$

(Toulouse, (1977)). *In zero field* the transformation (7.1) leaves the Hamiltonian (2.11) invariant

$$H_{J'}\{S_i'\} = H_J\{S_i\} \tag{7.2}$$

and does not change the partition function

$$Z_{J'} = \sum_{\{S_i'\}=\pm 1} \exp\left[-\tfrac{1}{2}\beta \sum_{ij} S_i' J_{ij}' S_j' \right]$$

$$= \sum_{\{S_i\}=\pm 1} \exp\left[-\tfrac{1}{2}\beta \sum_{ij} S_i J_{ij} S_j \right] = Z_J \tag{7.3}$$

The invariance (7.3) means that despite the fact that we changed the model and considered a completely new set of bonds, all thermodynamic properties remain unchanged. We therefore consider the disorder to be 'trivial' if it can be eliminated by a transformation of the type (7.1). An example is the

Mattis model (2.76) where this transformation leads for $\tau_i = \xi_i$ to a model with only ferromagnetic interactions in which all disorder is eliminated.

The transformation (7.1) does not eliminate 'nontrivial' disorder, which is described by the frustrated plaquettes of Fig. 2.4b. For the $\pm J$ model, this disorder is conveniently defined by the 'frustration'

$$\phi_{ijkl} = J_{ij}J_{jk}J_{kl}J_{li}/J^4 = \pm 1 \tag{7.4}$$

with $\phi_{ijkl} = -1$ for frustrated plaquettes. A natural generalization of (7.4) is

$$\phi_P = \prod_P \text{sgn}(J_{ij}) \tag{7.5}$$

for any loop P on a lattice and for arbitrary exchange. The frustration ϕ is obviously 'gauge invariant', i.e. invariant with respect to the transformation (7.1).

The term 'gauge transformation' is borrowed from electrodynamics and gauge theories of quantum field theory. In what follows, we show that there is indeed an analogy between spin glasses and these theories. We consider first a single charged particle with the wave function $\psi(\mathbf{r}, t)$ coupled to a static magnetic field $\mathbf{B} = \nabla \times \mathbf{A}$ in units $c = \hbar = 1$. The field $\mathbf{B}$ is invariant under the local transformation of the vector potential A_μ ($\mu = 1, 2, 3$)[1]

$$A'_\mu = A_\mu + \nabla_\mu \Lambda \tag{7.6}$$

for any function $\Lambda(\mathbf{r})$. It is coupled to the charged particle by making the substitution $\nabla_\mu \to \nabla_\mu - ieA_\mu$ in the kinetic part of the Hamiltonian. If the wavefunction $\psi(r)$ transforms as

$$\psi'(r) = e^{-ie\Lambda(r)}\psi(r) \equiv U(r)\psi(r), \tag{7.7}$$

then the combination $(\nabla_\mu - ieA_\mu)\psi$ will be invariant and therefore so will the Hamiltonian.

We consider the solution of (7.6) with (7.7) for the gauge $A'_\mu = 0$

$$U = \exp(-ie \int dx_\mu A_\mu) \tag{7.8}$$

For the loop indicated in Fig. 7.1 with the corners at the lattice points $\mathbf{i}, \mathbf{i} + a_0\mathbf{e}_\mu, \mathbf{i} + a_0\mathbf{e}_\mu + a_0\mathbf{e}_\nu, \mathbf{i} + a_0\mathbf{e}_\nu$, this leads to

$$U = \exp[-ie(A_\mu^{(1)}\Delta x_\mu + A_\nu^{(2)}\Delta x_\nu - A_\mu^{(3)}\Delta x_\mu - A_\nu^{(4)}\Delta x_\nu]$$

[1]Hitherto we have used Greek indices from the middle of the alphabet ($\mu, \nu, \cdots$) for spin component indices. In this chapter we depart from this practice and follow instead the standard convention in gauge field theories, letting $\mu, \nu, \cdots$ denote spatial or space-time coordinates and $r, s, t, \cdots$ label (internal) spin components.

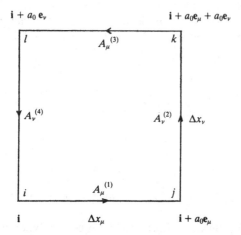

Figure 7.1: Loop in magnetostatics on a lattice; a_0 is the lattice constant and $\mathbf{e}_\mu$ the unit vector in direction μ; A_μ, A_ν are components of the vector potential.

$$= \exp[-iea_0^2 F_{\mu\nu} e_\mu e_\nu] \equiv U_{ij} U_{jk} U_{kl} U_{li} \qquad (7.9)$$

where a_0 is the lattice constant, $\mathbf{e}_\mu$ is the unit vector in direction μ and

$$F_{\mu\nu} = a_0^{-1}[A_\mu(\mathbf{i}) - A_\mu(\mathbf{i} + a_0\mathbf{e}_\nu) - (A_\nu(\mathbf{i}) - A_\nu(\mathbf{i} + a_0\mathbf{e}_\nu))] \qquad (7.10)$$

In the limit $a_0 \to 0$, the tensor $F_{\mu\nu}$ becomes

$$F_{\mu\nu} = \nabla_\mu A_\nu - \nabla_\nu A_\nu \qquad (7.11)$$

which is just the magnetic field or, if multiplied by the area a_0^2, the magnetic flux ϕ_M which goes through the loop. The quantities $U_{ij} = \exp(-ieA_\mu \Delta x_\mu)$ in (7.9) play the same role in defining the flux ϕ_M as the bonds J_{ij} in (7.4) for the frustration ϕ. In the Ising spin glass with $\pm J$ bonds, the phase factors U_{ij} can only assume the values ± 1. The transformation of the wave function (7.7) should be compared with the spin transformation in (7.1). However, the transformation (7.2) belongs to the discrete Abelian group Z_2 while (7.6) with (7.7) belongs to the one-dimensional Abelian continuous group $U(1)$.

There are still other important differences between the two gauge theories: In general, the vector potentials A_μ or U_{ij} and fields $F_{\mu\nu}$ are dynamic variables whereas the bonds J_{ij} and the frustration ϕ are quenched variables. In addition, an external field h breaks the gauge invariance (7.2) of the Ising model. The analogy would be much closer (and the calculations simpler) if one considered an annealed system in which the bonds were treated as dynamic variables (Toulouse and Vannimenus 1980). However, it seems that such a model does not yield very useful information about spin glasses.

Apart from trivial disorder, the ground state of a spin glass is then defined in two dimensions by the distribution of frustrated plaquettes and the strings between them (Fig. 2.5) (or, in higher dimensions, by a straightforward generalization (Fradkin et al, 1978)). Starting from the ideal ferromagnet, one considers first the energy of a pair of frustrated plaquettes in a ferromagnetic (unfrustrated) host lattice of Ising spins. The energy of such a pair increases linearly with the distance or with the number of broken bonds between the two plaquettes. This property is similar to 'confinement' in lattice gauge theories. Hence a single frustrated plaquette cannot exist below a characteristic temperature T_0. The confinement is lost above T_0 which represents some kind of 'melting temperature' (Fradkin et al, 1978).

Larger concentrations of frustrated plaquettes have been treated mostly numerically (Bieche et al, 1980; Barahona et al, 1982). For a very small concentration r of negative bonds, one observes a very small density of 'loose' spins (i.e. spins in zero internal field) and a still smaller concentration of minority spins in a ferromagnetic ground state. At larger concentrations $r = r_c \approx 0.15$, the ferromagnetic state breaks up into domains because the loose spins form closed loops and the nonfrustrated plaquettes (which support ferromagnetism) no longer percolate. Unfortunatley, it is not clear which phase one enters for $r > r_c$: For $2d$ Ising spins and $\pm J$ bonds ($r = 0.5$) transfer matrix analysis suggests a spin glass phase at $T = 0$ but a vanishing spin glass order parameter q_{EA}. This state differs from the paramagnetic phase in having power law rather than exponential decay of the spin glass correlation function $[\langle S_i S_j \rangle_{T=0}]_{av}$ (Morgenstern and Binder, 1980a, Morgenstern and Horner, 1982). It remains unclear whether or not this state persists down to $r = r_c$. Since the ferromagnetic correlation length ξ_F diverges for $r < r_c$, one presumably has, slightly above r_c, very large ferromagnetic domains separated by rather irregular walls (Barahona et al, 1982).

So far in this section we have only discussed Ising spins. In the XY model, one has continuous spin rotations $\mathbf{S}_i = (\cos \zeta_i, \sin \zeta_i)$ with $|S_i| = 1$. It is convenient to consider a slightly generalized model in which the random bonds J_{ij} are replaced by random difference angles b_{ij}

$$H = -\sum_{ij} J_{ij} \cos(\zeta_i - \zeta_i) \rightarrow -J_0 \sum_{ij} \cos(\zeta_i - \zeta_j - b_{ij}) \qquad (7.12)$$

$$\equiv -J_0 \sum_{ij} [e^{i(\zeta_i - \zeta_j)} V_{ij}^* + \text{h.c.}], \quad V_{ij} = e^{ib_{ij}}, \quad J_0 > 0 \qquad (7.13)$$

(Fradkin et al, 1978). In the original XY model, one has simply $b_{ij} = (0, \pi)$. A local gauge transformation now consists of the rotation of a spin S_i and the simultaneous rotation of the adjacent V_{ij}'s

$$S_i' = e^{i\chi_i} S_i, \quad V_{ij}' = e^{i\chi_i} V_{ij} e^{-i\chi_j} \qquad (7.14)$$

which again leaves the Hamiltonian invariant and which agrees with (2.82) for $U_j = \exp(-i\chi_j)$. For the conventional XY model ($V_{ij} = \pm 1$), the transformation angles χ_i are also restricted to $(0, \pi)$ and the gauge group is Z_2 as in the Ising case. Here, the angles b_{ij} are treated as continuous variables, which corresponds to complex bonds J_{ij}. The frustration ϕ_{ijkl} is defined by the loop of Fig. 7.2 with

$$\exp[2\pi i \phi_{ijkl}] = V_{ij} V_{jk} V_{kl} V_{li} \qquad (7.15)$$

or

$$2\pi \phi_{ijkl} = b_{ij} + b_{jk} + b_{kl} + b_{li} \quad (\text{mod } 2\pi) \qquad (7.16)$$

In order to compare with the loop of Fig. 7.1 for magnetostatics, we change the notation slightly, using

$$2\pi \phi_{ijkl} = b_\mu(\mathbf{i}) - b_\mu(\mathbf{i} + a_0 \mathbf{e}_\nu) - (b_\nu(\mathbf{i}) - b_\nu(\mathbf{i} + a_0 \mathbf{e}_\mu)) \qquad (7.17)$$

There is now a complete analogy between the frustration ϕ_{ijkl} and the magnetic field (7.10) and between the angles $b_\mu(\mathbf{i})$ and the vector potentials A_μ. Hence the frustration ϕ corresponds to the magnetic flux through the loop.

We return now to the conventional XY model (with real V_{ij}'s). A single frustrated plaquette produces only a spin rotation by an angle π, as indicated in Fig. 7.2 and mentioned already in Section 2.5. This is in contrast to the 2π-vortices in an XY model for ideal ferromagnets (Kosterlitz and Thouless, 1973). The half-vortices in spin glasses can have either positive or negative 'helicity' $\tau = \pm 1$. Hence a single plaquette leads, in addition to the continuous rotation degeneracy, to a two-fold degenerate ground state (Villain 1977a,b, 1978). For small spin deviations, the continuous variables lead to magnons. The discrete variables are associated with large spin deviations and can be envisaged as systems with two local minima. A change of helicity then is equivalent to a transition between these minima. These modes exist only in systems with frustration and act like Ising spins. It is tempting to attribute the slow relaxation processes in spin glasses to these 'intervalley' transitions. However, one has to keep in mind that in reality

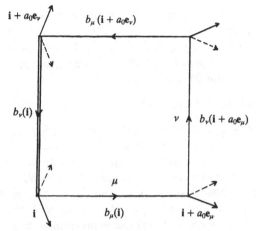

Figure 7.2: Loop in the XY model on a lattice; $b_{ij} = b_\mu(\mathbf{i})$ are difference angles which replace positive and negative bonds (single and double lines). The arrows at the corners indicate the two possible spin directions in the ground state.

frustrated plaquettes exist only in pairs and that one has a clear picture only for systems with a small concentration of frustated plaquettes.

A local gauge transformation in a generalized XY spin glass is defined by the single parameter χ_i in (7.14). We have a one-dimensional continuous Abelian gauge group $G = U(1) = SO(2)$ in which all elements commute. In the *Heisenberg model*, we have three spin rotation axes, and finite rotations do not commute. The rotations form a nonAbelian group, which is much more complicated. There is an additional difficulty: In the 2d Ising model, one has strings (Fig. 2.5), in the 3d Ising model closed loops (Fig. 2.6), in the 2d XY model half-vortices (Fig. 2.7), and in the 3d XY model half-vortex rings as elementary defects (Toulouse 1979). However, no 'topological' defects similar to frustration seem to exist in vector spin glasses. For this reason, we will discuss frustration for vector spins on a more phenomenological basis which necessarily has to be more macroscopic. This theory has some similarity with Yang–Mills theory (Yang and Mills, 1954) which is the basis of all nonAbelian gauge theories and which we will discuss first (see Kogut (1979) for an introduction to Yang–Mills theory). Yang and Mills considered particles with an internal degree of freedom ('isotopic spin'). The isotopic spin operates in the state spanned by the states

$$\psi = \left\{ \begin{pmatrix} 1 \\ 0 \end{pmatrix}, \begin{pmatrix} 0 \\ 1 \end{pmatrix} \right\} \tag{7.18}$$

of a particle which might be either a proton or a neutron. In an experiment, the two states are indistinguishable if one ignores the proton charge. Since the isotopic spin is conserved, all physical quantities have to be invariant against a local rotation by an angle $\Theta(\mathbf{r}, t)$ in the isotopic spin space. This is equivalent to the local gauge transformation

$$\psi' = U\psi, \quad U = \exp[-2i\Theta(\mathbf{r}, t) \cdot \vec{\tau}] \tag{7.19}$$

with the isotopic spin $\vec{\tau} = \frac{1}{2}\vec{\sigma}$, where $\vec{\sigma}$ are the Pauli matrices and where U is a 2×2 matrix. For a charged particle in an electrodynamic field, we had to replace ∇_μ by $\nabla_\mu - ieA_\mu$ in the kinetic part of the Hamiltonian. In the Yang–Mills theory, one has to postulate gauge potentials B_μ which are 2×2 matrices and which appear in the combination $(\nabla_\mu - igB_\mu)\psi$. Here, g is an (unknown) coupling constant and $\nabla_\mu = \{\nabla_l, \partial/\partial it\}$ with the spatial components $l = x, y, z$ and $\hbar = c = 1$. Local gauge invariance then requires

$$U^{-1}(\nabla_\mu - igB'_\mu)\psi' = (\nabla_\mu - igB_\mu)\psi \tag{7.20}$$

If (7.19) and (7.20) are to hold, the gauge potentials have to transform as

$$B'_\mu = UB_\mu U^{-1} + (i/g)U\nabla_\mu U^{-1} \tag{7.21}$$

In the case of a charged particle in an electromagnetic field, we solved (7.6) with (7.7) for U on a loop and obtained the gauge invariant fields $F_{\mu\nu}$ (7.11). A similar procedure now leads to

$$F_{\mu\nu} = \nabla_\mu B_\nu - \nabla_\nu B_\mu + ig[B_\mu, B_\nu] \tag{7.22}$$

where we used the Baker–Hausdorff formula $e^x e^y = e^{x+y+\frac{1}{2}[x,y]+\cdots}$. There is an important difference between (7.11) and (7.22). Since the potentials B_μ do not commute, there is now a nonlinear relation between the gauge fields and potentials. The 2×2 matrices B_μ and $F_{\mu\nu}$ can be written in terms of vectors $\mathbf{b}_\mu$ and $\mathbf{f}_{\mu\nu}$

$$B_\mu = 2\vec{\tau} \cdot \mathbf{b}_\mu, \quad F_{\mu\nu} = 2\vec{\tau} \cdot \mathbf{f}_{\mu\nu} \tag{7.23}$$

which leads with (7.22) and $[\sigma^r, \sigma^s] = 2i\epsilon^{rst}\sigma^t$ for $r, s, t = 1 \ldots 3$ to

$$\mathbf{f}_{\mu\nu} = \nabla_\mu \mathbf{b}_\nu - \nabla_\nu \mathbf{b}_\mu - \mathbf{b}_\mu \times \mathbf{b}_\nu \tag{7.24}$$

For infinitesimal small angles $\Theta \to 0$, the potentials $\mathbf{b}_\mu$ and fields $\mathbf{f}_{\mu\nu}$ then transform as

$$\mathbf{b}'_\mu = \mathbf{b}_\mu + \Theta \times \mathbf{b}_\mu + g^{-1}\nabla_\mu \Theta \tag{7.25}$$

$$\mathbf{f}'_{\mu\nu} = U\mathbf{f}_{\mu\nu}U^{-1} = \mathbf{f}_{\mu\nu} + \mathbf{\Theta} \times \mathbf{f}_{\mu\nu} \tag{7.26}$$

An application of this formalism to spin glasses has been attempted by Dzyaloshinskii and Volovik (1978), Volovik and Dzyaloshinskii (1978) (see also Dzyaloshinskii 1980, 1983). There arise two serious problems: First, one would like to identify the gauge-invariant fields $\mathbf{f}_{\mu\nu}$ with frustration densities, the microscopic origin of which remains unknown, and second, the frustration has no inherent dynamics since the bonds are quenched. Fortunately, the frustration current which arises from time dependence of the fields $\mathbf{f}_{\mu\nu}$ vanishes for any Hamiltonian, as shown in the next section. A purely dissipative current due to thermal noise will still be possible and will be discussed in connection with a hydrodynamic order parameter in the next section. In addition, general rotations of the spin will only be possible for more general bonds, analogous to those assumed for the generalized XY model.

We consider now a local transformation or rotation of the spin density $\mathbf{S}(\mathbf{r}, t)$ as generated by the operator $\mathbf{L} = \{L^r\}$ with the commutation relations $[L^r, L^s] = i\epsilon^{rst}L^t$, $(r, s, t = 1, 2, 3)$

$$\mathbf{S}' = e^{i\mathbf{\Theta}\cdot\mathbf{L}}\mathbf{S} \tag{7.27}$$

Here the components L^r are 3×3 matrix representations of angular momenta, and components in spin space are labelled with upper indices. For infinitesimal transformations $\mathbf{\Theta} \to 0$ the spin density transforms as

$$\mathbf{S}' = \mathbf{S} + \mathbf{\Theta} \times \mathbf{S} \tag{7.28}$$

and one has instead of (7.23)

$$B_\mu = \mathbf{L}\cdot\mathbf{b}_\mu, \quad F_{\mu\nu} = \mathbf{L}\cdot\mathbf{f}_{\mu\nu} \tag{7.29}$$

The gauge potentials $\mathbf{b}_\mu = \{\mathbf{b}_l, \mathbf{a}/iv\}$ and $\mathbf{f}_{\mu\nu} = \{\mathbf{f}_{kl}, \mathbf{g}_l\}$ with $k, l = 1, 2, 3$ contain spatial and time components, where v replaces the light velocity. From (7.24) and (7.25) we have, for $\mathbf{\Theta} \to 0$ (with $\nabla_\mu = \{\nabla_l, \nabla_t/iv\}$ and $g = 1$)

$$\mathbf{b}'_l = \mathbf{b}_l + \mathbf{\Theta} \times \mathbf{b}_l + \nabla_l\mathbf{\Theta}, \quad \mathbf{a}' = \mathbf{a} + \mathbf{\Theta} \times \mathbf{a} \tag{7.30}$$

$$\mathbf{f}_{kl} \equiv \epsilon_{klm}\vec{\rho}_m^F = \nabla_k\mathbf{b}_l - \nabla_l\mathbf{b}_k - \mathbf{b}_k \times \mathbf{b}_l \tag{7.31}$$

$$\mathbf{g}_l \equiv \mathbf{j}_l^F = \nabla_t\mathbf{b}_l - \nabla_l\mathbf{a} - \mathbf{a} \times \mathbf{b}_l \tag{7.32}$$

The identification of $\vec{\rho}_m^F$ with a frustration density and of $\mathbf{j}_m^F$ with a frustration current density is suggested by the conservation law for the covariant derivatives $D_\mu = \nabla_\mu - iB_\mu$

$$D_t \vec{\rho}_m^F + \sum_{lm} \epsilon_{klm} D_l \vec{j}_m^F = 0 \qquad (7.33)$$

which follows from the Jacobi identity

$$[D_t, [D_k, D_l]] + [D_l, [D_t, D_k]] + [D_k, [D_l, D_t]] = 0 \qquad (7.34)$$

The frustration density $\vec{\rho}_m^F$ fulfils the transversality condition

$$\sum_{lm} \epsilon_{klm} \nabla_l \vec{\rho}_m^F = 0 \qquad (7.35)$$

in analogy to $\nabla \cdot B = 0$ in electrodynamics. This agrees with the fact that in an Ising system in three dimensions, frustration lines always have to be closed (see Section 2.5).

This concept of local gauge invariance has been applied to vector spin glasses in two different ways. In a strictly static approach ($\mathbf{a} = 0, \mathbf{j}^F = 0$), one replaces ∇_k by the covariant derivative $\nabla_k - iB_k = \nabla_k - i\mathbf{L} \cdot \mathbf{b}_k$ in the Landau–Ginzburg expression (2.14) for the free energy. For vector spins, this gives the effective Hamiltonian

$$H_{eff} =$$

$$\tfrac{1}{2} \int d^d x [r_0(\mathbf{S}(x))^2 + \tfrac{1}{2}u(\mathbf{S}^2(x))^2 + \sum_l |(\nabla_l - i\mathbf{L} \cdot \mathbf{b}_l(x))\mathbf{S}(x)|^2] \quad (7.36)$$

where the coefficients r_0 and u are assumed to be independent of disorder. For $\mathbf{b}_l = 0$ the model (7.36) reduces to the Landau–Ginzburg Heisenberg ferromagnet. For a fixed $\mathbf{b}_l \neq 0$ it describes an antiferromagnet with a spin density wave of wavevector $\mathbf{b}$. One can model a kind of spin glass by assuming a random distribution of (gauge-invariant) fields $\mathbf{f}_{kl}$ connected with the potentials $\mathbf{b}_l$ by (7.31) with a distribution

$$P[\mathbf{f}_{kl}] \propto \exp\left[-\frac{1}{2f} \int d^d x \sum_{kl} \mathbf{f}_{kl}(x) \cdot \mathbf{f}_{kl}(x)\right] \qquad (7.37)$$

of width f. This model is in fact the Landau–Ginzburg description of a ferromagnet with random Dzyaloshinskii–Moriya anisotropy. It has been studied for small frustration density f by the renormalization group formalism described in Section 8.1. The ferromagnetic fixed point becomes unstable but near $d = 4$ no new fixed point appears (Hertz, 1978).

In a second approach, one treats the frustration on a macroscopic scale as a dynamic variable (Dzyaloshinskii and Volovik, 1978) despite the fact that the exchange interactions J_{ij} are quenched variables. We will show in the next section that this leads to an interesting connection between a rotation angle $\Theta(\mathbf{r}, t)$ (which will be introduced as a hydrodynamic order parameter) and the gauge fields $\mathbf{f}_{kl}$.

7.2 Hydrodynamics and spin waves

In hydrodynamics one considers modes which are slow compared to all microscopic modes. The hydrodynamic modes are determined by conservation laws and (below the phase transition) by the broken symmetry. In the latter case, there might be 'Goldstone' modes: The Goldstone theorem states that if the ground state of a system has a lower continuous symmetry than the Hamiltonian, there must exist excitations whose spectrum extends in the long-wavelength limit to zero without a gap (see Forster (1975) for an introduction to hydrodynamics and Goldstone modes). Examples of Goldstone modes are spin waves in ferro- and antiferromagnets. In both systems, the total magnetization and energy are conserved or constants of motion. In a ferromagnet, the magnetization is also the order parameter whereas in antiferromagnets the order parameter, the staggered magnetization, is not conserved. In both cases, the rotation symmetry in one direction is spontaneously broken but states created by rotations perpendicular to the (staggered) magnetization are still degenerate (see Halperin and Hohenberg 1969).

In the spin glass phase, no macroscopic axis is distinguished but locally the rotation symmetry is completely broken and we expect the hydrodynamics of spin glasses to be similar to those of a complicated antiferromagnet. Specifically, we also expect in spin glasses spin waves as Goldstone modes. However, hydrodynamic theories hold only for frequencies ω with $\omega\tau \ll 1$ where τ is the smallest 'microscopic' relaxation time of the system. Experiments and the MFT both indicate that there are extremely long relaxation times in spin glasses. So we do not know whether or not the spin waves we derive below really can be measured, for instance, by neutron scattering, or if the condition $\omega\tau \ll 1$ is too restrictive. Following conventional theories, we show how they are damped by energy-dissipating mechanisms such as spin diffusion or order parameter relaxation (Halperin and Saslow, 1977). We also show how this hydrodynamic theory is connected with the gauge theory discussed in the preceding section.

Hydrodynamic modes exist only for long wavelengths ($\mathbf{k} \to 0$) for which the underlying lattice becomes unimportant. We construct the corresponding continuum theory with a 'coarse-grained' spin density $\mathbf{S}(\mathbf{r})$ by considering blocks of N_R spins in a volume V_R and assume that $\mathbf{S}(\mathbf{r})$ varies slowly on the length scale $V_R^{1/3}$. A block spin (operator) is then defined by

$$\mathbf{S}_R = \frac{1}{N_R} \sum_{j \in R} \mathbf{S}_j = \frac{1}{N_R} \int_{V_R} d^3r \, \mathbf{S}(\mathbf{r}) \tag{7.38}$$

The spin commutators $[S_i^r, S_j^s] = i\epsilon^{rst} S_i^t \delta_{ij}$ $(r,s,t = 1,2,3)$ lead to

$$[S^r(\mathbf{r}), S^s(\mathbf{r}')] = i\epsilon^{rst} S^t(\mathbf{r})\delta(\mathbf{r} - \mathbf{r}') \tag{7.39}$$

In order to find a suitable order parameter for the spin glass, we consider the microscopic staggered magnetization of an ideal antiferromagnet at $T = 0$

$$\mathbf{M}_{stagg}(\mathbf{r}) = \frac{1}{N}\sum_j \eta_j \mathbf{S}_j \delta(\mathbf{r} - \mathbf{R}_j), \quad \eta_j = \langle S_j^z \rangle_g = \pm 1 \tag{7.40}$$

where $\langle S_j^z \rangle_g$ is the expectation value of the spin j in the ground state. A coarse graining procedure similar to (7.38) leads with (7.39) to

$$[M_{stagg}^r(\mathbf{r}), M_{stagg}^s(\mathbf{r}')] = i\epsilon^{rst} S^t(\mathbf{r})\delta(\mathbf{r} - \mathbf{r}') \tag{7.41}$$

This suggests as a spin glass order parameter the coarse-grained tensor

$$t^{rs}(\mathbf{r}) = \frac{1}{N_R}\sum_{j \in R} \langle S_j^r \rangle_g S_j^s \tag{7.42}$$

with $\langle S_j^r \rangle_g \neq 0$ but $\sum_{j \in R} \langle S_j^r \rangle_g = 0$ in the spin glass phase. This parameter is defined for a single ground state or 'valley'. If applied to finite temperatures, this theory does not take into account any transition between different valleys or the 'intervalley' contributions discussed below (2.42).

We consider small rotations with the angle $\mathbf{\Theta}_j$ of the spin j out of the ground state $\langle S_j^r \rangle_g$ $(r = x, y, z)$ with $\mathbf{\Theta}_j = \text{const}$ within each block R. One has with

$$U = \exp[i\sum_j \mathbf{\Theta}_j \cdot \mathbf{S}_j] \tag{7.43}$$

the spin expectation value in the rotated state

$$\langle S^r \rangle = Z^{-1}Tr(\rho S_j^r) = Z^{-1}Tr(U^{-1}\rho_g U S_j^r)$$

$$= \langle S_j^r \rangle_g + \epsilon^{rst}\Theta_j^s \langle S_j^t \rangle_g \quad (\mathbf{\Theta} \to 0) \tag{7.44}$$

Here, ρ is the density operator of the rotated state and ρ_g of the ground state (we use the summation convention). Of course, (7.44) is a special case of (7.28). Inserting (7.44) into (7.42), with the EA spin glass order parameter (2.39) at $T = 0$ (in the form $q_{EA} = N_R^{-1}\sum_{j \in R}\langle S_j^r \rangle_g^2$ for a single valley and component r), leads to

$$\langle t^{rs}(\mathbf{r})\rangle = q_{EA}(\delta^{rs} + \epsilon^{rst}\Theta^t(\mathbf{r})) \tag{7.45}$$

We solve (7.45) for $\Theta^t(\mathbf{r})$ by multiplying by ϵ^{rsp} and with $\epsilon^{rst}\epsilon^{rsp} = 2\delta^{tp}$

$$\Theta^r(\mathbf{r}) = (2q_{EA})^{-1}\epsilon^{rst}t^{st}(\mathbf{r}) = (2q_{EA})^{-1}\epsilon^{rst}N_R^{-1}\sum_{j \in R}\langle S_j^s \rangle_g S_j^t \tag{7.46}$$

This equation holds for any infinitesimal rotation and can therefore be read as an operator equation. The operator $\Theta(\mathbf{r})$ then becomes a dynamic variable which has the meaning of an order parameter. From (7.39) and (7.46) one derives the commutators

$$[\Theta^r(\mathbf{r}), \Theta^s(\mathbf{r}')] = \frac{i}{4q_{EA}\rho^2}\epsilon^{rst}S^t(\mathbf{r})\delta(\mathbf{r} - \mathbf{r}') \qquad (7.47)$$

$$[S^r(\mathbf{r}), \Theta^s(\mathbf{r}')] = -i(\delta^{rs} + \tfrac{1}{2}\epsilon^{rst}\Theta^t(\mathbf{r}))\delta(\mathbf{r} - \mathbf{r}') \qquad (7.48)$$

where $\rho = N/V = N_R/V_R$ is the density. The dynamics (7.39), (7.47) and (7.48) are restricted to small angles and do not include topological excitations such as the change of helicity as discussed for the XY model in the preceding section.

To clarify the way we introduced the angle Θ as a new dynamic variable, let us consider the simpler case of an antiferromagnet with an easy plane of magnetization. If the staggered magnetization (the order parameter) in the ground state points in the z-direction, the system is still degenerate with respect to rotations in the xy-plane. The variable Θ describes fluctuations of the phase of this order parameter in the $x - y$ plane, writing $M^\perp_{stagg} = M^x_{stagg} + iM^y_{stagg} = |M^\perp_{stagg}|e^{i\Theta}$. In hydrodynamics, one then has to take into account fluctuations of Θ and of the magnetization density M_z (since the total magnetization is conserved). (See Halperin and Hohenberg (1969) for details).

Since the expectation value of the spin density $\mathbf{S}(\mathbf{r})$, averaged over the block volume V_R, is assumed to be zero, the commutators (7.39), (7.47) and (7.48) simplify to

$$[S^r(\mathbf{r}), \Theta^s(\mathbf{r}')] = -i\delta^{rs}\delta(\mathbf{r} - \mathbf{r}') \qquad (7.49)$$

$$[S^r(\mathbf{r}), S^s(\mathbf{r}')] = [\Theta^r(\mathbf{r}), \Theta^s(\mathbf{r}')] = 0 \qquad (7.50)$$

and the angles Θ and the spin density $\mathbf{S}$ become canonically conjungate variables. The relations (7.49) and (7.50) hold only in the absence of remanence or of a magnetic field. Otherwise, one has to go back to (7.39) with (7.47) and (7.48).

Next we construct a coarse-grained Hamiltonian of the Landau–Ginzburg form, restricting ourselves to the most important terms. At temperatures far below T_f all powers in the spin density, except the term $\mathbf{S}^2(\mathbf{r})$, can be ignored. Terms of order Θ^n ($n = 2, 4, \ldots$) do not appear since in the absence of anisotropy the energy is independent of the simultaneous rotations of all spins by the same angle Θ. In the long-wavelength limit, the leading angle fluctuations are of the form $(\nabla_l\Theta)^2 \equiv \sum_{rl}(\nabla_l\Theta^r)^2$. This leads to the effective Hamiltonian

$$H_{eff} = \tfrac{1}{2} \int d^3r (\chi_0^{-1}|\mathbf{S}(\mathbf{r})|^2 + \rho_s|\nabla_l\Theta(\mathbf{r})|^2) \qquad (7.51)$$

with the uniform susceptibility χ_0 and the spin waves stiffness ρ_s as phenomenological constants. Terms of order $(\nabla_l\mathbf{S})^2$ lead, after Fourier transformation, to higher powers in the wavevector $\mathbf{k}$ and can be ignored.

In what follows, we ignore quantum fluctuations and replace the commutators (7.49) and (7.50) by Poisson brackets

$$[a,b] \longrightarrow \frac{\hbar}{i}\{a,b\} \qquad (7.52)$$

This leads to the canonical equations

$$\partial_t S^r = \{H_{eff}, S^r\} = -\frac{\delta H_{eff}}{\delta\Theta^r} = \rho_s\nabla^2\Theta^r \qquad (7.53)$$

$$\partial_t\Theta^r = \{H_{eff}, \Theta^r\} = \frac{\delta H_{eff}}{\delta S^r} = \chi_0^{-1}S^r \qquad (7.54)$$

or to

$$\chi_0\frac{\partial^2\Theta^r}{\partial t} = \rho_s\nabla^2\Theta^r \qquad (7.55)$$

The solutions of (7.55) are six-fold degenerate modes

$$\omega = \pm c_0|k|, \quad c_0^2 = \frac{\rho_s}{\chi_0} \qquad (7.56)$$

This degeneracy will be partly lifted by anisotropy, remanent magnetization, or a magnetic field and expresses the macroscopic isotropy of spin glasses (see Saslow (1980) and Fischer (1980) for details).

The spin wave equation (7.56) can also be derived microscopically (Ginzburg, 1978). The derivation is somewhat lengthy and we give only the result. One has for small, slowly varying deviations Θ from the ground state $\mathbf{S}_i^o \equiv \langle\mathbf{S}_i\rangle_g$ with the equation of motion

$$\partial_t\mathbf{S}_i = \sum_j J_{ij}\mathbf{S}_i \times \mathbf{S}_j \qquad (7.57)$$

and with

$$\mathbf{S}_i = \mathbf{S}_i^o + \Theta \times \mathbf{S}_i \qquad (7.58)$$

(see (7.28)) after some manipulations

$$\sum_j \chi_{ij}^{rs}\frac{\partial^2\Theta_j^s}{\partial t^2} = \sum_j A_{ij}^{rs}(\Theta_j^s - \Theta_i^s) \qquad (7.59)$$

Here $\chi_{ij}^{rs} = \partial M_i^r / \partial h_j^s$ is the susceptibility with $\mathbf{M}_i = \mathbf{S}_i - \mathbf{S}_i^o$, and

$$A_{ij}^{rs} = 2J_{ij}(\mathbf{S}_i^o \cdot \mathbf{S}_j^o \delta^{rs} - S_j^{or} S_i^{os}) \tag{7.60}$$

the generalized spin wave stiffness. The latter becomes small (or eventually zero) if $J_{ij}S_i^o S_j^o < 0$ for many spins or for large amount of frustration. It depends on the ground state and on the bonds. Numerical calculations indicate that the averaged spin wave stiffness ρ_s is finite for a symmetric Gaussian distribution of nearest neighbour interactions (Reed, 1978) and also for RKKY glasses (Walstedt and Walker 1981). The MFT yields $\rho_s > 0$ on finite time scales (which is relevant for spin waves) but $\rho_s = 0$ for equilibrium in the Parisi solution (Sompolinsky et al, 1984; Kotliar et al, 1987). This agrees with our assumption $\rho_s > 0$ for hydrodynamics, which we had to restrict to a single valley and which holds only in the absence of frustration (see below).

There have been various unsuccessful attempts to detect spin waves in spin glasses by means of neutron scattering. (Spin waves in 'reentrant' systems with both a ferromagnetic and a spin glass transition will be discussed in Section 10.3.) There are several possible reasons for this failure: As mentioned before, the basic assumption $\omega\tau \ll 1$ for all microscopic relaxation times τ might not be fulfilled. Secondly, the spin wave stiffness ρ_s and the spin wave energy ω might be too small in order to be observable in neutron scattering experiments. Thirdly, the hydrodynamic modes might be too strongly damped by faster modes. This damping will be considered below, again in the framework of a phenomenological theory. In metallic spin glasses, one has an additional damping mechanism (the 'Korringa relaxation') due to the interaction between the localized spins and the conduction electrons. Evidence for other modes in the hydrodynamic regime $(\omega, \mathbf{k})$ of energies and wave vectors comes from computer calculations. Walker and Walstedt (1977, 1980) consider dilute systems with RKKY interactions (e.g. for CuMn) and observe more or less localized modes rather than propagating ones such as spin waves. Krey (1980, 1981, 1982) observed that the eigenmodes of $Eu_x Sr_{1-x} S$, if projected onto plane waves with wavevector $\mathbf{k}$, become heavily damped. Nevertheless, one cannot exclude the existence of 'true' hydrodynamic modes which most likely are completely outnumbered by more localized ones.

The spin waves (7.56) can be damped by spin diffusion and the relaxation of the order parameter. To describe these effects, we use the phenomenological equations

$$\partial_t S^r = -\frac{\delta H_{eff}}{\delta \Theta^r} + \Gamma_s \nabla^2 \frac{\delta H_{eff}}{\delta S^r} + \xi_s^r \tag{7.61}$$

$$\partial_t \Theta^r = \frac{\delta H_{eff}}{\delta S^r} + \Gamma_\theta \frac{\delta H_{eff}}{\delta \Theta^r} + \xi_\theta^r \tag{7.62}$$

instead of (6.86) or (7.53) and (7.54) with the constants Γ_s and Γ_θ for spin diffusion and order parameter relaxation. The noise is assumed to be Gaussian

$$\langle \xi_s^r(\mathbf{r}, t) \xi_s^s(\mathbf{r}', t') \rangle = -2\Gamma_s \delta^{rs} \delta(t - t') \nabla^2 \delta(\mathbf{r} - \mathbf{r}') \qquad (7.63)$$

$$\langle \xi_\theta^r(\mathbf{r}, t) \xi_\theta^s(\mathbf{r}', t') \rangle = 2\Gamma_\theta \delta^{rs} \delta(t - t') \delta(\mathbf{r} - \mathbf{r}') \qquad (7.64)$$

Omitting the noise terms, we have now instead of (7.56)

$$\omega = \pm c_0 |k| - iDk^2 \qquad (7.65)$$

with the spin diffusion constant

$$D = \tfrac{1}{2}(\Gamma_\theta \rho_s + \chi_0^{-1}\Gamma_s) \qquad (7.66)$$

The damping of the spin waves to lowest order is proportional to k^2 and can be ignored for $\mathbf{k} \to 0$. Unfortunately, the constants Γ_s and Γ_θ are unknown and (7.61), (7.62) with the Hamiltonian (7.51) do not describe all the damping mechanisms discussed above. One of these additional damping mechanisms can be derived from the gauge theory discussed in the preceding section. First, we show that for *unfrustrated* systems the hydrodynamic theory and the gauge theory lead to identical results. In hydrodynamics, one considers only a single valley. The complicated multi-valley structure and hence also inter-valley transitions are due to frustration. We already mentioned as an example the additional degeneracy of the ground state of the XY model due to helicity which is not taken into account in hydrodynamics.

In the absence of frustration ($\vec{\rho}_m^F = \mathbf{j}_m^F = 0$), the field equations (7.31) and (7.32) have the solution in linear approximation

$$\mathbf{b}_l(\mathbf{r}, t) = \nabla_l \Theta(\mathbf{r}, t),$$

$$\mathbf{a}(\mathbf{r}, t) = \partial_t \Theta(\mathbf{r}, t) = \vec{\omega}_0(\mathbf{r}, t) = \chi_0^{-1} \mathbf{S}(\mathbf{r}, t) \qquad (7.67)$$

Here, $\vec{\omega}_0$ is the local angular velocity which can be interpreted as the Larmor precession produced by the local internal field $\mathbf{h} = \chi_0^{-1}\mathbf{S}$. The simplest Hamiltonian in terms of the gauge potentials $\mathbf{a}$, $\mathbf{b}_l$ reads

$$H_{eff} = \tfrac{1}{2} \int d^3r (\chi_0 \mathbf{a}^2 + \rho_s \mathbf{b}_l \cdot \mathbf{b}_l) \qquad (7.68)$$

and agrees with the Hamiltonian (7.51) in the absence of frustration (i.e. with (7.67)). In contrast to (7.36), this Hamiltonian is no longer gauge invariant. Evidently, the breaking of rotation invariance in hydrodynamics is equivalent to breaking local gauge invariance. The physical reason for this is obvious: The coarse-grained spin density $\mathbf{S}(\mathbf{r}, t)$ or magnetization acts like a magnetic field which breaks gauge symmetry. With (7.67) the second term in (7.68) describes the restoring force against small spin rotations which should be zero in a gauge- or rotation-invariant system. In a spin glass, a gauge transformation is practically the same as a rotation in spin space: In the first case, one rotates the frame, in the second case the spin themselves. The comparison with hydrodynamics suggests that all gauge invariant models are fairly irrelevant for spin glasses.

If frustration is present, the rotation angle $\mathbf{\Theta}$ is no longer well defined, and one has to describe the dynamics by the variables $\mathbf{a}$ (or $\mathbf{S}$) and $\mathbf{b}_l$. Writing all components in spin space with upper indices r, s, t and all components in real space with lower indices l, m, n, we have from (7.30) in linear approximation

$$\frac{\delta b_l^r(\mathbf{r})}{\delta\Theta^s(\mathbf{r}')} = \nabla_l \delta(\mathbf{r} - \mathbf{r}')\delta^{rs} + \epsilon^{rst} b_l^t \tag{7.69}$$

which leads to

$$-\frac{\delta H_{eff}}{\delta\Theta^s} = -\frac{\delta H_{eff}}{\delta b_l^r}\frac{\delta b_l^r}{\delta\Theta^s} = \nabla_l \frac{\delta H_{eff}}{\delta b_l^s} + \epsilon^{srt}\frac{\delta H_{eff}}{\delta b_l^r} b_l^t \tag{7.70}$$

The equations of motion (7.61) and (7.62) are now replaced by

$$\partial_t S^r = \nabla_l \frac{\delta H_{eff}}{\delta b_l^r} + \Gamma_s \nabla^2 \frac{\delta H_{eff}}{\delta S^r} + \xi_s^r \tag{7.71}$$

$$\partial_t b_l^r = \nabla_l \frac{\delta H_{eff}}{\delta S^r} - 2\kappa\rho_s^{-1}\frac{\delta H_{eff}}{\delta b_l^r} + \xi_b^r \tag{7.72}$$

The second term on the right-hand side of (7.72) replaces the order parameter relaxation in (7.62) and can be identified with the frustration current density

$$j_l^{F,r} = \partial_t b_l^r - \nabla_l \frac{\delta H_{eff}}{\delta S^r} = -2\kappa\rho_s^{-1}\frac{\delta H_{eff}}{\delta b_l^r} \tag{7.73}$$

which is purely diffusive. (In (7.73) we used $a^r = \delta H_{eff}/\delta S^r$ and assumed $\xi_s^r = \xi_b^r = 0$.) This result can be proved for any Hamiltonian (Volovik and Dzyaloshinskii, 1978). With the Hamiltonian (7.68), the equations (7.71) and (7.72) lead to the modes

$$\omega_{1...6} = -i\kappa(\mathbf{k}) \pm [\tilde{c}^2 k^2 - \kappa^2(\mathbf{k})]^{\frac{1}{2}} \tag{7.74}$$

with $\kappa(\mathbf{k}) = \kappa + \frac{1}{2}k^2\Gamma_s/\chi_0$, $\tilde{c}^2 = c_0^2 + 2\kappa\Gamma_s/\chi_0$ and $c_0^2 = \rho_s/\chi_0$. For $\kappa = 0$ we recover the propagating modes $\omega = \pm c_0|k|$ (7.56) of hydrodynamics.

However, these modes become overdamped for $\kappa(\mathbf{k}) \gg \tilde{c}|k|$ and are relaxing or diffusive modes

$$\omega_{1\ldots3} = -2i\kappa(\mathbf{k}), \quad \omega_{4\ldots6} = -iDk^2 \tag{7.75}$$

with the diffusion constant $D = \tilde{c}^2/2\kappa$. Owing to the frustration network, spin waves no longer exist in the long-wavelength limit.

7.3   Effects of anisotropy: hysteresis, remanence, torque, ESR and NMR

In the limit $\mathbf{k} \to 0$ the modes discussed in the preceding section depend strongly on anisotropy effects. The DM interaction (6.55) between two spins i, j depends on their position vectors $\mathbf{R}_i, \mathbf{R}_j$ in the form $\mathbf{R}_i \times \mathbf{R}_j$ and is *unidirectional*, i.e. it defines a preferred direction in space. It leads to an anisotropy energy $E_a(\theta)$: The simultaneous rotation of all spins by an angle θ from their equilibrium position costs a finite amount of energy. This energy can be determined, for instance, by a torque experiment. However, on the macroscopic scale there is no preferred direction, since the local microscopic preferred directions $\hat{\mathbf{R}}_i \times \hat{\mathbf{R}}_j$ are randomly distributed. The system is macroscopically isotropic: its physical properties do not depend on any preferred directions in the underlying lattice. In particular, the anisotropy energy $E(\theta)$ is the same for all rotation axes.

The most obvious effect of anisotropy is the remanent magnetization which is often observed after switching off a field. This field fixes a direction of the spins, and the spins remain in this direction since any rotation out of it would cost some energy. The remanence is observed in the form of hysteresis loops if one sweeps through a second field between positive and negative values. In the example shown in Fig. 7.3c, the sample first was cooled down below T_f in a very large field (18 k Oe). After switching off this field, a smaller field was applied in the same and opposite directions. The resulting hysteresis loop exhibits two types of anisotropy, as shown schematically in Figs. 7.3a and 7.3b. The *unidirectional* anisotropy of Fig. 7.3a

$$E_a^{(1)} = -K_1 \cos\theta \tag{7.76}$$

changes sign if the spin system is rotated by an angle $\theta = \pi$ and shows up as a displacement of the hysteresis loop. Such an anisotropy can exist only in noncollinear spin sytems. In a system with parallel spins, a rotation $\theta = \pi$ is equivalent to an inversion, which does not change its energy in zero field. An anisotropy energy of the type (7.76) is expected from the DM

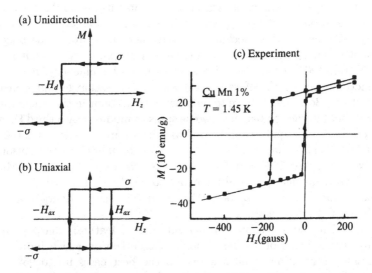

Figure 7.3:   Hysteresis cycles for (a) unidirectional anisotropy $E_a^{(1)} = -K_1 \cos\theta$, and (b) uniaxial anisotropy $E_a^{(2)} = -\frac{1}{2}K_2\cos^2\theta$, with $H_d = K_1/M_R$ and $H_{ax} = K_2/M_R$, where M_R is the remanent magnetization. (c) Typical experimenal data for C̲uMn 1% at $T = 1.45$ K show usually a mixture of these both anisotropies (from Préjean et al, 1980; and Alloul and Hippert, 1983).

interaction, and the displacement of the loop indeed is strongly enhanced by additional nonmagnetic impurities with large spin–orbit coupling such as Au or Pt. Fig. 7.4 shows this effect for different concentrations of Au in C̲uMn. It turns out that the anisotropy constant K_1 varies linearly with the impurity concentration.

Fig. 7.3 shows a second anisotropy contribution of the form

$$E_a^{(2)} = -\tfrac{1}{2}K_2\cos^2\theta \qquad\qquad (7.77)$$

which is uniaxial and determines the width of the hysteresis loop. In ideal ferromagnets such a term would depend on the lattice and define an 'easy' axis, and would be due to crystal field effects. In spin glasses, this term again is independent of the lattice directions and its origin is still unclear. A term which is quadratic in the spin–orbit coupling constant λ in (6.58) would lead to the anisotropy (7.77), but most likely this effect is too small to explain the data (see Henley et al, 1982).

In Section 6.3, we showed how the Dzyaloshinskii–Moriya (DM) interaction could be derived microscopically for a spin glass containing additional nonmagnetic impurities with orbital moments. A similar mechanism also exists in pure $\underline{Au}$Fe or $\underline{Cu}$Mn, leading to the unidirectional anisotropy (7.76). We discuss this mechanism briefly here for $\underline{Au}$Fe. The magnetic moment of an Fe atom in Au is due to a spin-split 'virtual bound state' (see Kittel, 1964) which has also a small orbital moment. An incoming conduction electron therefore is scattered by the exchange and spin–orbit interaction. Both effects have to be taken into account in the calculation of the indirect interaction between two Fe spins as modified by a third Fe atom. The calculation is quite complicated and it is hard to assess whether the effect is sufficiently large in order to explain the unidirectional anisotropy observed in $\underline{Au}$Fe. This is certainly not the case for $\underline{Cu}$Mn, where Mn has a completely quenched orbital moment. So far, there is no satisfactory explanation for the anisotropy of $\underline{Cu}$Mn (for details see Goldberg et al, 1986; Goldberg and Levy 1986).

Figs. 7.3 and 7.4 show another interesting feature: sharp jumps in the hysteresis loops of the field-cooled magnetization which indicate the simultaneous rotation of most spins (in the best cases up to 95% of all spins). These jumps are independent of the crystallographic directions: This is another proof that a spin glass is macroscopically isotropic, though it possesses macroscopic anisotropy at low temperatures. However, the spin system is not completely 'rigid' with respect to rotations by an angle π. If one varies the field very slowly, one observes that each jump consists in reality of a large number of smaller ones. This suggests the motion of some kind of defects similar to Bloch walls in ideal ferromagnets instead of the simultaneous flip of most spins (Felten and Schwink, 1984; Guy, 1982). The change of the remanence observed in sweeping through the hysteresis loop is reversible. However, the state with a given magnetization is metastable, and one has in addition irreversible decay of the remanence which extends over long times and will be discussed in Section 9.2. So far, no magnetization jumps have been observed in $\underline{Au}$Fe. This system has an extremely large anisotropy field, and possible jumps would be hard to observe.

In a hysteresis loop one measures only spin rotations with an angle $\theta = \pi$. Smaller rotation angles can be tested by means of the ESR, NMR, the transverse ac susceptibility and torque experiments. (In a torque experiment, one rotates the remanent magnetization M_R by rotating the external field. This leads to a change of the anisotropy energy and is seen in the (mechanical) torsion if the sample hangs on a thin thread.) For small angles, the assumption of rigid spins turns out to be well justified. In ESR and NMR after field cooling one observes rather sharp resonance lines at the frequency

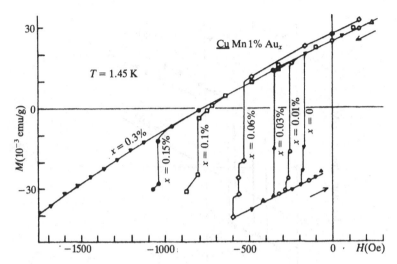

Figure 7.4: Hysteresis cycles of various ternary $\underline{Cu}$Mn 1%-Au$_x$ spin glasses in a small field. The remanent magnetization is obtained after removing a field of 18 kOe in which the sample was cooled down to 1.45 K (from Préjean et al, 1980).

$$\omega = \gamma(H_z + K_1/M_R) \tag{7.78}$$

where $\gamma = g\mu_B/\hbar$ is the gyromagnetic ratio and M_R the remanent magnetization. Fig. 7.5 shows ESR data for $\underline{Cu}$Mn 1.35 at%. After the cooling field $\mathbf{H}_c$ is switched off, a small field $\mathbf{H}_z \parallel \mathbf{H}_c$ and a rf or ac field $\mathbf{H}_1$ are applied. By field cooling, the system is trapped in a metastable state and by applying the field $\mathbf{H}_1$ in the y-direction the remanent magnetization $\mathbf{M}_R$ is rotated by an angle θ out of the direction of $\mathbf{H}_c$. This rotation angle depends on the angle θ_H between the total field $\mathbf{H} = \mathbf{H}_z + \mathbf{H}_1$ and $\mathbf{H}_c$. One has $\theta < \theta_H$ since the spins follow the field only partly.

We describe these effects in a model which is a simple extension of the hydrodynamics discussed in the preceding section. Assuming completely rigid spins, we now need only consider macroscopic variables and replace the coarse-grained spin density $\mathbf{S}(\mathbf{r})$ (7.52) by the uniform magnetization $\mathbf{M}$ and the angle $\Theta(\mathbf{r})$ by the uniform angle Θ. Both are assumed to be thermal averages. Instead of the effective Hamiltonian (7.68), we write down a Landau–Ginzburg expression for the free energy

$$F(\mathbf{M}, \Theta) = \tfrac{1}{2}\chi_0^{-1}(\mathbf{M} - \mathbf{M}_r)^2 - \mathbf{M} \cdot \mathbf{H} - E_a(\Theta) \tag{7.79}$$

in which one considers deviations of the magnetization from the remanent

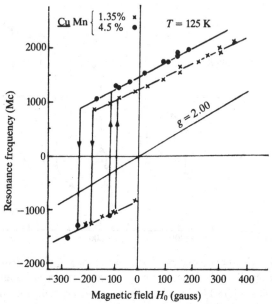

Figure 7.5: Electron spin resonance frequency around zero field for C̲uMn 1.35% (×) and 4.7% (•) alloys field–cooled down to 1.35 K (from Monod and Berthier, 1980).

magnetization $\mathbf{M}_r$, the field $\mathbf{H}$ and the anisotropy $E_a(\mathbf{\Theta})$. Since the spin glass is macroscopically isotropic (with the exceptions discussed in Section 6.2), the anisotropy E_a depends only on $|\mathbf{\Theta}|$ and is independent of the axis of rotation. In the absence of the field $\mathbf{H}$, the free energy (7.79) is then also independent of the direction of $\mathbf{M}_r$. As in Fig. 7.1, we define the direction of $\mathbf{M}_r$ for $\mathbf{H} = 0$ by the cooling field $\mathbf{H}_c$ and consider small deviations of $\mathbf{M}_r$ from this direction due to the field $\mathbf{H}$. Variations of $F(\mathbf{M}, \mathbf{\Theta})$ with respect to $\mathbf{M}$ and $|\mathbf{\Theta}|$ then lead to the equilibrium values M_0 and θ_0

$$\mathbf{M}_0 = \chi_0 H + M_r^0 \tag{7.80}$$

$$\sin \theta_0 = \frac{H_R H}{H_1^2} \sin(\theta_H - \theta_0) \tag{7.81}$$

with $H_R = M_r/\chi_0$, $H_1 = (K_1/\chi_0)^{\frac{1}{2}}$ and with the angle θ_H between the cooling field $\mathbf{H}_c$ and the applied field $\mathbf{H}$. In the case of the transverse susceptibility $\chi_\perp$, one has $\theta_0 \ll 1$, and the sum of the anisotropy energies

(7.76) and (7.77) becomes, apart from a constant,

$$E(\theta) = \tfrac{1}{2}K\theta^2, \quad K = K_1 + K_2 \tag{7.82}$$

In this limit, one derives from (7.79)

$$\chi_\perp = \chi_0 + \frac{M_r}{H_z + K/M_r} \tag{7.83}$$

For $H_z = 0$, the second term on the right hand side of (7.83) describes the response M_r/H_a induced by the anisotropy field $H_a = K/M_r$. The anisotropy constant K turns out to be independent of the remanent magnetization and is therefore an intrinsic property. Equation (7.81) agrees well with experimental data on CuMn 7.7 at% (Alloul and Hippert 1983).

The description of a spin glass state by $\mathbf{M}$ and a single scalar angle θ is not sufficient for all experiments. To understand a torque experiment with arbitrary angles, we have to remember that in a spin glass state the full rotation invariance ($SO(3)$) is broken. The effect of random anisotropy is to break the degeneracy of states as created by the transformation (7.27) or (7.44). (A macroscopic spin glass state is then defined by a basis 'triad' $(\hat{\mathbf{n}}, \hat{\mathbf{p}}, \hat{\mathbf{q}})$ of unit vectors and the rotation matrix R($\mathbf{\Theta}$), in addition to the magnetization $\mathbf{M}$.) The rotation matrix R($\mathbf{\Theta}$) can also be represented by an 'anisotropy triad' $(\hat{\mathbf{N}}, \hat{\mathbf{P}}, \hat{\mathbf{Q}})$ of unit vectors (Saslow 1982a,b, 1983). The anisotropy term in the free energy (7.79) then becomes

$$E_a(\mathbf{\Theta}) = -K_1 Tr\mathsf{R}(\mathbf{\Theta}) = \tfrac{1}{2}K_1((\hat{\mathbf{n}} \cdot \hat{\mathbf{N}} + \hat{\mathbf{p}} \cdot \hat{\mathbf{P}} + \hat{\mathbf{q}} \cdot \hat{\mathbf{Q}}) - 1) \tag{7.84}$$

where one assumes only unidirectional anisotropy (7.76). For a single axis, the second term would simplify to $-\tfrac{1}{2}K_1(\hat{\mathbf{n}} \cdot \hat{\mathbf{N}})$ where $\hat{\mathbf{N}}$ and $\hat{\mathbf{n}}$ are the directions of the cooling field $\mathbf{H}_c$ and of the remanent magnetization $\mathbf{M}_r$, respectively. This 'vector model' would apply to a system with a single easy axis (due to crystal fields) but is incorrect for spin glasses (Henley et al, 1982).

The triadic character of the anisotropy can be observed in a torque experiment. First, one rotates the field from $\theta_H = 0$ to $\theta_H = \pi$ around the x-axis, changing the remanence from $\mathbf{M}_r = (0, 0, M_r)$ to $\mathbf{M}_r = (0, 0, -M_r)$. Hereupon, one applies either a torque $\Gamma_y = K\theta_y$ around the y-axis or $\Gamma_x = K\theta_x$ around the x-axis. These torques turn out to produce completely different responses. An initial torque $\Gamma_x = -dE_a/D\theta_x$ leads, with $E_a = -K_1 \cos\theta$ near $\theta = \pi$, to a linear relation between Γ_x and θ, whereas $\Gamma_y(\theta)$ measures the torque perpendicular to the x-axis and is very small.

A second verification of the triadic character of the spin glass anisotropy comes from ESR. The hydrodynamic equations (7.53) and (7.54) are now replaced by (Henley et al, 1982)

$$\gamma^{-1}\partial_t\delta\mathbf{M} = -\frac{\partial F}{\partial\Theta} = \delta\mathbf{M}\times\mathbf{H} + \tfrac{1}{2}K_1[R(\Theta_0) - Tr R(\Theta_o)1]\delta\Theta \qquad (7.85)$$

$$\gamma^{-1}\partial_t\Theta = \frac{\partial F}{\partial\mathbf{M}} = \chi_0^{-1}(\delta\mathbf{M} - \delta\Theta\times\mathbf{M}_r^0) \qquad (7.86)$$

where we inserted the free energy (7.79) and considered small deviations $\delta\mathbf{M} = \mathbf{M} - \mathbf{M}_0$ and $\delta\Theta = \Theta - \Theta_0$ with $R(\Theta) = R(\delta\Theta)R(\Theta_0)$ from the equilibrium values $\mathbf{M}_0, \Theta_0$ (7.80) and (7.81). The first term on the right-hand side of (7.85) is the usual precessional term and the static limit of (7.86) leads back to (7.58) in the form $\delta\mathbf{S} = \delta\Theta\times\mathbf{S}^0$. The solution of (7.85) and (7.86) depends on the remanence field $H_r = M_r/\chi_0$, the anisotropy field $H_1 = (K_1/\chi_0)^{1/2}$, and the angle θ_r between the cooling field $\mathbf{H}_c$ and $\mathbf{M}_r^0$. One finds three modes, which in general have both transverse and longitudinal components. These modes decouple for $\theta_r = 0$ and are

$$\gamma^{-1}\omega_\pm = \tfrac{1}{2}(H - H_r) \pm \tfrac{1}{2}[(H + H_r)^2 + 4H_1^2]^{\frac{1}{2}} \qquad (7.87)$$

$$\gamma^{-1}\omega_L = H_1 \qquad (7.88)$$

They have been observed in an ESR experiment (Gullikson et al, 1983, 1985). However, the rigid spin model seems to break down for angles $\theta_H > 90^o$ (Hoekstra et al, 1985). The transverse modes can also be measured in zero-field-cooled samples with $\mathbf{M}_r = 0$, $\mathbf{H}\to 0$. (7.86) then reduces to

$$\gamma^{-1}\omega_\pm = H_1 \pm \tfrac{1}{2}H \qquad (7.89)$$

In the opposite limit, $K_1/M_r^0 \equiv H_1^2/H_r \ll 1$ we recover the result (7.78) for ω_+ as shown in Fig. 7.5.

These experiments show convincingly that for small deviations from the initial state, the anisotropy is unidirectional and of the form (7.76). For large-angle deviations, one has in addition uniaxial anisotropy and, moreover, a breakdown of the rigid spin model. The modes (7.87) and (7.88) (and also the more complicated modes for $\theta_r \neq 0$) can also be derived from the microscopic model (6.55) with exchange and DM anisotropy, including a magnetic field. The phenomenological anisotropy constant K_1 in (7.53) can be expressed in terms of the ground state spin configuration $\{\mathbf{S}_i\}$ and the anisotropy constants $\mathbf{D}_{ij}$

$$K_1 = \chi_0 H_1^2 = -E_a = -\sum_{ij}\mathbf{D}_{ij}\mathbf{S}_i^o\times\mathbf{S}_j^o \qquad (7.90)$$

where E_a is the anisotropy energy for $\theta = 0$ (Beton and Moore, 1984).

The remanence, torque, ESR, and NMR experiments discussed so far have all been performed at low temperatures ($T \ll T_f$). Does the anisotropy (7.76) or (7.77) also exist above the freezing temperature? For the crystal-field anisotropy of an ideal ferromagnet this is, of course, the case. In a spin glass, (7.90) suggests that the anisotropy K_1 vanishes for $T \geq T_f$. The data for the transverse susceptibility as given by (7.83) and for the torque agree with $K_1 = 0$ for $T \geq T_f$ (Youm and Schultz, 1986; de Courtenay et al, 1984) whereas the ESR data indicate a nonvanishing anisotropy slightly above the freezing temperature. Most likely, this is a dynamic effect: Dynamically, the spins still appear to be fairly rigid slightly above T_f and the metastable states still exist for sufficiently short times.

8

Short-range interactions:
beyond mean field theory

8.1 Renormalization in pure systems

In Section 2.2 we reviewed the basic facts about broken symmetry and phase transitions in pure systems, within mean field theory. This helped set the stage for the mean field description of spin glasses that we studied in Chapters 3–6. We will now want to study the phase transition and the nature of the low-temperature phase for spin glasses with fairly short-ranged interactions in low dimensionality (2 or 3). This situation is far from the region where mean field theory is a good guide, so we will have to learn new methods. Again, we start by learning about them in pure systems.

Renormalization is a very general approach to problems with many strongly interacting degrees of freedom. In a statistical mechanical problem, the basic idea is to carry out the trace in the partition function over some of the variables, leaving a new problem with fewer degrees of freedom. One follows how the parameters in the Hamiltonian of the system change as this procedure is iterated many times. Generally, the partial trace corresponds to removing the short-distance degrees of freedom, so each successive effective Hamiltonian has a larger lattice constant (or equivalent microscopic length) than its predecessor: each renormalization step corresponds to a change of scale. Different thermodynamic phases are identified with flows to different fixed points in the parameter space of the Hamiltonian under this semigroup.

In general, this transformation cannot be carried out exactly except for special models, but one can learn a great deal simply from the qualitative nature of the flows in parameter space: the fixed points and their basins of attraction. These determine the phase diagram of the system, and linearization of the renormalization group equations around the fixed points gives the macroscopic properties of the corresponding phases.

There are several formal ways to implement the renormalization procedure in practice, leading to different natural approximation schemes. One of these is based on the most direct way one could imagine trying to do it: Consider an Ising spin lattice; the partition function is a trace over the allowed values of all N spins. Now imagine carrying out the trace over some fraction of the spins — to make it simple, consider a one-dimensional system and trace over every other spin. We have to perform a series of calculations like

$$\sum_{S_2=\pm 1} e^{\beta J S_1 S_2} e^{\beta J S_2 S_3}$$

$$= \cosh^2 \beta J \sum_{S_2} (1 + S_1 S_2 \tanh \beta J)(1 + S_2 S_3 \tanh \beta J)$$

$$= 2\cosh^2 \beta J (1 + S_1 S_3 \tanh^2 \beta J) \tag{8.1}$$

Thus the elimination of the second spin leads to a coupling J' between the first and the third given by

$$\tanh \beta J' = \tanh^2 \beta J \tag{8.2}$$

In this one-dimensional example, the new interactions are still between nearest neighbours; the new problem is still of the same form as the old one and the transformation is exact. Successive iteration is sufficient to solve the problem completely: We continue eliminating half the remaining spins until the problem is no longer a strongly correlated one, i.e until $J = O(T)$. The lattice spacing we have reached at this point (expressed in terms of the original units, of course) is then (by definition) the correlation length. It is easy to check that this leads to a correlation length $\xi \propto e^{2\beta J}$ for low temperatures ($T \ll J$). (A way to see this result without using the renormalization group argument starts with the observation that the long-range order of the Ising chain is destroyed at low temperatures by occasional broken bonds or 'kinks' of energy $2J$. These occur with Boltzmann probability $e^{-2\beta J}$, so we can identify the correlation length with the reciprocal of this kink density.)

In higher dimensions, new couplings (e.g. next-nearest neighbour interactions) are generated in each renormalization step, and approximations have to be made, truncating the paramater space of the Hamiltonian. But this simple example illustrates the main idea.

In this one-dimensional problem, the problem always flowed toward weak coupling (i.e. high-temperature behaviour) with successive iterations of the renormalization transformation (except for T exactly zero). In higher dimensions, one can find flow toward strong coupling if the initial coupling is sufficiently strong (i.e. the temperature is sufficiently low). For a special intermediate value of the temperature, the flow is neither toward weak nor strong coupling; rather the parameters of the Hamiltonian go asymptotically to finite fixed point values. This special temperature is the critical temperature of the system.

We learned in mean field theory how various quantities (the susceptibility (2.18), the correlation length (2.20) or the order parameter (2.15)) diverged or behaved singularly at a second order phase transition. More generally, these and other quantities are found experimentally to be singular, but with critical exponents different from mean field ones. The experimentally measurable critical exponents are defined by the divergence of the specific heat

$$C_{sing} \propto |T - T_c|^{-\alpha}, \tag{8.3}$$

the way the order parameter goes to zero at T_c

$$M \propto (T_c - T)^{\beta}, \tag{8.4}$$

the divergence of the susceptibility

$$\chi \propto |T - T_c|^{-\gamma}, \tag{8.5}$$

the behaviour of the order parameter as a function of the conjugate field right at T_c

$$M \propto h^{1/\delta}, \tag{8.6}$$

the divergence of the correlation length

$$\xi \propto |T - T_c|^{-\nu}, \tag{8.7}$$

and the decay of correlations right at T_c

$$\langle S(0)S(r) \rangle \propto r^{-(d-2+\eta)} \tag{8.8}$$

(or, in k–space,

$$G(k) \propto k^{-(2-\eta)}). \tag{8.9}$$

These exponents are not all independent; from the assumption of scale invariance (lack of any characteristic length longer than the lattice spacing) at the critical point one can derive the following *scaling laws* relating them:

$$\alpha + 2\beta + \gamma = 2 \tag{8.10}$$

$$d\nu = 2 - \alpha \tag{8.11}$$

$$(2 - \eta)\nu = \gamma \tag{8.12}$$

$$\beta = \tfrac{1}{2}\nu(d - 2 + \eta) \tag{8.13}$$

$$\delta = \frac{d + 2 - \eta}{d - 2 + \eta} \tag{8.14}$$

These equations reduce the number of independent critical exponents to 2. Note the critical role of the dimensionality d, which did not appear in the mean field values of the exponents.

These scaling laws emerge as a general consequence of the structure of the renormalization group if the critical temperature (or, more generally, critical point) is associated with a fixed point of the renormalization transformation which is unstable against small changes in temperature or conjugate field. For a few problems, one can carry out the renormalization group transformation exactly and find the remaining two independent critical exponents without approximation. More often, one must resort to approximations.

One simple kind of approximation is based on the direct scheme described above of tracing over a fraction of the spins in the lattice to generate a new effective Hamiltonian on a larger length scale. As we mentioned, this can only be carried out exactly for one dimension, and in higher dimensions it is necessary to truncate the new Hamiltonian after a few terms. For example, one can work within a restricted parameter space in which only nearest-neighbour and next-nearest-neighbour interactions are kept and all the longer-range terms are discarded. We will in fact use the simplest such scheme, in which only nearest-neighbour couplings are retained, later in this chapter to illustrate this kind of method. In view of its crudity, it will turn out to work surprisingly well for describing both the critical point and the low-temperature phase of Ising spin glasses in 2 or 3 dimensions.

The other common way to carry out the renormalization idea is to write the relevant fluctuating variables (here, the spins) in terms of Fourier components and get rid of the components with large wavenumber. This is best illustrated in the context of the Wilson–Fisher calculation of critical exponents just below four dimensions, which is carried out in a framework which starts with the Landau–Ginzburg formulation we used for mean field theory in Section 2.2. We now regard the Landau–Ginzburg free energy (2.14) as an effective spin Hamiltonian to be traced over to obtain thermodynamical quantities:

$$Z = Tr_S \exp(-\beta H[\mathbf{S}]) \tag{8.15}$$

where (for zero external field)

$$\beta H[\mathbf{S}] = \tfrac{1}{2} \int d^d x [r_0 \mathbf{S}^2(\mathbf{x}) + c_0 (\nabla \mathbf{S}(\mathbf{x}))^2 + \tfrac{1}{2} u (\mathbf{S}^2(\mathbf{x}))^2] \tag{8.16}$$

(We have generalized to an m-component vector $\mathbf{S}(\mathbf{x})$. Higher-order terms in $\mathbf{S}(\mathbf{x})$ and its gradients are omitted, so this model describes small and slowly-varying (i.e. semimacroscopic) magnetization fluctuations.)

As $\mathbf{S}$ is a continuous variable, the trace (8.15) is an integral over all real values. Furthermore, since $\mathbf{S}(\mathbf{x})$ is a function of a continuous variable $\mathbf{x}$, the trace becomes a functional integration like the one we introduced in Chapter 4 for dynamics. We write

$$Z = \int D\mathbf{S} \exp(-\beta H[\mathbf{S}]) \tag{8.17}$$

(Analogously to that case, the functional integral is formally defined as the limit of a multidimensional conventional integral over a set of $\mathbf{S}(\mathbf{x}_i)$ defined on a discrete lattice of points, in the limit of vanishing lattice spacing.)

The mean field theory of Section 2.2 is obtained from (8.16) and (8.17) in the saddle-point approximation. The fluctuations at the level of (2.19) are obtained in Gaussian approximation, i.e. by expanding (8.16) to quadratic order around the stationary point and carrying out the resulting Gaussian integral.

Going beyond Gaussian or mean field approximation clearly means doing perturbation theory in u. This is a standard problem in field theory (we are just dealing with 'ϕ^4' field theory). The structure of the perturbation theory is very similar to what we used for the corresponding problem in the field theory of soft-spin dynamics in Section 4.3. The diagrams are built from free correlation functions or propagators (cf. (2.17–19))

$$G_0(k) = \langle S_{k\mu} S_{-k\mu} \rangle_0 = \frac{1}{r_0 + c_0 k^2} \tag{8.18}$$

with interactions represented by wavy lines standing for factors of $-u$. In addition there is a factor $m/2$ for each closed loop (recall that m is the number of components of $\mathbf{S}$). Examining the form of the diagrams qualitatively, we find that typically when we go from nth to $n+1$st order in perturbation theory we add two new spin correlation functions G to be integrated over. Thus we get corrections like

$$u \int d^d k \left(\frac{1}{r + ck^2} \right)^2 \propto r^{d/2-2} \tag{8.19}$$

where r and c differ from the bare parameters r_0 and c_0 in including self-energy corrections. Thus, above $d = 4$, we get finite corrections at the critical point (i.e. in the limit $r \to 0$), so we can hope that perturbation theory will converge, with only quantitative corrections to mean field theory (i.e. no change in critical exponents). Below $d = 4$, in contrast, terms of all orders in perturbation theory diverge as we approach the critical point. Thus mean field theory must break down completely, and renormalization group methods are called for to find the critical behaviour.

Each iteration of the renormalization group on this model consists of three steps:

1. Carry out the part of the functional integration over S_k with k lying in an outer shell in momentum space: $\Lambda > k > \Lambda/b$, where Λ is a cutoff momentum inverse, say, several lattice spacings in the original problem we are trying to model. This is done to a particular order in perturbation theory in u (here, first order).

2. Rescale length (i.e. wavevectors) so that the remaining Fourier components S_k once again have wavenumbers between 0 and Λ: $k' = bk$.

3. Rescale the spin fields themselves by a factor (conventionally defined as $b^{1-\eta/2}$); η will be chosen later in order to find a fixed point.

The perturbation theory graphs which give the corrections to r_0 and u are just the zero-external-momentum limits of the irreducible 2- and 4-point vertices, respectively. They are shown in Fig. 8.1. In these diagrams the internal G lines must have momenta in the shell $\Lambda/b < k/\Lambda$, so we have

$$r' = r + (m + 2)u \int_{\Lambda/b < k/\Lambda} \frac{d^d k}{(2\pi)^d} \left(\frac{1}{r + ck^2} \right) \qquad (8.20)$$

$$u' = u - (m + 8)u^2 \int_{\Lambda/b < k/\Lambda} \frac{d^d k}{(2\pi)^d} \left(\frac{1}{r + ck^2} \right)^2 \qquad (8.21)$$

Corrections to c come from the k-dependence (as $k \to 0$) of the self-energy diagrams; to order u this is zero.

The rescaling of wavevectors puts the rS^2 term into its old form, but we want to write the uS^4 term

$$\frac{u}{4N} \sum_{k'_i} (\mathbf{S}_{k'_1} \cdot \mathbf{S}_{k'_2})(\mathbf{S}_{k'_3} \cdot \mathbf{S}_{-k'_1-k'_2-k'_3}) \qquad (8.22)$$

in terms of $N' = N/b^d$, the new number of sites in the lattice. This divides u by a factor b^d. In the $c(\nabla S)^2$ term,

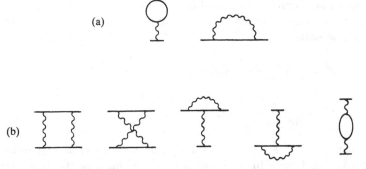

Figure 8.1: Perturbation theory graphs for correction to r_0 (a) and u (b).

$$c \sum_{k<\Lambda/b} k^2 |\mathbf{S}_k|^2 = cb^{-2} \sum_{k'<\Lambda} (k')^2 |\mathbf{S}_{k'}|^2 \tag{8.23}$$

we get a factor b^{-2}. The field rescaling gives a factor $b^{2-\eta}$ in quadratic terms and $b^{4-2\eta}$ in quartic ones, so, putting everything together, we get

$$r' = b^{2-\eta} \left[r + (m+2)u \int_{\Lambda/b<k<\Lambda} \frac{d^d k}{(2\pi)^d} \left(\frac{1}{r+ck^2} \right) \right] \tag{8.24}$$

$$u' = b^{4-d-2\eta} \left[u - (m+8)u^2 \int_{\Lambda/b<k<\Lambda} \frac{d^d k}{(2\pi)^d} \left(\frac{1}{r+ck^2} \right)^2 \right] \tag{8.25}$$

$$c' = b^{-\eta} c \tag{8.26}$$

To obtain a fixed point we will need to take $\eta = 0$ so that c remains fixed. Doing this, we can take $c = 1$ henceforth and omit c from subsequent equations.

It is easiest to proceed by taking $b = e^l$ and going over to differential equations

$$\frac{dr}{dl} = 2r + (m+2)u \frac{K_d}{1+r} \tag{8.27}$$

$$\frac{du}{dl} = \epsilon u - (m+8)u^2 \frac{K_d}{(1+r)^2} \tag{8.28}$$

where $K_d = \Omega_d/(2\pi)^d$, with Ω_d the solid angle in d dimensions, and where we have defined $\epsilon = 4-d$. For $d > 4$, then, the effective nonlinearity goes to zero, asymptotically things are described by the Gaussian approximation. There should therefore be no qualitative corrections to mean field theory, i.e. mean field exponents should be correct.

For $d < 4$ however, the Gaussian fixed point $u^* = r^* = 0$ is unstable; the stable point is

$$r^* = \tfrac{1}{2}\epsilon \left(\frac{m+2}{m+8}\right) + O(\epsilon^2) \tag{8.29}$$

$$u^* = \frac{\epsilon}{(m+8)K_d} + O(\epsilon^2) \tag{8.30}$$

The fact that u^* is of $O(\epsilon)$ is what justifies stopping at first order in the perturbation theory in u for small ϵ. If we linearize the renormalization group equations around this fixed point, we find

$$\frac{d(r - r^*)}{dl} = \left[2 - \left(\frac{m+2}{m+8}\right)\epsilon\right](r - r^*) \tag{8.31}$$

And by the same kind of argument that we used to find the correlation length in the problem of the one-dimensional chain, we argue that when we have reached a scale where r is of order unity, we begin to have a weakly-coupled problem, i.e. we have reached the correlation length. Since $r - r^* \propto T - T_c$, this gives

$$\xi \propto \left(\frac{T_c}{T - T_c}\right)^{\nu} \tag{8.32}$$

with the critical exponent ν given, to first order in ϵ, by

$$\nu = \tfrac{1}{2}\left[1 + \tfrac{1}{2}\epsilon\left(\frac{m+2}{m+8}\right)\right] + O(\epsilon^2) \tag{8.33}$$

We see that we get a correction to the mean field result $\nu = \tfrac{1}{2}$ in (2.20) proportional to ϵ. Other critical exponents can be computed by similar arguments.

We say that 4 is the *upper critical dimensionality* for this problem. We recall that there is generally another critical dimensionality, the *lower critical dimensionality* discussed at the end of Section 2.2, below which order is completely destroyed. One can also make expansions of critical exponents just above the lower critical dimensionality, as well as in the reciprocal of the number of spin components m. The reader is referred to the books of Ma, Toulouse and Pfeuty, and Amit for details.

We stress that the renormalization group is quite a general tool, and not merely a way to compute critical exponents. For example, the properties of the low-temperature phase may be inferred from the way the effective couplings grow toward infinity under the renormalization transformation. We will use such an argument to construct a theory for the spin glass phase of short-ranged Ising systems in the next section.

8.2   Scaling theory of the spin glass phase

Unlike the mean field theory of previous chapters, the theory of spin glasses with short-range interactions has not been thoroughly worked out yet. The description given in the next two sections must be regarded as tentative and incomplete. In particular, it assumes that the ergodicity breaking in the spin glass phase is of the trivial sort, with a single thermodynamic phase, up to an overall spin flip. This has not been proved.

In this section we use a renormalization group picture to describe the low-temperature phase of Ising spin glasses. Our treatment follows Bray and Moore (1986). Essentially the same arguments in a slightly different language were given by Fisher and Huse (1986, 1988b).

The basic idea is to study the flow of the couplings in the range where they are very strong and getting stronger. For long enough distances, this flow determines the form of the correlation functions everywhere in the low-temperature phase. The theory is grounded in both numerical calculations which, given sufficient computer time, could in principle determine the relevant renormalization group flows to any desired accuracy, and in a simple approximate renormalization scheme which we will describe.

The picture that emerges is rather different from that of mean field theory. At the same time, it is also very different from that of conventional ordered phases.

Here is the basic idea: We work with a nearest-neighbour Ising model (2.11) *of finite size L in each dimension* on a generalized cubic lattice in d dimensions with a symmetric bond distribution and, for now, zero external field. Suppose we consider the free energy of such a block (with free boundary conditions) and compare this with the free energy obtained if the L^{d-1} spins on one end of the block are constrained to point in the opposite direction (with those on the other end held fixed in their previous orientations). The difference $\Delta F(L)$ is something like an effective bond strength on length scale L. That is, if we perform a renormalization group calculation in the style of (8.1), tracing over spins until only a fraction L^{-d} of the original ones remain, the effective coupling in the resulting problem should scale with L like $\Delta F(L)$.

Calculating $\Delta F(L)$ is very simple at low temperature in an Ising ferromagnet. Then the states obtained with the two boundary conditions just differ from each other in that the second one has a domain wall across the sample, implying a $\Delta F(L) \propto L^{d-1}$ and, in particular, a lower critical dimensionality of unity. In a spin glass, frustration invalidates this simple result, since the domain wall can wander and adjust its shape to minimize its free energy. The result is a much lower exponent than $d-1$ and a larger lower critical dimensionality.

Essentially everything we want to know about equilibrium properties can be obtained from $\Delta F(L)$. If it decreases toward zero (as in (8.2)) with increasing L the system is in a disordered (paramagnetic) phase, while if it grows to infinity the system is in the spin glass phase. The borderline case where $\Delta F(L)$ settles at a particular finite value occurs just at the spin glass transition temperature T_f.

In a spin glass, the $\Delta F(L)$ are random quantities, and we have to study their distribution $P(\Delta F(L))$, which we write as

$$P(\Delta F(L)) = \frac{1}{L^y} f_L(\Delta F(L)/L^y) \tag{8.34}$$

where f_L is normalized such that $\int f_L(x)dx = 1$. Asymptotically for large L, $f_L \to f_\infty$, a so-called *scaling function* function independent of L. Then the paramagnetic phase is characterized by negative y, the spin glass phase by positive y, and T_f by $y = 0$, i.e. a fixed distribution $P(\Delta F) = f_\infty(\Delta F)$.

Thus to know whether there is a spin glass phase in a particular system, we have only to find whether the exponent y at zero temperature is positive. This was done numerically by Bray and Moore (1984) for the Gaussian-bond-distribution Ising model. They simply evaluated the ground state energy difference $\Delta F(L) \to \Delta E(L)$ numerically for samples of several sizes and plotted the width of their distribution against L on log–log paper (Fig. 8.2). For $d = 2$, y is clearly negative (they find $y = -0.291 \pm 0.002$) so there is no finite-temperature transition. For $d = 3$ there are only three data points, but they all fall nicely on a line with $y = +0.19 \pm 0.01$, suggesting that T_f is nonzero in this case. Hence the lower critical dimensionality lies between 2 and 3, according to these results.

Fig. 8.3 shows the scaling functions f_∞ for both dimensionalities. The important feature in them is the fact that they are continuous in the limit where their arguments go to zero. We will see that this is responsible for many anomalous properties of spin glasses which are not found in ferromagnets or other pure magnetic systems.

A very simple approximate renormalization group theory gives essentially the same results as these numerical calculations. This approach was introduced by Migdal (1975) and reformulated by Kadanoff (1976). The idea is that, at each stage of the renormalization, one approximates the problem by moving bonds in the lattice so that the trace over some frac-

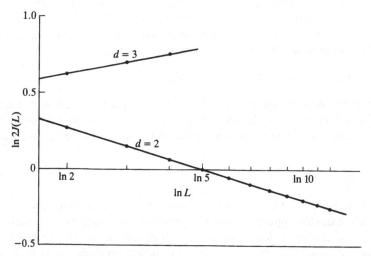

Figure 8.2:   Scale dependence of the block coupling $J(L)$ for two-dimensional ($d = 2$) and three-dimensional ($d = 3$) Ising spin glasses (from Bray and Moore, 1984).

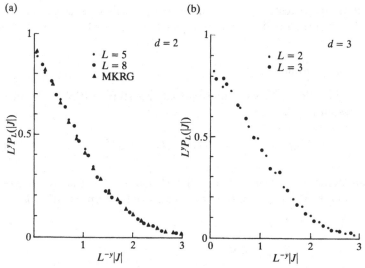

Figure 8.3: Scaling function f_∞ (from Bray and Moore, 1984).

tion of the spins can be carried out exactly. This is illustrated for $d = 2$ in Fig. 8.4. After moving vertical bonds as shown in Fig. 8.4b, we can trace over every other spin as in the one-dimensional renormalization (8.1) to combine each successive pair of horizontal bonds into a new bond. This gives (a simple generalization of the one-dimensional result (8.2))

$$\tanh \beta J_{AC} = \tanh \beta J_{AB} \tanh \beta J_{BC} \tag{8.35}$$

(and correspondingly for J_{DF}). This has changed the horizontal lattice scale by a factor 2. In the next step, we make the scale change in the vertical direction. Then every other horizontal bond is moved up and added to another one:

$$J'_{AC} = J_{AC} + J_{DF} \tag{8.36}$$

In d dimensions, steps like this are repeated in all $d-1$ nonhorizontal directions. Thus in this approximation the problem remains a single-parameter one (no higher-spin or longer-range couplings are introduced by the trace over the eliminated spins) at every renormalization step. It is therefore straightforward to calculate the way the distribution of bond strengths changes with length scale (although it does require some numerical computation to get the distributions).

We can get some insight into the problem, following Southern and Young (1977), who made the further approximation of forcing the bond distribution into a Gaussian of the correct variance at each renormalization step. This allows us to restrict our attention to the scale of a single parameter, the variance of the bond distribution, which we denote by $\tilde{J}^2$. In the $T \to 0$ limit, (8.35) reduces simply to

$$J_{AC} = \min(J_{AB}, J_{BC}) \tag{8.37}$$

We find by an elementary calculation that this step multiplies the variance of a Gaussian by a factor

$$\lambda = 8 \int r\,dr \int_0^{\pi/4} \frac{d\theta}{2\pi} e^{-r^2/2} r^2 \sin^2 \theta = 1 - \frac{2}{\pi} \tag{8.38}$$

The remaining $d - 1$ steps like (8.36) each just double the variance of the distribution, so the approximate renormalization group equation is

$$\tilde{J}^2_{l+1} = 2^{d-1} \lambda \tilde{J}^2_l \tag{8.39}$$

Thus $\tilde{J}_l \propto 2^{yl} \tilde{J}_0 = L^y \tilde{J}_0$, with

$$y = \tfrac{1}{2}[(d-1) + \log_2 \lambda] \tag{8.40}$$

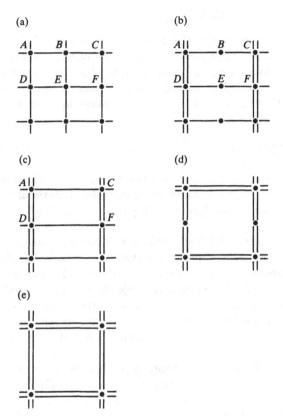

Figure 8.4: Migdal–Kadanoff renormalization: (a) the original bonds. (b) moving every other vertical bond. (c) result after tracing over spins B and E. (d) moving every other horizontal bond. (e) final result.

This gives $y = -0.23$ in $d = 2$ and $y = +0.26$ in $d = 3$, in reasonable agreement with the numerical results cited above. Keeping the correct form of the distribution instead of forcing it into a Gaussian (Bray and Moore, 1984,1985) also gives similar agreement.

Equation (8.40) implies that y cannot exceed $\frac{1}{2}(d-1)$, since λ must be less than unity (as a consequence of (8.37)). Fisher and Huse (1986, 1988b) argue that this is a rigorous bound on y generally.

We now concentrate on the case of positive y, i.e. above the lower critical dimensionality, as in $d = 3$. We can use the results on the low-temperature scaling of the bond distribution to find the long-distance behaviour of the spin glass correlation function

$$G(r_{ij}) = [(\langle S_i S_j \rangle - \langle S_i \rangle \langle S_j \rangle)^2]_{av} = T^2 \chi_{SG}(r_{ij}) \qquad (8.41)$$

(cf. (2.64)). At zero temperature, this goes to zero because all the spins
are frozen, making $\langle S_i S_j \rangle = \langle S_i \rangle \langle S_j \rangle$. At a small nonzero temperature,
this equality can be violated for the small fraction of spin pairs for which
a domain wall can be thermally excited between them. From the above
arguments we have that the typical domain wall energy for length scale r
is proportional to r^y; hence

$$G(r) \propto T P(\Delta F(r) = 0) \propto \frac{T}{Jr^y} \qquad (8.42)$$

where J is of the order of the microscopic bond energies. (This is where
the assumption of a single phase enters the theory. For nontrivial broken
ergodicity $G(r_{ij})$, like the related quantity $G_1(r_{ij})$ discussed below, has a
nonzero limit as $r \to \infty$.)

The power law behaviour (8.42) is a little surprising for an Ising system:
We would naively expect that its fluctuations would decay exponentially like
those of the Ising ferromagnet (although we saw in Section 2.2 that systems
with a continuous symmetry have power law behaviour in their correlation
functions in their ordered phases). It is clear from the nature of our argument that this behaviour in the present model is due to the continuous
nature of the effective coupling or domain wall energy distribution $P(J(L))$
near $J(L) = 0$.

In our discussion of the mean field theory in previous chapters, we made
a great deal of the fact that the low-temperature state exhibited broken
ergodicity, as manifested formally by replica symmetry breaking, with ultrametricity and lack of self averaging as consequences. Do any such exotic
effects occur for the short-range system?

To answer this question, consider the correlation function

$$G_1(r_{ij}) = [\langle S_i S_j \rangle^2]_{av} - [\langle S_i \rangle^2]_{av}[\langle S_j \rangle^2]_{av} \qquad (8.43)$$

From (2.52–53) we see that *for the SK model* in the limit $r_{ij} \to \infty$

$$G_1(r_{ij}) \to q^{(2)} - q^2 = \int P(q) q^2 dq - \left(\int P(q) q \, dq \right)^2 \qquad (8.44)$$

which is nonzero when $P(q)$ is not a delta-function, i.e. below the AT line.
By contrast, the same scaling arguments that led to the power law decay
(8.41) can be applied to $G_1(r_{ij})$ in the present description. That is, $G_1(r_{ij})$
decays to zero at infinite separation, implying that $P(q)$ must be a delta-
function. There is therefore no nontrivial broken ergodicity in the scaling
theory of the short-range system.

A remnant of such effects persists in a finite system, however: In a system of size L we have $G_1(L) \sim L^{-y}$, and for the small value of y in $d = 3$ this can be rather noticeable. It is much larger, for example, than the corresponding effect for correlation functions in the ordered phase of a Heisenberg ferromagnet, where, using (2.23), they behave like L^{-1} in three dimensions.

There is another way in which the spin glass phase obtained in this picture differs significantly from conventional ordered phases. In the conventional situation the spatial pattern of broken symmetry is essentially temperature independent below the transition temperature; only the magnitude changes with T. This is not true in a spin glass. Here, the spatial pattern of local equilibrium magnetizations changes nearly completely if the temperature is changed by any finite amount, no matter how small, as the following argument (Fisher and Huse, 1986, 1988b; Bray and Moore, 1987) shows.

At zero temperature, consider a low-energy excitation of the system, a 'droplet' of reversed spins of size L and typical energy $\sim JL^y$. The energy comes from the domain wall, which is expected to have some fractal dimensionality d_s. d_s is an independent exponent between $d - 1$ and d. Now consider what happens when we introduce small random changes in the bond strengths of typical size δJ. This will change the wall energy by a random amount of order $\delta J L^{d_s/2}$. If $d_s/2 > y$, this will (for large enough L) be much larger than the original wall energy JL^y, implying that the original ground state is not stable against these small perturbations in the bond strengths. Thus the ground state will become completely rearranged on scales longer than

$$L^* \propto (J/\delta J)^{1/(d_s/2-y)} \qquad (8.45)$$

Using $d - 1 < d_s < d$ and the inequality $y \leq (d - 1)/2$ that we argued for above, we see that this instability will always occur for spin glasses. The response to a small change in temperature δT gives a change in wall free energy which scales in the same way, with δT in place of δJ. We recall that in mean field theory there was also an extreme sensitivity of the spin glass phase (in that case, the local magnetization configurations of the many phases characteristic of the broken ergodicity). So here too the short-range system appears to have a similar property, despite the absence of strict broken ergodicity.

So far we have only discussed the static properties of the spin glass phase. The dynamics are also different from what we are accustomed to in pure systems. We approach this problem again in the renormalization group formulation first given by McMillan (1984). Now, in addition to calculating how the trace over spins which are eliminated within a cell at each renormalization step determines a new effective bond strength, we also examine how the local equilibrium implied by this calculation is reached. This means calculating the relaxation time for the degrees of freedom inside the cell, which is determined by the time it takes for a domain wall to propagate across a cell.

We assume the low-temperature limit everywhere. That is, we will ignore all but the fastest (lowest-barrier) allowed activated processes, since ones with higher barriers will be exponentially negligible in comparison for $T \rightarrow 0$. The motivation for doing this is that in 3 dimensions the critical temperature is fairly small because the system is just above its lower critical dimensionality. Thus there should be a reasonable range of temperatures, including T_f where such an approximation is reasonable.

An explicit and exact calculation in this limit can be carried out, but it is somewhat complicated, so here we only give the qualitative arguments, as McMillan did. For each stage of the renormalization, the barrier hindering the relaxation process will be of the same order as the typical bond strengths at this scale:

$$\delta V_l \propto \tilde{J}_l \sim 2^{yl}, \tag{8.46}$$

and the time it takes to move the domain wall across the cell is

$$\tau_l = \tau_{l-1} \exp(\beta \delta V_l) \tag{8.47}$$

since τ_{l-1} was the effective characteristic time for flipping spins within the elementary cell at this stage of renormalization. Thus at each stage the effective time unit is multiplied by a factor like that in (8.47), or, equivalently, the total effective barrier is increased by δV_l.

Note the implicit assumption that each length scale has an associated timescale, and that equilibrium is reached on one length scale long before it is reached on the next one, allowing the internal degrees of freedom at a given length scale to be treated solely in terms of their renormalized bonds obtained from an equilibrium calculation. This requires that $\tau_{l+1} \gg \tau_l$, which will be true at low enough temperatures, provided only that the barrier increment δV_l is not zero. We will assume that the exceptional cases where it does vanish do not affect the following arguments qualitatively.

The total barrier V_l after l steps is obtained by summing up the δV_j for all the steps up to this point:

$$V_l = \sum_{j=1}^{l} \delta V_j \tag{8.48}$$

For positive y, this sum is essentially dominated by the last term, $j = l$. Then the barrier V_l scales with the same exponent y as $\tilde{J}_l$. Of course, we have only shown this in Migdal approximation. The barriers might actually scale with some other exponent. In their phenomenological treatment, Fisher and Huse (1986, 1988b), allowed for this possibility with the ansatz

$$V_l \sim 2^{\psi l} \tag{8.49}$$

We can now calculate the time dependence of the spin autocorrelation function $\tilde{C}(t) = [\langle S_i(0)S_i(t)\rangle]_{av} - q$ as follows. Blocks of size L will have a relaxation time $\sim \exp(\text{const } \beta L^{\psi})$, or, equivalently, relaxation processes at time t involve blocks or domains of size

$$L(t) \propto [T\ln(t/\tau_0)]^{1/\psi} \tag{8.50}$$

Now from the same kind of argument we made about static correlations we find a $\tilde{C}(t)$ at time t proportional to the number of bonds at scale $L(t)$ which are weaker than T:

$$\tilde{C}(t) \propto \frac{1}{L(t)^y} \propto \frac{1}{[T\ln(t/\tau_0)]^{y/\psi}} \tag{8.51}$$

This extremely slow decay makes the observation of the equilibrium spin glass state effectively impossible on accessible experimental timescales except right in the neighbourhood of T_f (see the next section). This makes it important to generalize these dynamical arguments to nonequilibrium (effectively broken-ergodicity) states (Fisher and Huse, 1988a; see Section 9.2).

The Fourier transform of (8.51) (the noise spectrum of the spin fluctuations) is

$$\tilde{C}(\omega) \propto \frac{1}{|\omega \ln \omega|^{1+y/\psi}} \tag{8.52}$$

Except for the logarithmic corrections, this is an example of so-called $1/f$ noise, which occurs in a very wide variety of complex physical systems.

We can extend our analysis to include the effects of an external field. The argument (Bray, 1988) is a simple generalization of an old argument due to Imry and Ma (1975) about the effect of a random field on the ordering in a ferromagnet. A version of their argument is as follows: We consider how the exchange interaction and the (random) field scale under renormalization at low temperatures. For an ordered ferromagnetic state this is simple: We have $J(L) \propto \Delta F(L) \propto L^{d-1}$ (the domain wall energy), while the width of the random field distribution scales like $L^{d/2}$. Therefore, for $d-1 > d/2$, i.e. for $d > 2$, the exchange will grow faster and dominate at large distances, and the ferromagnetic order will be stable. For $d < 2$, the random field grows larger at long enough length scales, so ferromagnetic order will be destroyed, even at $T = 0$. The ordered state will break up into domains: The crossover length scale $L \propto (J/h)^{2/(2-d)}$, where the two energies are of the same order, gives the typical size of the ferromagnetic domains.

The only difference in the spin glass case is that the domain wall energy is proportional to L^y instead of L^{d-1}. The uniform field scales the same way for a spin glass that the random field does in a ferromagnet. Thus for $d/2 > y$ a magnetic field will destabilize the spin glass state; this condition is *always* satisfied in the cases we have met so far. Indeed, the inequality $y < (d-1)/2$ mentioned above implies that this will happen in any dimensionality. This instability is similar to the one we found above to a small change in temperature or in the random bond strengths. Again, the extreme sensitivity to field is reminiscent of the mean field behaviour, despite the lack of broken ergodicity in the present situation.

The state induced by the field consists of domains of the two zero-field spin glass states. The size of these domains is again set by the crossover length where $\tilde{J}(L) \sim h(L)$:

$$L_h \sim (J/h)^{2/(d-2y)} \tag{8.53}$$

If we compare this with the length scale (8.45) on which small temperature perturbations change the equilibrium spin arrangement, we see that since $d > d_s$ the field is a more serious perturbation of the order than the temperature change.

We can calculate the magnetization per site as

$$m \sim L^{-d/2} \sim (h/J)^{d/(d-2y)} \tag{8.54}$$

This shows that the magnetization depends nonanalytically on the field. There is of course also an analytic term in m proportional to h which comes from the reorientation of single spins and small clusters by the field; the nonanalytic term (8.54) comes from the reorientation of large domains of spins.

So far we have only shown that the external field breaks up the *zero-field* spin glass state. We have not proved anything about whether the finite-field phase goes over smoothly into the high-temperature paramagnetic phase as the temperature is raised, or whether instead it is separated from the paramagnetic phase by a kind of AT line. We will return to this question in the next section.

This discussion has dealt only with Ising models. Numerical calculations of the domain wall energy $\Delta F(L)$ have also been carried out for vector spin glasses (Morris et al, 1986). The results are at least roughly consistent with $y \approx (d-4)/2$ for $2 < d < 4$, indicating a lower critical dimensionality of 4.

8.3 The spin glass transition

In this section we use scaling and renormalization group methods to study the spin glass transition. Different implementations of the renormalization group suggest themselves for different dimensionalities: a Landau–Ginzburg formulation in momentum space for high dimensionality analogous to the one we gave for ferromagnets in Section 8.1, and a real–space technique for low dimensionalities (including the experimentally relevant case of $d = 3$). An interesting special case of the latter that we will also examine is when d is below the lower critical dimensionality, so the transition occurs at $T = 0$. In all these cases, we start with the simpler situation where the external field vanishes.

The basic features of critical phenomena in pure systems that we reviewed in Section 8.1 carry over to spin glasses. In particular, the order parameter q behaves as

$$q \propto (T_f - T)^\beta, \tag{8.55}$$

while the associated order parameter susceptibility is

$$\chi_{SG} \propto |T - T_f|^{-\gamma}. \tag{8.56}$$

The correlation length

$$\xi \equiv \left(\frac{\int d^d r\, r^2 \chi_{SG}(r)}{\int d^d r \chi_{SG}(r)} \right)^{\frac{1}{2}} \propto |T - T_f|^{-\nu} \tag{8.57}$$

Right at T_f

$$\chi_{SG}(r) \propto r^{-(d-2+\eta)}, \tag{8.58}$$

$$\chi_{SG}(k) \propto k^{-(2-\eta)} \tag{8.59}$$

while q depends on the conjugate field h^2 at this point like

$$q \propto (h^2)^{1/\delta}. \tag{8.60}$$

Here h^2 can be taken either as the square of a uniform applied field or as the variance of a random applied field. (The remaining critical exponent α is defined by the singular behaviour of the specific heat, just as in (8.3).) The exponents obtained in mean field theory are $\alpha = -1$, $\beta = 1$, $\gamma = 1$, $\delta = 2$, $\nu = \frac{1}{2}$ and $\eta = 0$).

As in conventional phase transitions, the critical exponents for spin glasses obey the scaling laws (8.10–14). However, the values measured in experiments or simulations in three dimensions are very different from the mean field ones. Most of this section will be concerned with calculating these exponents.

We begin with the Landau–Ginzburg formulation, which begins with the expansion in powers of the replica-formalism order parameter $q^{\alpha\beta}$, just as in (3.66), augmented by a gradient term to describe the energy cost of the spatial variations of the order parameter (which we were not interested in in the mean field case):

$$\beta H_{eff}[\mathbf{q}] = \tfrac{1}{2} \int d^d x [\tfrac{1}{2} r \, tr\mathbf{q}^2 + \tfrac{1}{2} c \, tr(\nabla \mathbf{q})^2 - \tfrac{1}{3} w \, tr\mathbf{q}^3] \qquad (8.61)$$

We omit the quartic term in q in (3.66) which was responsible for replica symmetry breaking in mean field theory. This is because here we are not interested in the low-T phase, but only in the vicinity of the critical point, where replica symmetry breaking is unimportant.

This problem differs most significantly from the ferromagnetic example worked out in Section 8.1 in that the interaction term in the effective Hamiltonian is cubic rather than quartic. This has the consequence that when we try to do perturbation theory in w the corrections introduced at each successive order are now proportional to

$$w^2 \int \frac{d^d k}{(2\pi)^d} \left(\frac{1}{r + ck^2} \right)^3 \propto r^{d/2-3}, \qquad (8.62)$$

which diverges as $r \to 0$ for $d < 6$ instead of $d < 4$. This means that the upper critical dimensionality of this problem is 6. We can then develop expansions of the critical exponents in powers of $\epsilon = 6 - d$ in the same way we did in $4 - d$ for the ferromagnet. The only other technical detail to keep track of is the matrix nature of the order parameter.

The graphs for the contribution to the change in the parameters r, w, and c from integrating over momenta in the 'shell' $\Lambda/b < k < \Lambda$ are shown in Fig. 8.5. The renormalization group equations (Harris et al, 1976) are (for spin dimensionality m and replica dimensionality n)

$$\frac{dr}{dl} = (2 - \eta)r - K_6 m(n - 2) \left(\frac{w}{1+r} \right)^2 \qquad (8.63)$$

$$\frac{dw}{dl} = \tfrac{1}{2}(\epsilon - 3\eta)w + K_6[(n - 3)m + 1]w^3 \qquad (8.64)$$

$$\frac{dc}{dl} = -\eta c + \tfrac{1}{3}m(n - 2)K_6 w^2 c \qquad (8.65)$$

As in the ferromagnetic example, we choose η so that c can be fixed to have the value 1 when w is at its fixed point. Solving for this fixed point value,

(a)

(b)

Figure 8.5: Graphs which lead to the renormalization group equations of Harris et al (1976): (a) corrections to the quadratic vertex, which lead to changes in r and c. (b) corrections to w.

$$K_6(w^*)^2 = \frac{-\epsilon}{2 + m(n - 4)} \qquad (8.66)$$

we can solve for the critical exponents η and ν. Taking $n \to 0$ we obtain

$$\eta = \frac{-m\epsilon}{3(2m - 1)} + O(\epsilon^2) \qquad (8.67)$$

and

$$\nu = \tfrac{1}{2} + \frac{5m\epsilon}{12(2m - 1)} + O(\epsilon^2) \qquad (8.68)$$

The other exponents can be obtained from (8.67) and (8.68) from the scaling laws (8.10–14). Of course, the ϵ-expansion results are of little use in three dimensions, but they do help to illuminate the structure of the theory.

In ferromagnets and many other systems, we can gain further information about the nature of the transition as a function of dimensionality by expanding around the lower critical dimensionality. However, so far there is no corresponding systematic expansion about the lower critical dimensionality for spin glasses. Indeed, as we have seen, even the value of the lower critical dimensionality itself is not known exactly; its determination is not a simple matter of power counting the way it is for the ferromagnet (Section 2.2).

Nevertheless, we can hope that for low dimensionality the Migdal–Kadanoff real space renormalization that we used in studying the low-T phase in the preceding section will be reasonably accurate. This hope is also supported by the quite good agreement between the value of the low-T

exponent y obtained in this approximation and the numerical calculations we cited above.

Southern and Young (1977) also applied the Migdal–Kadanoff method outlined above, forcing the bond distribution into a Gaussian at each step, to the critical properties of the Ising spin glass transition in $d = 3$. They found a transition temperature of $T_f = 0.86$ (in units where the variance of the original bond distribution is unity). At this temperature, the variance $\tilde{J}_l^2$ of the renormalized bond distribution $P_l(J)$ approaches a constant $(J^*)^2$ after many iterations. Linearizing the recursion relation for $\tilde{J}_l^2$ around this fixed point value and seeing at what length scale the deviations $\delta J_l^2 = J_l^2 - (J^*)^2$ became of order unity (in analogy to what we did in obtaining (8.33) for the ferromagnet), they obtained the correlation length critical exponent $\nu \approx 2.8$.

They also checked on the importance of their forcing the renormalized bond distribution into a Gaussian by actually following the renormalization of the distribution $P_l(J)$ explicitly. They found that they got the same T_f, to within their numerical accuracy. Riera and Hertz (unpublished) have evaluated ν, keeping the full $P_l(J)$ at each step. Again, the value they obtain is the same, within numerical accuracy, as that obtained by Southern and Young in the approximation of forcing the $P_l(J)$ into Gaussians.

The value of T_f obtained in these approximations is actually rather close to that obtained in Monte Carlo simulations (Bhatt and Young, 1986b). The value of ν is in greater doubt. It agrees reasonably well with Bray and Moore (1984), but the more recent Monte Carlo results of Bhatt and Young (1985), Ogielski and Morgenstern (1985) and Ogielski (1985) give significantly smaller values. However, it is difficult to get really close to T_f in the simulations because the relaxation times of the system become so long, so one should not necessarily accept the Monte Carlo values, either.

McMillan (1984) extended the one-parameter renormalization theory to dynamics. The idea is as outlined for the low-T phase in the preceding section (8.49–50). At the critical point, where $\tilde{J}_l$ approaches a fixed point value J^*, the mean barrier increments $\overline{\delta V_l}$ must also approach a fixed point value δV^*, since the $\tilde{J}_l$ and the $\overline{\delta V_l}$ scale the same way in this approximation. Equivalently, the effective timescale τ_l is multiplied at each renormalization step by a constant factor:

$$\tau_{l+1}/\tau_l = \exp(\delta V^*/T_f) \tag{8.69}$$

This is a familiar situation in conventional critical dynamics: At the transition there is a constant ratio between the time and space rescaling factors. It is conventional to describe it by a dynamical critical exponent z, defined by

$$\tau'/\tau = (L'/L)^z \tag{8.70}$$

where L and L' are the length scales (i.e. effective block spin sizes) before and after the renormalization step, and τ and τ' are the corresponding time scales. In our present problem. $L' = 2L$, so

$$z = \frac{\delta V^*}{T_f \ln 2} \tag{8.71}$$

We now examine the case where the zero-temperature exponent y is negative, i.e. we are below the lower critical dimensionality. This picture applies to the Ising model in two dimensions and to vector spin models in both two and three dimensions. Here there is another interesting problem: The system clearly freezes with a nonzero q at exactly zero temperature (at least for continuous bond distributions; we comment on the discrete case later), so $T_f = 0$ in this case. Let us examine this zero-temperature transition.

A special feature of this transition is that since the order parameter jumps discontinuously to a finite value at T_f, the exponent $\beta = 0$. Thus instead of two independent critical exponents there is just one to be calculated. Again it is ν which is easiest to obtain. To find it we note first that temperature always enters into the renormalization group equations in the combination T/J, where J is the bond strength. Thus the scaling of J with length scale L as L^y (at fixed T) is equivalent to scaling of T like L^{-y}. Thus if we start scaling at an initial temperature T, we will scale to $T = O(1)$ when $L \sim T^{-1/y}$. This identifies the correlation length

$$\xi \propto T^{-1/y} \tag{8.72}$$

i.e.

$$\nu = -1/y = 1/|y|. \tag{8.73}$$

When the bond distribution is discrete (e.g. the $\pm J$ model (2.7)) the situation is different. There can be exact cancellations between competing bonds, leading to a vanishing renormalized bond strength. Then, for positive y, with each successive renormalization step more and more of the weight in $P(J(L))$ will fall in a single delta-function peak at $J(L) = 0$. As a consequence, the degeneracy of the ground state will grow with the sample size, and the spin glass correlation function $[\langle S_0 S_r \rangle]_{av}$ (where $\langle \cdots \rangle$ indicates averaging over all the ground states) will fall off for large r.

In this situation the bond distribution will not scale like (8.34), and our previous analysis does not apply. Instead we argue as follows: Let $p(L)$ be the weight in the bond distribution $P(J(L))$ at length scale L at $J \neq 0$, i.e. the probability that the two ends of a block of size L are coupled. That is,

$$|\langle S_0 S_L \rangle| = 1 \tag{8.74}$$

with probability $p(L)$. At each rescaling step, this probability will be reduced by the same factor (call it $2^{-\eta}$). This implies that

$$[\langle S_0 S_L \rangle^2]_{av} = p(L) = L^{-\eta}, \tag{8.75}$$

i.e. the power law decay (8.58). (The same decay occurs for *all* moments of $|\langle S_0 S_L \rangle|$.) Numerical computations (Bray and Moore, 1986) give a value $\eta = 0.20 \pm 0.02$

Thus the zero-temperature transition for the two-dimensional $\pm J$ model has two independent critical exponents, like the usual case of finite-T transitions, and unlike that of the two-dimensional Gaussian-distributed model.

We now turn to the situation in the presence of an external field. In the preceding section we already argued that a field destabilized the zero-T state. Now whether this means that there is a new kind of phase or that the system is simply paramagnetic for all temperatures in a finite field depends on where the flow of the parameters under the renormalization group leads. If it is always to the weak-coupling (high-T) fixed point, the former possibility will occur. If it is to a new fixed point for suffiently low T and/or h, there can be a new phase, separated from the paramagnetic one by a line in the $h - T$ plane. We would identify this line as the counterpart of the AT line of mean field theory.

How do we describe a potential transition in finite field? We start with the Landau–Ginzburg expansion (8.61), augmented by a term

$$\delta \beta H_{eff} = -\tfrac{1}{2} h^2 \int d^d x \sum_{\alpha \neq \beta} q^{\alpha\beta} \tag{8.76}$$

representing the (replicated) effect of the external field h. This term changes the symmetry of the effective Hamiltonian (it can not be written as a trace in replica indices). This changes the nature of the transition: Instead of all $\tfrac{1}{2} n(n-1)$ modes going soft as they do in zero field, only $\tfrac{1}{2} n(n-3)$ do so. The remaining n-dimensional subspace of modes is 'stiffened up' against the instability by the new term in H_{eff}.

Green et al (1983) found a fixed point of the resulting renormalization group equations for $d > 6$, and for $d > 8$ the resulting phase boundary in the $h - T$ plane had just the form (3.28 or 5.36) we found for the AT line in the SK model. For $6 < d < 8$, the exponent Θ which describes the form of the line near $h = 0$ and $T = T_f(h = 0)$ is modified. For $d < 6$, however, no new fixed point can be found (Bray and Roberts, 1980). They therefore conclude that broken ergodicity of the sort found in mean field theory persists in finite-range systems down to six dimensions below this AT line, although the exponent Θ is modified below $d = 8$. Below six dimensions, they suggest that there is no AT line and no broken ergodicity, just a crossover as a function of field and reduced temperature.

Another relevant piece of evidence here is the finding (Temesvári et al, 1988) that in zero field Parisi-type order cannot be stable below six dimensions. They find that assuming such order leads to a contradiction where a positive-definite quantity comes out negative (like χ_{SG} in the AT instability). This result is not just an expansion in fluctuations around mean field theory; it is generally valid for *any* state with Parisi order, analogous to the proof (at the end of Section 2.2) of the breakdown of ferromagnetic order at or below two dimensions in Heisenberg ferromagnets.

These conclusions raise a number of interesting questions:

1. Why is the form of the AT line changed from the mean field form for $6 < d < 8$, when 6 is supposed to be the upper critical dimension?

2. How do we reconcile the apparent finding of broken ergodicity above $d = 6$ with our analysis of the preceding section, which seemed to rule out ergodicity breaking quite generally, irrespective of dimensionality?

3. How do we explain the apparent AT lines seen in experiments in three-dimensional systems?

The first question was answered quite elegantly by Fisher and Sompolinsky (1985). The peculiar behaviour comes about because field-dependent quantities (e.g. $T_f(h)$) turn out to depend singularly on the coefficient of the quartic term $\sum_{\alpha\beta}(q^{\alpha\beta})^4$ in the Landau–Ginzburg expansion (3.66). So even though we know that this coefficient decays away to zero under the renormalization group for $d > 6$, we have to know just how it decays in order to calculate field-dependent quantities. (This never entered our previous analysis of the spin glass transition near $d = 6$, since there we worked strictly at $h = 0$.) Such dependence on parameters in the Hamiltonian that scale away to zero ('irrelevant variables') can lead to violations of the scaling laws; in this case the variables are said to be 'dangerously irrelevant'.

The simplest example of dangerous irrelevance is that of the quartic term in the Landau–Ginzburg effective Hamiltonian (8.16) above four dimensions (see (8.25) and (8.28)). Despite the fact that it scales to zero with increasing renormalization parameter l like $\exp(4 - d)l$, it affects the singular part of the free energy, which is proportional to u^{-1}. This leads to a violation of the so-called *hyperscaling* law $d\nu = 2 - \alpha$ (8.11) above four dimensions.

In the spin glass, the mean field value $\frac{2}{3}$ of the AT exponent Θ is another example of such an effect. From simple scaling, we have (8.55) $q \propto (\Delta T)^{\beta}$ at $h = 0$ and (8.60) $q \propto (h^2)^{1/\delta}$ at $T = T_f$, so ΔT scales like $(h^2)^{1/\beta\delta}$. This means that $\Theta = 2/\beta\delta$. For the mean field values of the exponents $\beta = 1$ and $\delta = 2$, $\Theta = 1$. This value is correct for the exponent describing the crossover between the critical behaviours (8.60) and (8.55) *above* the transition, and we would naively expect the AT line to obey the same scaling. That it does not is a result of dangerous irrelevance.

Let us then see how this happens. We take the model

$$\beta H_{eff} =$$

$$\frac{1}{2}\int d^dx[\frac{1}{2}r\,Tr\mathsf{q})^2) + \frac{1}{2}(Tr(\nabla\mathsf{q})^2) - \frac{1}{3}w\,Tr\mathsf{q}^3 - \frac{1}{6}y\sum_{\alpha\beta}(q^{\alpha\beta})^4] \quad (8.77)$$

The coefficients w and y are both positive with this sign convention. The renormalization group equation for the quartic term y has the form

$$\frac{dy}{dl} = (4-d)y - Aw^4 \quad (8.78)$$

where the first term comes from simple power counting as in (8.25) and (8.28) for the ferromagnet, and the second one comes from the elimination of modes with wavevectors in the shell $\Lambda/b < k < \Lambda$. The coefficient A is positive. Above six dimensions (after initial transient terms have died away) the dominant behaviour of the renormalization of w (8.64) is simply governed by $dw/dl = \frac{1}{2}(6-d)w$, so we can just substitute $w(l) \propto \exp\frac{1}{2}(6-d)l$ in (8.78). We then find that $y \propto -A\exp 2(6-d)l$ for large l. The critical behaviour of quantities can then be found by integrating the renormalization group equations until a value $l = l^*$ defined by $r(l^*) \approx 1$.

To find the way the shape of the AT line is affected by this renormalization, we then want to know how an external field scales. This is given by

$$\frac{dh^2}{dl} = (1 + \frac{1}{2}d)h^2, \quad (8.79)$$

using the same kind of simple power-counting arguments that led to the first terms in (8.27) and (8.28) in the ferromagnet and to (8.63), (8.64) and (8.65) for the spin glass. (Here we set $\eta = 0$ since we are above 6 dimensions.) Thus

$$h_{AT}^2 \sim \exp[-(1 + \frac{1}{2}d)l^*]h_{AT,MFT}^2(r(l^*), w(l^*), y(l^*))$$

$$\sim |r|^{\frac{1}{2}+\frac{1}{4}d}\frac{y(l^*)}{w^3(l^*)}r^3(l^*) \sim |r|^{\frac{1}{2}d-1} \quad (8.80)$$

Thus our AT exponent Θ defined by $T_f(h) - T_f(0) \propto h^\Theta$ is given by

$$\Theta = \frac{4}{d-2} \tag{8.81}$$

This goes over to the mean field result (5.36) at $d = 8$. This derivation thus shows explicitly how the particular way y scales can affect the critical behaviour, even though it is 'irrelevant'.

As $d \to 6$, $\Theta \to 1$, which is the value expected from scaling. Below $d = 6$, scaling is obeyed ($\Theta = 2/\beta\delta$), but, as we have noted, there is apparently no fixed point corresponding to an AT transition line, so the exponent Θ will just describe a continuous crossover rather than a sharp transition.

We turn now to our second question, about the apparent conflict between the finding of a broken ergodicity phase above $d = 6$ and our previous argument against such a state, without reference to dimensionality. The point here is that the scaling description which formed the basis of the argument against broken ergodicity was not made in the replica formalism, and so cannot distinguish between a transition where all the $q^{\alpha\beta}$ modes go soft and one in which only the $\frac{1}{2}n(n-3)$ in the subspace satisfying $\sum_{\alpha\beta} q^{\alpha\beta} = 0$ do so. Thus it cannot hope to capture the AT transition which Green et al found for $d > 6$. On the other hand, the apparent absence of any such transition for $d < 6$ suggests that a correct description need not take this kind of exotic possibility into account for such dimensionalities; a replica-symmetric description should be sufficient.

This brings us to our third question of how we are to account for the 'experimental AT lines' that we described in Section 5.2 (Figs. 5.2 and 5.3) if there really is no transition. We try to give a qualitative answer, describing the observed onset of irreversible behaviour as a dynamical effect due to the failure of the system to reach equilibrium at sufficiently small fields. (Remember that all experimental evidence for an 'AT line' was based on dynamical properties such as the vanishing of the magnetic viscosity or the slope of $\chi''(\omega)$.) The basic point is that the size (8.53) of the domains that have to flip in order to equilibrate is very large for small field, so their equilibration time is very large (exponentially so, in our dynamical picture).

We have to begin by generalizing the argument leading to our domain size estimate (8.53) to the region of the transition, where we have to include the critical temperature dependence of the prefactors of the L-dependent parameters $\tilde{J}(L)$ and $h(L)$. The former generalizes to $\tilde{J}(L/\xi)^y$, where ξ is the correlation length $\propto \Delta T^{-\nu}$. Similarly, the field energy for the same block is proportional to $q^{\frac{1}{2}}L^{d/2} \propto \xi^{-\beta/2\nu}$. Incorporating these factors into the determination of L_h, we obtain

$$L_h(\xi) \propto (J\xi^{-y+\beta/2\nu}/h)^{2/(d-2y)} \tag{8.82}$$

Now from our dynamical theory, the characteristic energy barrier at length scale L is proportional to L^{ψ}, and this now generalizes to $(L/\xi)^{\psi}$. Thus the relaxation time of the domain structure induced by a field h is

$$\tau_{\xi}(h) = \tau_0 \exp[\text{const.}(L_{\xi}(h)/\xi)^{\psi}] \qquad (8.83)$$

If the experimental time is small compared to (8.83), the system will not have a chance to equilibrate on all the length scales less than $L_{\xi}(h)$ and will appear (partially) frozen. Or, for a given t there is a characteristic field $h(t,\xi)$ below which the system will appear frozen, given by setting $t = \tau_{\xi}(h)$, which yields

$$\left(\frac{h(t,\xi)}{J}\right)^2 \propto \frac{\xi^{-d+\beta/\nu}}{[\ln(t/t_0)]^{(d-2y)/\psi}} \qquad (8.84)$$

or, in terms of the reduced temperature ΔT,

$$\left(\frac{h(t,\Delta T)}{J}\right)^2 \propto \frac{\Delta T^{\beta\delta}}{[\ln(t/t_0)]^{(d-2y)/\psi}} \qquad (8.85)$$

where we have used the scaling laws to write the exponent of ΔT in this way. This exponent is just what we argued above we would expect from scaling (i.e. $\Theta = 2/\beta\delta$). The prefactor depends only logarithmically on time (because of the exponential dependence of relaxation times on length scale), so an experiment with only a narrow window of observation timescales may not show significant time-dependence, and the transition will appear to be a static one.

From the current estimates of critical exponents (see Section 8.5) the exponent $\beta\delta$ should be around 3.3 or $\Theta \approx 0.61$. This is not far from the mean field value of $2/3$, but this is of course just a coincidence. From (8.80) we can see that Θ rises from $2/3$ to unity as d varies from 8 to 6, and then falls again as d falls below 6, presumably crossing the mean field value somewhere just above $d = 3$.

As discussed in Chapter 6, most experimentally interesting spin glasses are approximately isotropic Heisenberg systems, but the small anisotropy that is inevitably present, from either dipole–dipole or Dzyaloshinskii–Moriya interactions, affects the possibility and nature of the spin glass transition decisively. We now show how a small amount of random anisotropy can lead to an Ising-like transition in an otherwise isotropic spin glass. The anisotropic interaction has the form

$$H_D = -D \sum_{ij,\mu\nu} K_{ij}^{\mu\nu} S_i^{\mu} S_j^{\nu} \qquad (8.86)$$

(with $\sum_{i\mu} K_{ii}^{\mu\mu} = 0$, where i and j label lattice sites and μ and ν label spin components). Now under a renormalization group transformation with a scale factor of L, D will scale like

$$D(L) \sim L^{d/2}D \tag{8.87}$$

Meanwhile the isotropic exchange scales away like L^y with $y < 0$ (for $d = 3$). So even if we start with $D/J \ll 1$, we will come to a length scale L^* where the anisotropy takes over. After that, the problem will essentially scale like the Ising one. The condition to be below the transition temperature is roughly that $J(L^*) > 1$. This leads to

$$T_f \propto J(D/J)^{-2y/(d-2y)} \tag{8.88}$$

Using the result (Morris et al, 1986) $y \approx \frac{1}{2}$ we find $T_f \propto J(D/J)^{\frac{1}{4}}$ in $d = 3$.

Metallic spin glasses with RKKY interactions require special consideration. They represent a limiting case of systems with long-range forces with different critical properties from short-range interaction systems. to see how this comes about, return to the simpler case of ferromagnets. If there are forces which fall off like $r^{-(d+\sigma)}$, the Landau–Ginzburg effective Hamiltonian (8.16) acquires a new term which can be written

$$\beta H_{LR} = \tfrac{1}{2} \sum_k ak^\sigma |S_k|^2 \tag{8.89}$$

This term differs from the conventional gradient term in containing the factor k^σ iinstead of k^2. Under the rescalings of length and spins as done in Section 8.1, it rescales by a factor $b^{2-\eta-\sigma}$, where b is the length rescaling factor. Now the perturbative elimination of the modes with wavevectors in the shell $\Lambda/b < k < \Lambda$, can make further corrections to the rescaling of terms in the effective Hamiltonian as in (8.24–25) *only if these terms are analytic functions of wavevector*, since the perturbation theory diagrams will always be analytic at small external wavevectors. Therefore there is *no* other correction to the long-range part of the effective Hamiltonian; we have exactly

$$a(L) = aL^{2-\eta-\sigma} \tag{8.90}$$

The same considerations apply to random long-range forces where

$$[J_{ij}^2]_{av} \sim \frac{1}{r_{ij}^{d+\sigma}} \tag{8.91}$$

RKKY forces correspond to the case $d = \sigma$. We obtain

$$J_{LR}^2(L) \propto J_{LR}^2 L^{2-\eta-\sigma} \tag{8.92}$$

Thus for a zero-T fixed point, with $2 - \eta = d$, the long-range force will scale like

$$J_{LR} \propto JL^{(d-\sigma)/2} \tag{8.93}$$

and will be the dominant one at large length scales if $(d - \sigma)/2 > y$. This will not happen for the 3-dimensional Ising case with RKKY-like long-range interactions ($d = \sigma = 3$), but it *will* happen for the experimentally relevant case of 3-dimensional vector models, since there $y < 0$. Thus such systems are at their lower critical dimensionality.

We can also generalize our previous analysis of the effect of anisotropy on a system that would not otherwise have a finite T_f to this case. The result is (Bray, Moore and Young, 1986)

$$T_f \propto \frac{J_{LR}}{[\ln(J_{LR}/D)]^{\frac{1}{2}}} \tag{8.94}$$

Thus even for fairly small anisotropy (e.g. $D/J_{LR} = O(10^{-2})$) the transition temperature will not be much smaller than that for the corresponding Ising model and will not be terribly sensitive to changes in the strength of the anisotropy.

This section and the preceding one have been essentially phenomenological scaling theories, bolstered occasionally by a Migdal–Kadanoff-style real space renormalization calculation. While the results are physically appealing, it would be much more satisfactory to have a systematic expansion in a replica Landau–Ginzburg model which gave a stable account of a spin glass phase. Here we have assumed that such a theory would not have broken replica symmetry, but this has not been proved. Although we know from results cited above (Temesvári et al, 1988) that broken replica symmetry of the Parisi sort apparently becomes unstable below six dimensions, there remains the possibility of a different kind of replica symmetry breaking.

In very recent work De Dominicis and Kondor (1989) have suggested that such a theory could actually be constructed via a suitable loop expansion around Parisi theory. They give arguments that one-loop corrections should modify the small-x behaviour of the order function $q(x)$ in zero field, changing the linear x-dependence to x^ρ, where $\rho = 3/(d - 3)$. They then show that such a modified Parisi theory would be stable against the instability identified by Temesvári et al. Much remains to be done to carry out this programme, particularly the extension to finite field, but their results, however preliminary, indicate that we should not hastily dismiss the possibility of replica symmetry breaking below six dimensions. It is also worth noting some recent simulation results (Sourlas, 1988) which also suggest nontrivial broken ergodicity (a nontrivial $P(q)$) in three dimensions.

Any real theory (as opposed to a scaling one), with or without replica symmetry breaking, would be a very significant advance and would essentially complete the equilibrium description of short-range spin glasses.

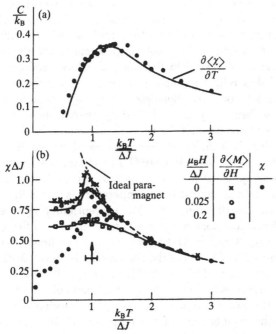

Figure 8.6: Specific heat $C(T)$ and susceptibility $\chi(T)$ of a two-dimensional spin glass model with a Gaussian distribution of bonds from MC simulations (from Binder and Schröder, 1976).

8.4 Monte Carlo Calculations

Monte Carlo (MC) simulations are an important link between theory and experiment. They can be used to show that a model yields physical properties which agree more or less with the experimental data, and also to test theoretical predictions (which may be based on certain hypotheses or approximations). In the case of spin glasses, the EA model with a Gaussian or $\pm J$ distribution of nearest neighbour interactions has been investigated rather thoroughly by simulations where the specific heat, susceptibility, field-dependent magnetization, correlation function $C(t)$ and other physical quantities are measured. MC simulations are also useful for the direct visualization of the microscopic spin structure at a given temperature. They have the further advantage that they can be applied to the same model in different dimensions. Since in Ising spin glasses the critical dimension most likely lies between two and three, one can compare systems with and without a phase transition. However, MC simulations are conventionally based on Glauber dynamics (see Section 4.2), and at low temperatures one has the same difficulties in obtaining thermal equilibrium as in the experiments.

MC simulations have been an important tool in order to determine critical exponents of 3d Ising spin glasses and to check the dynamics in the spin glass phase as discussed in Section 8.2. Here, again, one runs into the problem of slow relaxation processes, and a test of the correlation function (8.51) seems to be fairly hopeless, except very near to T_f. MC simulations have also been very useful in testing predictions of the MFT, as we discussed in Sections 3.5 and 5.2.

Many of the earlier MC simulations were performed on two-dimensional lattices. Fig. 8.6 shows the specific heat $C(T)$ (a) and the susceptibility $\chi(T)$ (b) for 2d Ising spins on a 80×80 lattice with a Gaussian distribution of nearest-neighbour bonds. The temperature is normalized to the width $\Delta_{ij} = \Delta$ of the distribution (2.6). The susceptibility $\chi(T)$ is calculated either from the fluctuations of the magnetization or by differentiating the (field-cooled) magnetization $M(h)$. We denote the first of these by $\tilde{\chi}(T)$ and the second by $\chi(T)$. Similarly, the specific heat is calculated either from energy fluctuations $C/k_B = [\langle H^2 \rangle - \langle H \rangle^2]_{av}/N(k_B T)^2$ or from $\partial[\langle H \rangle]_{av}/\partial T$. Here, the thermal averages $\langle \ldots \rangle$ are actually are time averages. These quantities are found to differ below the freezing temperature T_f (as defined by the temperature of the cusp), which indicates that at low temperatures thermal equilibrium is not completely reached. The two susceptibilities calculated this way behave (at least approximately) like experimental quantities: The experimental field-cooled susceptibility, which is constant below the freezing temperature (see Fig. 1.4) can be identified with $\chi(t)$, whereas the ac susceptibilty in the limit of vanishing frequency, which exhibits a more or less pronounced cusp, behaves like $\tilde{\chi}(T)$. Actually, in the MC simulation of Fig. 8.6 both the plateau in $\chi(T)$ and the cusp in $\tilde{\chi}(T)$ have to be nonequilibrium properties, since we know from Section 8.2 that there is no phase transition in a 2d Ising system (Morgenstern and Binder 1979, 1980a,b). Above the freezing temperature, the susceptibility for the smallest field shows Curie behaviour, in contrast to most experimental data. This is an indication that the EA model with a symmetric bond distribution is insufficient to describe this detail (most probably a nonsymmetric distribution will correct this deficiency).

The specific heat $C(T)$ shows a broad maximum well above the freezing temperature which agrees with the experimental data (see Fig. 1.8). This maximum indicates the formation of short range order which leads already above T_f to a large change of the entropy or a freezing-in of degrees of freedom. At T_f one expects only a small change of entropy, which makes it plausible that no anomaly of the specific heat is observed at this

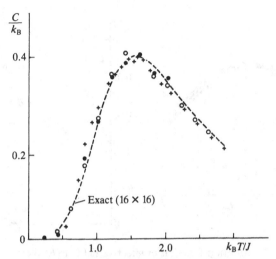

Figure 8.7: Specific heat vs temperature of a two-dimensional spin glass model with $\pm J$ interactions and $N = 16 \times 16$ spins. The broken line represents the result of an exact calculation based on a recursive method; the circles represent results from MC simulations starting either with a random spin configuration ($\bullet$) or with a ground state configuration ($\circ$) (from Morgenstern and Binder, 1980a).

temperature. Far below T_f these data certainly become incorrect since the relaxation times of the system become longer than the computer time (typically 2000 Monte Carlo steps per spin). Their accuracy can be checked by comparing with the specific heat as calculated by means of an essentially exact calculation of the partition function (for small systems). Fig. 8.7 shows the specific heat as calculated in this way for a spin glass with $\pm J$ interactions. A comparison of Figs. 8.6a and 8.7 shows that the broad maximum of $C(T)$ above T_f is indeed a equilibrium property, in contrast to the low temperature behaviour in Fig. 8.6a.

Another property of interest is the distribution $P(\tilde{h})$ of internal fields at lattice site i:

$$\tilde{h} = \left[\sum_j J_{ij} m_j \right]_{av} \tag{8.95}$$

which has also been calculated from the TAP equations (see Section 3.5) at zero temperature (Palmer and Pond, 1979). For infinite-range interac-

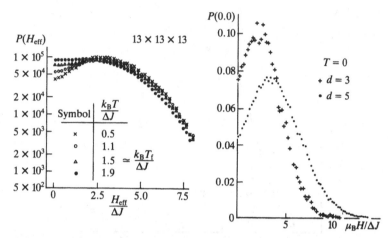

Figure 8.8: Distribution function of effective fields $P(\tilde{h})$ plotted vs $\tilde{h}$ at various temperatures as obtained from MC simulations for a spin glass model with a Gaussian distribution of nearest neighbour bonds (a) for $d = 3$ (from Binder, 1978), (b) for $d = 3$ and $d = 5$ (from Stauffer and Binder, 1979).

tions this distribution vanishes linearly for $\tilde{h} \to 0$ (see (3.97)). In systems with short-range interactions, this is not the case. Fig. 8.8a shows $P(\tilde{h})$ for a three-dimensional Ising spin glass at various temperatures. $P(\tilde{h} = 0)$ decreases with decreasing temperature but never becomes zero. The distribution at $T = 0$ and for dimensions $d = 3$ and $d = 5$ are seen in Fig. 8.8b: $P(0,0)$ decreases with increasing dimension and vanishes in infinite dimensions (which is equivalent to the SK model).

More complicated properties can also be measured in simulations. The results for the remanent magnetization as obtained by cooling in a field (thermo-remanent magnetization (TRM)) and by cooling in zero field and applying a field for a short time (isothermal remanent magnetization (IRM)) as a function of the initially applied field are shown in Fig. 8.9 for the EA model in two dimensions. The data have considerable similarity with the TRM and IRM measured on AuFe (Fig. 1.5). The maximum of the TRM seems to be a transient effect which depends on the cooling rate in both simulations and experiment.

Since most MC simulations are based on Glauber dynamics, the investigation of dynamical spin glass properties is straightforward. In general, one starts with some initial condition (for instance, the ferromagnetic state

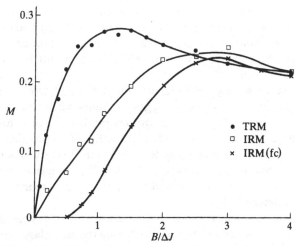

Figure 8.9: Monte Carlo results for the remanent magnetizations M_{IRM} and M_{TRM} of a square lattice Ising spin glass of size 50×50 with periodic boundary conditions at a temperature $T = \Delta J/4k_B$; ΔJ is the width of a Gaussian distribution of nearest neighbour bonds, M_{TRM} is obtained by cooling in a field, M_{IRM} is shown as a function of the initially applied field; see Kinzel (1979) for more details.

with all spins parallel) and investigates the relaxation of the system towards thermal equilibrium. We already discussed in connection with Fig. 4.1 the attempts to find the infinite-time limit q of the correlation function $C(t)$. (There is no reason here to distinguish between the thermal equilibrium value q and the long-time limit q_{EA} as we did in the MFT.) The function $C(t)$ has been investigated by Ogielski (1985), using a special-purpose computer. Fig. 8.10 shows typical data for a three-dimensional Ising spin glass with a $\pm J$–distribution of nearest neighbour bonds for 64^3 lattice sites and temperatures ranging from $T/J = 2.5$ to 1.30 where $T_f = 1.175J$. Here, $C(t)$ is measured in some cases over more than 10^7 steps! Ogielski fits his data by a 'stretched exponential' multiplied by a power function

$$C(t) = ct^{-x} \exp[-(t/\tau)^\beta] \qquad (8.96)$$

with temperature dependent exponents $x(T)$ and $\beta(T)$ and the constants c and τ (full lines in Fig. 8.10). A similar law is also observed in the relaxation rate $dM/d\ln t$ of the magnetization of various systems (Svedlindh et al, 1987). The relaxation of $C(t)$ for $T > T_f$ is considerably slower.

For $0.6 < T/T_f < 1$ Ogielski fits it by a power law $C(t) \propto ct^{-x}$ with a different temperature dependence of the exponent $x(T)$. This fit assumes that $C(t \to \infty) = 0$, i.e. $q = 0$. However, even the MC runs up to 10^8 steps are not sufficiently long to establish that this is actually the case. Nor are they sufficiently accurate to decide whether one has algebraic decay (as predicted by the MFT) or an inverse power of $\log t$ (as predicted by the scaling theory in Section 8.2).

At the longest times, Fig. 8.10 shows fluctuations of $C(t)$. The slow relaxation and the fluctuations both suggest a many-valley picture similar to Fig. 2.3 in which at least some of the free energy barriers are finite. One can also learn from Fig. 8.10 that a MC simulation of a spin glass with a conventional number of MC steps (say 2000 per spin) never yields thermal equilibrium at low temperatures. Near and below T_f, one has an extremely broad spectrum of relaxation times, and so far the longest relaxation time has not yet been determined.

MC simulations also can be used to determine critical exponents. The static exponents η and ν can be obtained by fitting the spin glass correlation function to the scaling form

$$G(r) \propto r^{-(d-2+\eta)} f(r/\xi) \tag{8.97}$$

One determines ν from the temperature dependence of the correlation length $\xi(T) \propto |T - T_f|^{-\nu}$ and η from the scaling plot of Fig. 8.11.

In this way Ogielski and Morgenstern (1985) obtained the values for η and ν indicated in Table 8.1. Improved values for ν can be obtained by means of finite size scaling, which has also been discussed in Section 8.2: Since L/ξ is the only relevant parameter for $L \to \infty$ and $T \to T_f$, one has $(\xi/T) \propto (T - T_f)^{-\nu}$ and one can introduce the scaling function $\overline{\chi}(L^{1/\nu}(T - T_f))$. Here L is the linear size of the system. The spin glass susceptibility then has the finite size scaling form

$$\chi_{SG} = L^{\gamma/\nu} \overline{\chi}(L^{1/\nu}(T - T_f)) \tag{8.98}$$

with $\chi_{SG} \propto L^{\gamma/\nu} = L^{2-\eta}$ from (8.10) at $T = T_f$. In this way Ogielski (1985) obtained

$$\nu = 1.3 \pm 0.1, \quad \eta = -0.22 \pm 0.05,$$
$$\gamma = 2.9 \pm 0.3, \quad \alpha = -1.9 \pm 0.3 \tag{8.99}$$

with the critical exponent α from (8.11). He claims that the value $\beta \approx 0.5$ as obtained from (8.13) is fairly inaccurate because of scaling corrections which might be large but have been ignored. The critical exponents by Bhatt and Young (1986b) shown in Table 8.1 are also based on finite size scaling. The good agreement of all these data yields convincing evidence that there

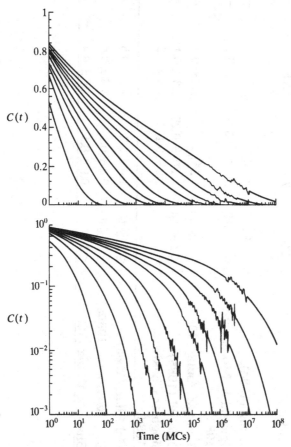

Figure 8.10: spin autocorrelation function $C(t)$ above T_f of the three–dimensional $\pm J$ model plotted vs time for lattice size 64^3 and temperatures (from left to right) $T/J = 2.5$, 2.0, 1.8, 1.7, 1.5, 1.45, 1.40, 1.35, and 1.3. Solid curves represent fit of the data to equation (8.96) (MC data from Ogielski, 1985).

Table 8.1

MODEL	REFERENCE	SIZES	T_f	EXPONENTS		
				ν	η	z
±J	Young (1984)	$L = 64$	$< 1.2^{**}$	1.2	-0.4	
	Ogielski & Morgenstern (1985)	$L = 32, 64$	1.2 ± 0.05	1.2 ± 0.1	-0.1 ± 0.1	5
	Bhatt & Young (1985)	$4 \leq L \leq 20$	$1.2^{+0.1}_{-0.2}$	1.3 ± 0.3	-0.3 ± 0.2	5
	Ogielski (1985)	$4 \leq L \leq 64$	1.175 ± 0.25	1.3 ± 0.1	-0.22 ± 0.05	5.5 ± 07
	Singh & Chakravarty (1986)	$\ell = 17^*$	1.2 ± 0.1	1.3 ± 0.2	-0.25 ± 0.2	
Gaussian	McMillan (1985)	$3 \leq L \leq 6$	1.0 ± 0.2	1.8 ± 0.5		
	Bray & Moore (1985a)	$2 \leq L \leq 4$	0.8 ± 0.1	3.3 ± 0.6		
	Bhatt & Young (1988)	$3 \leq L \leq 6$	0.9 ± 0.1	1.6 ± 0.4	-0.4 ± 0.2	

* length of series
** exponent estimates using $T_f = 1.2$

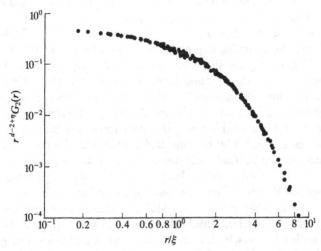

Figure 8.11: Scaling plot of the spin glass correlation function $G(r)$ (8.43) for the Ising spin glass with $\pm J$ nearest neighbour interactions in three dimensions. The plot demonstrates that $r^{d-2+\eta}G(r)$ vs r/ξ is a universal curve for temperatures in the range $1.32 \leq T/J \leq 1.8$ in the paramagnetic phase (from Ogielski and Morgenstern, 1985).

is indeed a finite temperature phase transition in three-dimensional short-range Ising spin glasses.

The dynamic critical exponent z indicated in Table 8.1 describes critical slowing down as T_f is approached from above. One has the characteristic relaxation time

$$\tau \propto \xi^z \propto (T - T_f)^{-z\nu} \tag{8.100}$$

The dynamical exponent depends on the way τ is defined. For the values quoted in Table 8.1, it is defined by

$$\tau = \int_0^\infty dt\, C(t) \tag{8.101}$$

Ogielski defines a second relaxation time $\bar{\tau}$ from

$$\bar{\tau} = \frac{\int_0^\infty dt\, t\, C(t)}{\int_0^\infty dt\, C(t)} \tag{8.102}$$

which scales differently: For a scaling form of $C(t)$ like (8.96) one has $\tau \propto \bar{\tau}^{1-x}$, or an effective dynamical exponent $\bar{z} = z/(1-x)$. Ogielski's measurements yield a value $\bar{z} = 6.0 \pm 0.8$.

Another numerical technique which gives useful information about critical exponents is high-temperature series. Here one makes a formal expansion of quantities of interest (e.g. χ_{SG}) in powers of βJ or, more usefully, $\tanh \beta J$. The terms in the series are then calculated to rather high order (nearly 20), and the resulting expression is fit to a form with the expected singularity. Optimizing the fit yields the critical temperature and exponents. In pure systems this method gives the most accurate critical exponents known for three-dimensional systems. In spin glasses it has played a less prominent role because the expansions were not carried out to high enough order until recently. However, the most recent calculations of exponents for the 3d $\pm J$ model by this method (Singh and Chakravarty, 1987) (which are also shown in Table 8.1) agree well with those found by MC.

So far, we have considered only Ising spins. The phase space for vector spins is considerably larger. Hence, smaller systems have to be studied and the data are less reliable. Some early MC simulations were performed for the specific heat, susceptibility and correlation function $C(t)$ of Heisenberg spins in three dimensions with a Gaussian distribution of nearest neighbour bonds (Binder, 1977; Ching and Huber, 1977a,b). The susceptibility shows a peak at $T/J \approx 0.3$, but there is now considerable evidence from the scaling theory (Section 8.3) and other MC simulations that there is no finite temperature phase transition in this system. More recent MC simulations by Olive et al (1986) yield a $\chi_{SG} \propto T^{-\gamma}$ and relaxation times which increase like a power of T as $T \to 0$, indicating that the lower critical dimension must be larger than 3.

As we remarked in the preceding section, RKKY spin glasses, where the magnitude of the interactions falls off with the distance R between spins like R^{-3}, should be at their lower critical dimensionality. This has been tested by Reger and Young (1988) who indeed found instead of a power law an essential singularity at $T = 0$ with $\chi_{SG} \propto \exp(C/T^{\sigma})$, with a new critical exponent $\sigma \approx 2.2$ and with the constant $C \approx 1.1$.

Real spin glasses presumably always contain a small amount of anisotropy which strongly modifies the scaling properties, and makes T_f positive as in the Ising case, as discussed at the end of the preceding section. MC simulations on spin glasses with RKKY interactions and additional anisotropy indeed indicate a finite-temperature phase transition with critical exponents which do not differ much from those indicated in (8.99) for 3d Ising spin glasses (Chakrabarti and Dasgupta, 1988).

8.5 Experimental evidence

We mentioned already in the Introduction that the cusp observed in the ac susceptibility of various systems (Fig. 1.1) suggests a second-order phase transition in spin glasses and indeed played a key role in arousing the interest in this field. However, the cusp turned out to be rounded (Fig. 1.2) and today we know that the linear susceptibility is not the best quantity to study for determining whether or not a spin glass has a sharp phase transition. One should rather investigate the spin glass susceptibility χ_{SG} (2.64) which plays the same role in a spin glass as the uniform susceptibility in a ferromagnet. For small external fields, χ_{SG} is proportional to the nonlinear susceptibility χ_{nl} in (2.67) which is measurable and diverges at T_f with the critical exponent γ if a second-order phase transition exists. In this case, χ_{nl} should scale as a function of the field h and the temperature difference $T - T_f$ as

$$\chi_{nl} = h^{2/\delta} F \left(\frac{T - T_f}{h^{2/\phi}} \right) \tag{8.103}$$

with the critical exponents δ and ϕ. The exponent ϕ can be expressed in terms of other critical exponents using scaling, $\phi = \beta + \gamma$, and the exponent δ can be determined either from the field dependence of the order parameter q at T_f

$$q(T_f, h) \propto h^{2/\delta} \tag{8.104}$$

(Chalupa, 1977) or from the nonlinear magnetization

$$\Delta M = M - \chi h \tag{8.105}$$

at $T = T_f$ and for $h \to 0$. Note that the nonlinear susceptibility $\Delta M/h$ differs for finite field from χ_{nl}, since it includes also higher order terms in the expansion (2.67). The exponent γ as defined by $\chi_{nl} \propto (T - T_f)^{-\gamma}$ is determined by plotting $\Delta M/h$ against h^2 for various temperatures. The initial slope of these curves then yields $\chi_{nl}(T)$ and for $T \to T_f$ the exponent γ. From δ and γ all other static critical exponents follow from the scaling laws (8.10)–(8.14). There remain two major problems: First, one does not know the critical region in which scaling should hold; and second, one rarely reaches thermal equilibrium below T_f. For the latter reason, practically all static measurements of critical exponents (including MC simulations for $d = 3$) have been performed at temperatures $T \geq T_f$. Of course, the extension of these investigations to $T < T_f$ would be extremely desirable since the theory of Section 8.2 predicts nontrivial static and dynamic properties for all temperatures $T \leq T_f$.

	δ	γ	β	ν	η	α	ϕ	range of T $\Delta T/T_f$	range of H M_{nt}^{max}/M	Reference
AgMn (0.4-20.5%)	3.1±0.2	2.2±0.2	1.0±0.1	1.4±0.1	0.46±0.2	-2.2	3.2±0.1	0.1	0.1	Bouchiat (1986)
AgMn (150 ppm)	6.6	3.8	0.7	1.7	-0.2	-3.2	4.5			Novak et al (1986)
AgMn (0.2, 0.5%)	3.3±0.2	2.1±0.1	0.9±0.2	1.3±0.2	0.4	-1.9±0.3	0±0.2			Lévy and Ogielski (1986)
CuMn (0.25%)	4.5	3.6±0.3	1	1.8	0.1	-3.4	4.5	0.7		Berton et al (1982)
CuMn (1%)	4.4	3.25±0.1	0.75±0.25	1.7	0.1	-2.7	4.2	2	0.5	Omari et al (1983)
AuFe (1.5%)	2.0±0.2	1.1±0.2	0.9	1.0	1.0	-0.9	2.0	0.1	0.01	Taniguchi et al (1983)
Eu$_x$Sr$_{1-x}$ ($x = 0.15, 0.25, 0.3, 0.4$)	4.1				0.2					Maletta and Felsch (1979)
CdCr$_{1.7}$In$_{0.3}$S$_4$	4.1±0.4	2.3±0.4	0.75±0.1	1.26±0.2		-1.9	3.1±0.5	0.16		Vincent and Hamann (1987)
a - AlGd (37%)	5.7±0.2	2.7±0.1	0.9	1.3	-0.1	-1.9	3.3± 0.4	0.3	0.3	Malozemoff et al (1982)
a - Fe$_{10}$Ni$_{70}$P$_{20}$	5.2±0.5	2.3±0.2	0.55	1.2	0.0	-1.5	2.9		0.15	Taniguchi et al (1985)
a - FeNiPBAl	10	3.4	0.38	1.39		-2.2	3.8			Lundgren et al (1986)
a - AlMnSi	5±1	0.9	0.2	2.0±0.1	0.0	+0.7	1.1±0.2	0.4	0.05	Beauvillain et al (1984a)
a - AlMnSi	3.2	3.1±0.1	1.4	2.0±0.1	0.4	-3.9	4.5			Beauvillain et al (1984b)
a - AlMnSi	3.4	3.4±0.1	1.4		0.4	-4.0	4.8			Beauvillain et al (1986)
a - FeMnP	5.5	3.6±0.15	0.8	1.4±0.1	-0.1	-3.2	4.4			Beauvillain et al (1986)
a - FeMnP	5.5	3.6±0.2	0.8	1.4±0.1	-0.1	-3.2	4.4			Beauvillain et al (1986)
simulations		2.9±0.3	0.5	1.3±0.1	-0.22±0.05	-1.9±0.3	3.4			Ogielski (1985)
series expansions		2.9±0.2			1.3±0.2					Singh and Chakravarty (1987b)

Table 8.2. Critical exponents of various spin glasses. Some of the values are deduced from scaling laws (e.g. the exponent α for the specific heat from $\alpha = 2(1 - \phi) + \gamma$)

The results presented in Table 8.2 strongly suggest that there is indeed a phase transition at a finite temperature T_f. In most cases, the critical exponents δ and γ (or δ and ϕ) have been measured and the remaining exponents determined from the scaling laws. Obviously, fairly different values have been obtained, and some of the discrepancies are due to experimental uncertainties. These uncertainties are largest for the exponents α and η (which have been determined from $\phi = \beta + \gamma$ and γ by means of (8.10) and from δ by means of (8.14)), but there is fair agreement between most values of γ and the value $\gamma = 2.9 \pm 0.3$ from Ogielski's MC simulations. The value of ν (as determined from ϕ and γ by means of (8.10) and (8.11)) compares favourably in most cases with the values obtained by various MC simulations and listed in Table 8.2.

Apart from experimental uncertainties, the discrepancy between the measured values of the critical exponents might be due to too large fields and temperature differences $T - T_f$. This problem is illustrated in Fig. 8.12, which shows the nonlinear magnetization ΔM vs h for AgMn 0.5 at% on a log–log plot (Bouchiat, 1986). The critical exponent $\delta = 3.1$ obtained from the low-field data ($h < 500$ Oe) definitely differs from the high-field value $\delta = 4.5$ found by Omari et al (1983) from data on CuMn in fairly high fields. One can ask why scaling up to 50 kOe works so well (Yeshurun and Sompolinsky, 1986), though one obtains a 'wrong' critical exponent. One possibility is that the critical region of these spin glass systems is rather large and that one has a crossover from an Ising system in low fields to a Heisenberg system in high fields, similarly to what we found in MFT in Section 6.4. Such a crossover can be due to random anisotropy and will be discussed at the end of this section.

A typical scaling plot of the nonlinear susceptibility χ_{nl} is shown in Fig. 8.13 for amorphous manganese aluminosilicate with 15 at%Mn and for rather small fields ($0 \leq h \leq 500$ Oe). Unfortunately, similar scaling plots can also be produced by MC simulations on a two-dimensional Gaussian Ising spin glass which has no finite-temperature phase transition (Kinzel and Binder, 1984). Again, fairly high fields have been used in the simulation. Hence these scaling plots do not prove or disprove a phase transition.

A less conventional method for measuring χ_{nl} is based on filtering out the higher harmonics of the response to an ac driving field (Taniguchi et al, 1983, 1985). For $T \geq T_f$, the ac and dc susceptibilities agree and both measure thermal equilibrium properties. Critical exponents determined in this way are also included in Table 8.2. If one varies the frequency, these experiments also yield the dynamic critical exponent $z\nu$. One has the response to the field $h = h_0 \cos \omega t$

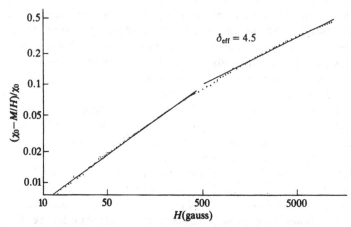

Figure 8.12: Log–log plot of the nonlinear magnetization vs magnetic field at T_f for $\underline{Ag}Mn$ 0.5% in the range of fields where $M_{nl}/M < 0.5$ (from Bouchiat, 1986).

$$M(\omega, t) = \sum_{k=1}^{n} [m'_k \cos(k\omega t) + m''_k \sin(k\omega t)] \qquad (8.106)$$

where, to lowest order in the field, $m'_1 = \chi'_1 h$, $m'_3 \propto \chi'_3 h^3$, $m'_5 \propto \chi'_5 h^5$, etc. and where χ'_1 and χ'_3, $\chi'_5 \ldots$ are the real parts of the linear and nonlinear susceptibilities. If a characteristic time τ of the system scales with the correlation length ξ, one has from (8.100) $\tau \propto |T-T_f|^{-z\nu}$ and the frequency dependence of some quantity scales as $\omega^{-1/z\nu}$. The nonlinear susceptibility χ'_3 then scales as

$$\chi'_3(\omega, T-T_f) = \omega^{-\gamma/z\nu} g_3(\omega\tau) = \omega^{-\gamma/z\nu} \tilde{g}_3(\omega^{\gamma/z\nu}|T-T_f|^{-\gamma})(8.107)$$

with $\tilde{g}_3(x) \to x$ and $\chi'_3(0, T-T_f) \propto |T-T_f|^{-\gamma}$ for $x \to 0$. Fig. 8.14a shows the linear susceptibility $\chi'_1(\omega, T)$ at ultralow frequencies ($10^{-3} \leq \omega \leq 10$ Hz) for $\underline{Ag}Mn$ 0.2 and 0.5 at%. One observes a sharp cusp in $\chi'_1(T)$ at T_f and a nearly logarithmic decrease of χ'_1 as a function of ω for $T < T_f$ and $\omega > 10^{-2}$ Hz. The latter agrees with (1.5). Fig. 8.14b indicates the frequency dependence of the nonlinear magnetization $m'_3 \propto \chi_3 h^3$ for $T > T_f$ on a log–log plot. The slope of this plot yields for $\omega > 10^{-2}$ Hz from (8.107) $-\gamma/z\nu = -0.30 \pm 0.02$ or with $\gamma = 2.1 \pm 0.1$ from $\chi'_3(0, T-T_f)$ the exponent $z\nu = 7.0 \pm 0.6$. This result agrees fairly well with the value $z\nu = 7.2$ obtained by Ogielski (1985) from his MC simulations on a 3d Ising spin glass. It suggests that, at least in small fields, $\underline{Ag}Mn$ is effectively an

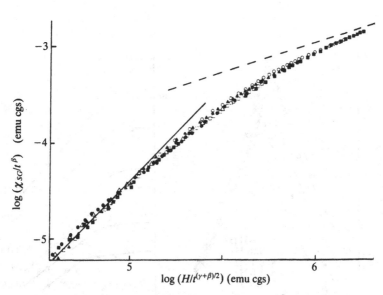

Figure 8.13: Scaling function $g(x) = \chi_{SG}/|T - T_f|^{\beta}$ plotted against $x = H/|T-T_f|^{(\gamma+\beta)/2}$ on log–log scales for an amorphous aluminosilicate with 15% Mn. The full line shows the initial slope of $g(x)$, the broken line shows the asymptotic slope $2\beta/(\gamma + \beta)$ for $x \to \infty$ (from Beauvillain et al, 1984).

Ising system. Possibly, this is due to anisotropy, as discussed in Section 8.3. The other static critical exponents, some of which have been obtained by means of the scaling laws (8.10)–(8.14), are listed in Table 8.2.

Fig. 8.14b shows another interesting effect: At the lowest frequencies and sufficiently far from $T = T_f$, the nonlinear susceptibility saturates as a function of frequency. Low frequencies mean a large length scale L. Above T_f, the coherence length ξ is finite and, if $L > \xi$, one moves out of the critical region for which (8.107) holds. The broken line in Fig. 8.14b shows the frequency ω_c for which $\omega \propto \xi^{-z}$.

The critical exponent z also has been obtained from scaling by including an external field h. In the presence of h, (8.100) generalizes to

$$\tau = |T - T_f|^{-z\nu} \tilde{F}\left(\frac{T - T_f}{h^{2/\phi}}\right) \tag{8.108}$$

with $\phi = \beta + \gamma$. One measures the onset of irreversibility as a function of the field for a given frequency ω or characteristic relaxation time τ either

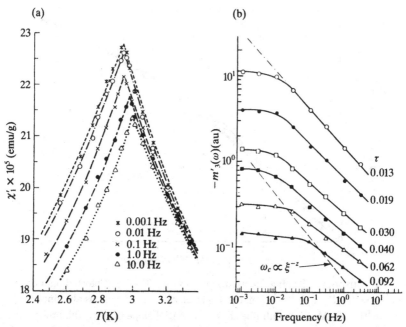

Figure 8.14: (a) Frequency dependence of the coefficient χ'_1 above and below T_f. Below T_f, the decrease of χ'_1 is nearly logarithmic at frequencies above 10^{-2} Hz.

(b) Frequency dependence of the nonlinear field term $-m'_3 \propto -\chi'_3 h^3$ of eq. (8.106) above T_f for several reduced temperatures $\tau = |T - T_f|/T_f$. The slope above 10^{-2} Hz is $-\gamma/(z\nu)$. The broken line shows qualitatively the shift of the frequency $\omega_c \propto \xi^{-z}$ as a function of τ from dynamic scaling (from Lévy and Olgielski, 1986).

by the onset of the imaginary part of the susceptibility $\chi''(\omega, T, h)$ or by its inflection point. If scaled in the form (8.108), the lines of constant τ or ω all merge into a single straight line in the $h - T$ plane. The critical exponents ϕ obtained thereby agree well with those indicated in Table 8.2 (ranging for the various systems between 3.3 and 4.0) and the values for $z\nu$ ($7.0 \leq z\nu \leq 8.2$) with the value $z\nu = 7.2$ of Olgieski (1985) (Bontemps et al, 1986; Vincent et al, 1986; Hamida and Williamson, 1986; Svedlindh et al, 1987). These lines in the $h - T$ plane were originally interpreted as evidence for an AT line as discussed in Chapter 5 (Paulsen et al, 1984; Rajchenbach and Bontemps, 1983) but according to our discussion of Section 8.3 are in reality crossover lines $|T - T_f| \propto h^{2/\phi}$ between a weak-field and a strong-field region for a fixed measuring time τ or frequency ω.

Finally, we discuss the possible crossover from Ising to Heisenberg behaviour due to random anisotropy. We already found such a crossover in the MFT with random Dzyaloshinskii–Moriya anisotropy added to the ordinary exchange interactions (Section 6.4). In MFT, this crossover manifests itself in a transition from an AT line to a GT line with increasing field; both lines are defined by a singularity in the spin glass susceptibility χ_{EA} and the onset of nonergodicity (see Fig. 6.4). In spin glasses with short-range interactions, there is only a crossover without any singularity or nonergodic behaviour. For weak anisotropy, one expects a scaling relation of the form

$$\chi_{nl}(T, h, D) = h^{2/\delta} F\left(\frac{T - T_f}{h^{2/\phi}}, \frac{D^{2/\phi_A}}{h^{2/\phi}}\right) \qquad (8.109)$$

which generalizes (8.103). Here D is the strength of the anisotropy energy defined in Section 6.4 (in units of T_f) and ϕ_A is another crossover exponent. This anisotropy can be strongly enhanced by additional nonmagnetic impurities with orbital moments, as we discussed in Section 6.4. Indeed, additional Au impurities lead to strong deviations from a single-parameter fit of χ_{nl} based on (8.109) (Yeshurun and Sompolinsky, 1982). A universal scaling for various AgMnAu and CuMnAu alloys has been obtained by de Courtenay et al (1986). Again, one observes different critical exponents δ_I and δ_H for small and large fields, which are now interpreted as Ising and Heisenberg values of δ. In addition, the crossover exponent ϕ_A in (8.109) is estimated to be $\phi_A = 0.8 \pm 0.1$, and adding Au impurities shifts the crossover toward higher fields and lower temperatures.

9

Dynamics on many time scales

Spin glasses differ from most magnetic materials in having dynamics on many timescales. That is, for example, if one measures the susceptibility of a spin glass to fields oscillating with frequencies $\omega = 10^{12}$, 10^8. 10^4, 1 and $10^{-4}\mathrm{sec}^{-1}$ one may get different results at all the measuring frequencies, indicating that the system has characteristic excitation and relaxation times over (at least) 16 orders of magnitude, ranging from the typical times associated with microscopic spin motions in solids to macroscopic times. For sufficiently low temperatures, this spectrum can extend at the long-time end to geological timescales. (A practical limit is set by the length of time society considers reasonable for a graduate student to complete a Ph.D.) This situation is completely different from that of conventional materials, where for practical purposes measurements at all frequencies less than about $10^{10}\mathrm{sec}^{-1}$ would give indistinguishable results.

How are we to understand such properties? In this chapter we review the various kinds of experiments that can be done to probe the dynamical behaviour on timescales ranging from microscopic ones to the longest ones experimenters have the patience to study, and we try to sketch a theoretical picture that enables us to get some insight into what the experimental results mean.

We have actually met this theoretical framework before (often) in the preceding chapters of this book. It is nothing other than our old friend *broken ergodicity* (introduced in Section 2.4), adapted to the kinds of experimental situations we meet here. In most of our discussion of static mean field theory (Chapter 3) we took 'broken ergodicity' to mean the existence of many equilibrium thermodynamic phases. We thought of these as valleys in the system's configuration space which were separated from each other by infinite free energy barriers, so that finite-time experiments could only probe one such phase at a time. Here we take a somewhat broader view, thinking in the context of a particular experiment of 'infinite' barriers as defined by the timescale of the experiment. Thus if there is a reasonable distribution of barrier heights, as the experimental time is (adiabatically) increased, the system will have the opportunity to explore larger and larger regions of configuration space. It may never get to true equilibrium, but here we are interested in characterizing this dynamical process rather than the ultimate equilibrium state.

We actually made use of this broader form of broken ergodicity in the picture we drew in trying to motivate the dynamical treatment of mean field theory (the Sompolinsky theory presented in Section 5.3). There each value of the Parisi parameter x was associated with a timescale, and $q(x)$ was an effective order parameter for this timescale. In a system of finite size, the barriers would be finite and the evolution through the sequence of timescales or values of x observable in principle. However, we never actually studied the finite-N problem for the SK model: The theory gave us a way to characterize the different degrees of broken-ergodicity states that could exist, but not a description of the dynamical processes by which one might evolve into another. This was because the theory was constructed for infinite N, where no such processes occur.

Here we are concerned about real spin glasses where (we believe) strict broken ergodicity in the sense of many equilibrium phases does not exist, but that there are many metastable configurations (or metastable regions of configuration space, which we can call 'metastable states'). We now want to adddress just the aspects of the problem ignored in Sompolinsky dynamics — the real dynamical properties.

Recalling our analysis of the equilibrium dynamics of short–range spin glasses in the preceding chapter, where we found that correlation functions decayed below T_f like an inverse power of $\log t$, we can see immediately that that theory can only have academic relevance in the spin glass phase (except maybe in the immediate neighbourhood of T_f): *The system will simply never equilibrate in the time available to do an experiment trying to measure such a decay.* Moreover, this pessimistic conclusion is not grounded only in the admittedly sketchy theoretical picture of Chapter 8: Independent of any theory, Every experimenter who has tried to do a measurement on a spin glass below T_f knows that it does not reach equilibrium in practical measuring times. Thus we need a theoretical formulation which takes explicit account of the nonequilibrium nature of the experimental situation.

The remainder of this chapter is divided into three sections. In the first of them (Section 9.1) we review the evidence for the existence of many time scales, describing the different kinds of experiments that have been done on various kinds of spin glass systems at characteristic timescales ranging from picoseconds to days. In the second (9.2) we sketch a theoretical framework for the explicit description of nonequilibrium effects in the rather general context of what are called *ageing* experiments, where one studies the dependence of the measured response of the system on the time since the external thermodynamic parameters (field and temperature) were last fixed. A generalization of the scaling theory of Section 8.2 gives a natural framework for such a description. Finally, in the last Section (9.3) we outline several kinds of more general phenomenological models that have been proposed for describing the dynamics of many-time-scale systems. These models incorporate naturally some general features which must be essential to such dynamics, though they are all constructed at a level which is several steps away from that of the lattice EA model (2.5). It is difficult in general to connect the levels, but in the case of Ising glasses we will see that the dynamical renormalization theory of Section 8.2 provides a link.

9.1 Experimental windows

Any experiment has a characteristic timescale (or characteristic frequency) associated with it. For a static magnetization measurement it is just the length of time over which the experiment was carried out, while for a Mössbauer experiment, to take just one example, it is the lifetime of the Mössbauer resonance (typically 10^{-7}sec). These experimental parameters can typically be varied over a range of 1 to 3 decades for a given kind of experiment. Then each measuring technique gives us a window a few decades wide within which we can see the characteristic excitation or relaxation processes of the system. As we noted above, spin glasses exhibit interesting dynamics over many decades of timescales, so there are many kinds of experiments we can use to study them.

The windows opened by several kinds of experiments are shown in Fig. 9.1. The timescales range from 10^{-13}sec for conventional neutron scattering to days or months for magnetization relaxation or remanence experiments carried out by patient experimenters. In between lie windows opened by neutron spin echo experiments, muon spin resonance, ESR and NMR, Mössbauer, and conventional ac susceptibility measurements.

There is a crude form of universality characterizing the susceptibility measured in these experiments for frequencies from around 1 to 10^{10} sec^{-1}. The susceptibility is always essentially Curie-like at high temperatures, has a (rounded) 'cusp' or maximum at an effective 'spin glass temperature' $T_f(\omega)$, and falls off well below the extrapolation of the high-temperature

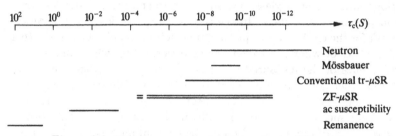

Figure 9.1: Windows on spin glass dynamics in different time or frequency ranges (Uemura, 1981).

form for $T < T_f(\omega)$. The temperature for which the maximum occurs varies slowly with the measuring frequency ω (so slowly that the variation is easy to miss if the experimenter does not vary ω over several decades). On the low-temperature side of the maximum, the susceptibility itself is also slowly frequency-dependent. These features are seen in *every* spin glass material, including the 'pseudo-spin glasses' mentioned in Chapter 2.

ac susceptibility measurements

This phenomenological universality is exhibited most cleanly in ac susceptibility measurements, which work in a window from around 1 to $10^4 \sec^{-1}$. Here it can even be formalized in the description we gave in the introduction: The effective freezing temperature $T_f(\omega)$ and the (real part of the) susceptibility for $T < T_f(\omega)$ vary approximately logarithmically with frequency:

$$\chi'(\omega) = \chi_0 + a \ln \omega \qquad (9.1)$$

and the consequent imaginary part

$$\chi''(\omega) = \frac{\pi a}{2} \operatorname{sgn} \omega \qquad (9.2)$$

implied by analyticity. Figs. 9.2 and 9.3 show how nicely (9.1) and (9.2) fit the data in some cases.

At a phenomenological level, this kind of frequency dependence finds a natural explanation in terms of a smooth and broad distribution of barrier heights in the system. That is, if we model the system as an ensemble of independent modes λ, each of which relaxes with a simple exponential decay $\propto \exp(-t/\tau_\lambda)$, with an activated relaxation time

$$\tau_\lambda = \tau_0 \exp(\beta \Delta_\lambda) \qquad (9.3)$$

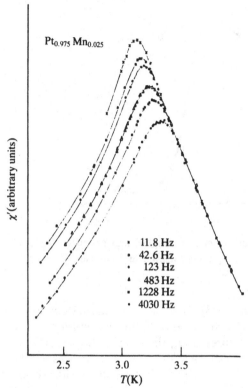

Figure 9.2: The frequency dependence of $\chi'(\omega)$ for $Pt_{1-x}Mn_x$ with $x = 0.025$ at frequencies from 11.8 to 4030 Hz (Lecomte et al, 1983).

with barrier Δ_λ, we get a susceptibility

$$\chi(\omega) = \int d\ln\tau\, g(\ln\tau)\frac{\chi_0}{1 - i\omega\tau} \qquad (9.4)$$

If the distribution g of $\ln\tau$ (i.e. of the dimensionless barriers $\beta\Delta$) is approximately constant in a range $(\ln\tau_{min}, \ln\tau_{max})$, then (9.4) reduces to (9.1) and (9.2) for $\tau_{max}^{-1} \ll \omega \ll \tau_{min}^{-1}$. Indeed, any reasonably smooth and broad barrier distribution will lead to a $\chi(\omega)$ very close to (9.4) over several decades of frequency. The point is that a spread of Δ's of order, say, a few times T_f will lead at resonably low temperatures to τ's extending over many orders of magnitude.

By doing very careful experiments and fitting to (9.4) one can even

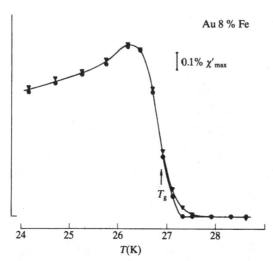

Figure 9.3: χ'' full inverted triangles) and $(\pi/2)\partial\chi'/\partial\ln\omega$ versus temperature near T_f in AuFe, measured at frequency $f = \omega/2\pi = 1.7\,\mathrm{Hz}$ (Lundgren et al, 1982a).

'measure' $g(\ln\tau)$ (Wenger, 1983). Now of course the system does not really consist of independently relaxing modes, so this distribution appears at this level just to be a way of representing the data. However, it is possible to connect it with theory at least one level deeper than this — the McMillan scaling theory of the dynamics given in Section 8.2 efffectively provides a way of calculating $g(\ln\tau)$, as we will see explicitly in Section 9.3. The very broad relaxation time distribution emerged naturally from the renormalization formulation, where the effective barriers grow with increasing lengthscale.

Furthermore, with the fluctuation-dissipation theorem (4.45), the logarithmic time dependence of correlation functions (8.51) obtained in the scaling theory gives

$$\chi''(\omega) \propto \frac{\mathrm{sgn}\,\omega}{|\ln\omega|^{1+y/\psi}} \tag{9.5}$$

which is very hard to distinguish experimentally from (9.4) with some arbitrary slowly-varying $g(\ln\tau)$. Thus, while they do not provide a severe test of the scaling theory, the available data are consistent with it.

Of course these measurements tell us virtually nothing about the existence or nonexistence of a phase transition at a nonzero temperature. Qualitatively similar curves are obtained for the one-dimensional Ising spin

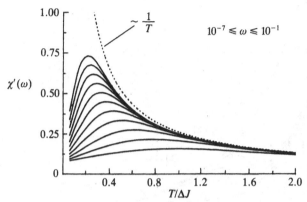

Figure 9.4: Real part of $\chi(\omega)$ for a finite-length 1-dimensional Gaussian Ising glass (6 spins) (Reger and Binder, 1985).

glass (Fig. 9.4), which we are sure has no transition. Similarly, the data shown in Figs. 2.11–12 for electric glasses in which we are also fairly sure there is no transition also resemble those of common magnetic spin glasses where we are now fairly confident that there is one. There is an apparent qualitative universality in the way all these systems relax to their equilibrium states, independent of whether those states have any sort of broken symmetry.

The McMillan dynamical renormalization theory (Section 8.2) appears to provide the best simple picture we have of how this generic behaviour arises in short range spin glasses. However, the observation of similar dynamical behaviour in a wide variety of systems whose degrees of freedom are at least superficially very different from 'spins' has prompted people to look for more general model frameworks for describing it. We will examine a few examples of such models in Section 9.3.

One should not take the logarithmic 'law' (9.1) too seriously; it is far from universal. For example, in the case of $Eu_{0.4}Sr_{0.6}S$ a power law gives a better fit to the data, as indicated in Fig. 5.3.

In very dilute $Eu_xSr_{1-x}S$, which has no spin glass transition, we can even see two kinds of dynamical freezing processes happening in the same system (Fig. 9.5). The high-temperature susceptibility maximum comes from exchange coupling within small Eu clusters and the low-temperature one from intercluster coupling via dipole–dipole forces (2.83). The former can be fit *quantitatively* by taking explicit account of only a few kinds of very small

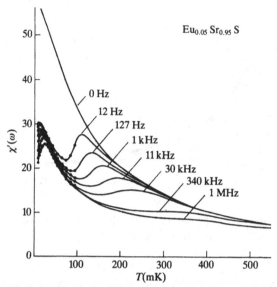

Figure 9.5: Susceptibility of $Eu_{0.05}Sr_{0.95}S$ for different measuring frequencies (Eiselt et al, 1979).

clusters, treating them as completely decoupled from the rest of the system, and adding up their calculated contributions to the response proportional to their probabilities of occurence. We notice, however, that the frequency dependence of the resulting susceptibility (the higher-temperature peaks in Fig. 9.5) is much stronger than that seen near the low-temperature peaks, where the real collective freezing occurs.

So far we have only mentioned small-field experiments here. Measurements have also been done in finite field to look for behaviour related to the AT line of mean field theory, as we discussed in Section 5.2, and the same sorts of dynamical features are observed here as well. In particular, the data shown in Fig. 5.4 indicate that the measured 'AT' lines depend strongly on frequency. Further evidence that the apparent AT line and possibly also the plateau in the field-cooled magnetization M_{fc} are actually dynamic effects comes from Monte Carlo simulations. The simulation of a two-dimensional Ising spin glass with nearest neighbour interactions appears to show both a plateau in M_{fc} and an AT line, though the model definitely has no phase transition (Morgenstern and Binder, 1979, 1980a,b). Here the 'AT' line is defined operationally by when the field-cooled and zero-field-cooled magnetizations begin to differ. This line depends on the time during which the

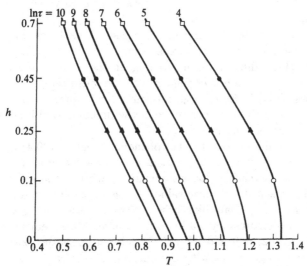

Figure 9.6: Contours of constant relaxation time in the two-dimensional nearest neighbour $\pm J$ model (Young, 1983a)

field was applied, but scales for different times (Fig. 9.6). With increasing time these lines shift to lower temperature and presumably tend to $h = 0$ for $t \to \infty$. The plateau in the field-cooled magnetization turns out to be a nonlinear effect: $M_{fc}(T = 0)/h$ varies like $h^{-1/\Delta}$ with an exponent $1/\Delta = 0.28 \pm 0.05$ at $T = 0$. It diverges as $h \to 0$, in agreement with the Curie susceptibility $\chi = 1/T$ which one expects for such a system without a phase transition (Kinzel and Binder, 1983, 1984).

Time-dependent magnetization measurements

We can open our window on spin glass dynamics to lower frequencies by going over to direct measurements of the time dependence of the magnetization at macroscopic times. This permits us to probe processes with relaxation rates as low as $10^{-5} \sec^{-1}$ (or even lower for very patient experimentalists).

Doing this, we discover that below the canonical frequency range of ac susceptibility measurements $1 \sec^{-1} < \omega < 10^4 \sec^{-1}$ the nice logarithmic behaviour of $\chi(\omega)$ and $T_f(\omega)$ breaks down. Qualitatively, the picture of χ' rising slowly (the natural parameter is again $\log t$, not t) as the measurement time increases is still valid up to times of the order of hours (Fig. 9.7), but the curves at equally spaced values of $\log t$ are no longer approximately equidistant. They appear to converge toward a curve which is definitely lower than the Curie value. There is a finite time $\tau_{max}(T)$ where the system does seem to have reached equilibrium, and just a little below T_f it is short enough to be measurable in these experiments.

That something different is happening at these very long times can also be seen in the frequency dependence of the 'cusp temperature' $T_f(\omega)$ Fig. 9.8 shows $T_f(\omega)$ in $Eu_{0.4}Sr_{0.6}S$ measured for frequencies from 10^{-2} to 10^6Hz. There is clearly some curvature in the function, and it seems to be leveling off for very low frequencies.

A natural speculation based on the phenomenology (9.1–9.4) would have been that there is no static phase transition in these materials at all: $\chi'(\omega)$ would just go on increasing (logarithmically slowly) as the frequency was lowered until it reached the Curie value $1/T$, as in one- or two-dimensional systems (e.g. Fig. 9.4). Data like those of Figs. 9.7 and 9.8 (as well as Fig. 5.3) offer evidence against such a hypothesis here. Measurements on other systems (e.g. Malozemoff and Imry, 1981) also support the existence of a nonzero $T_f(0)$, in agreement with the current theoretical ideas and other experiments we discussed in Chapter 8.

Other long-time magnetization measurements also yield interesting information. In the Introduction we showed results of measurements of the remanent magnetization (isothermal and thermoremanent). These data were fit in different systems by logarithmic, power law, or stretched exponential time dependence. Whatever the underlying fundamental dynamics, these data all share with the higher-frequency data reported in the previous section the feature that they cannot be explained in terms of a single relaxation time (or of a narrow distribution of them). Thus the broad spectrum of relaxation times must extend to the timescales ($O(10^4$ sec)) of these experiments, as we already inferred from data such as Fig. 9.7 closer to T_f.

A very interesting effect in similar experiments was first noticed by Lundgren et al (1983). This is the fact that the relaxation of the magnetization depends on how much time elapsed between the initial cooling of the system to a $T < T_f$ and the measurement time. In this situation there is thus an interplay between *two* timescales which can be varied independently. This phenomenon, called 'aging', can give valuable insight about the slow dynamics below T_f and has been studied intensively in recent years. Because of its conceptual importance we reserve the next section for it.

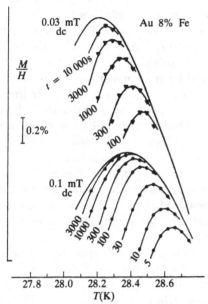

Figure 9.7: Long–time magnetization measurements in A̲u̲Fe (Lundgren et al, 1982b).

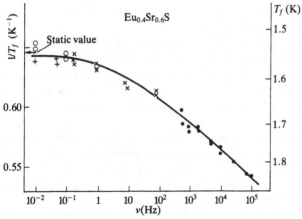

Figure 9.8: $T_f^{-1}(\omega)$ of $Eu_{0.4}Sr_{0.6}S$ (Ferré et al, 1981).

Neutron scattering (including spin echo)

At the other end of the frequency spectrum, in the range of characteristic rates of microscopic magnetic processes in solids, lies inelastic neutron scattering. The characteristic time of a conventional neutron scattering experiment is the reciprocal of the measured linewidth. Thus the neutron 'window' allows us to view processes in solids on frequency scales between about $10^{10} - 10^{13} \, \text{sec}^{-1}$. This is ideal for most magnetic materials, but in spin glasses it misses most of the action, which lies at much lower frequencies. A conventional neutron scattering experiment is unable to distinguish between degrees of freedom which are truly frozen and those which merely have relaxation times longer than 10^{-10} sec. To it, essentially all of the interesting degrees of freedom in the materials which have been studied appear static at any temperature of interest. (There is, however, some dependence of the neutron scattering cross section on the energy resolution of various spectrometers. This indicates that there is some spin dynamics on timescales even between 10^{-9} and 10^{-11} sec around the freezing temperature (Murani and Heidemann, 1978).)

Recently Huse (1989) has suggested that the *field dependence* of the neutron scattering cross section could be used to probe the spatial dependence of the generalized spin glass susceptibility $\chi_{SG}(\mathbf{r})$ (2.64) in something like the same way that the nonlinear susceptibility is used to measure the uniform component $\chi_{SG}(\mathbf{k} = 0)$. Such measurements would also be confined to the high-frequency window $10^9 < \omega < 10^{11} \text{sec}^{-1}$, however.

The one thing conventional neutron scattering is good for is establishing that what is going on in these materials is *not* conventional periodic magnetic order. This is especially relevant in the kinds of systems we will study in Chapter 12, where both spin glass freezing and conventional order occur and compete with each other.

It has also been used to look for spin waves. However, despite a lot of work, no one has found any spin waves associated with spin glass order (the Halperin–Saslow modes of Section 7.2). (Some short-wavelength excitations have been seen (Maletta et al, 1981), but they are probably due to short-range ferromagnetic order in the particular alloy systems where they have been observed.)

There is a clever kind of neutron scattering experiment invented by Mezei (1980) which enables one to open the observation window to much lower characteristic frequencies. Called 'neutron spin echo', it measures the evolution of the spin correlation function $\langle S_{\mathbf{q}}(0)S_{-\mathbf{q}}(t)\rangle$ directly as a fucntion of time, rather than through its Fourier transform as in conventional neutron scattering. Fig. 9.9a shows the results of such experiments on C̲uMn; at temperatures near and above T_f the curves extrapolate nicely

(a)

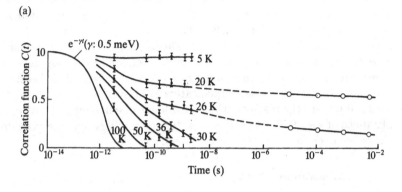

(b)

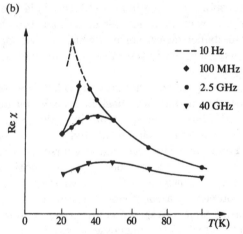

Figure 9.9: (a) Time dependence of the spin correlation function measured in C̲uMn by neutron spin echo (Mezei, 1981). The points at much longer times are obtained from ac susceptibility measurements (Tholence, 1981). (b) χ as a function of T for frequencies from 10 to 4×10^{10} Hz from ac susceptibility and neutron spin echo measurements.

to join the results obtained for timescales three to six orders of magnitude longer in ac susceptibility measurements like those discussed above (Tholence, 1980). Figure 9.9b shows the same data plotted another way: For a set of characteristic frequencies ranging from 10 Hz to 4×10^{10} Hz, the susceptibility is plotted as a function of temperature. (The low-frequency data come from Tholence's ac susceptibility measurements and the high-frequency ones from the neutron spin echo experiments on the same material.) This figure makes it clear that already at the highest frequency measured, some crudely spin-glass-like behaviour is going on, though with an effective $T_f(\omega)$ which is much higher than that seen in the low-frequency measurements. Of course the more quantitative sort of universal behaviour observed in the lower-frequency range does not hold over all these orders of magnitude, but it is still striking to see how the dynamical freezing phenomenon persists in the same qualitative form over 10 orders of magnitude (or 15, if we add the results of time-dependent magnetization measurements like those of Fig. 9.7) of frequency. A dynamics with even approximate self-similarity over so many time scales is remarkable.

Mössbauer effect

The gap in our frequency window between ac susceptibility ($< 10^5 \sec^{-1}$) and neutron spin echo experiments ($10^8 \sec^{-1}$) is filled, although not in the most satisfying way, by a number of resonance techniques. We start the list with Mössbauer measurements, which, as we noted above have a characteristic frequency set by the Mössbauer linewidth. This is $O(10^7 \sec^{-1})$, i.e. in the gap.

Mössbauer experiments actually provided some of the earliest evidence for spin glass behaviour. Way back in 1965 Violet and Borg measured the splitting of the Mössbauer spectrum in $\underline{Au}Fe$ (Fig. 9.10) which arose as a result of the internal fields generated by the frozen spin configuration of nearby Fe atoms. This freezing, of course, is that which occurs on the timescale of the inverse Mössbauer resonance width. All degrees of freedom which relax more slowly than this appear static. Thus, although for most solid-state processes, a Mössbauer experiment is essentially a static one (i.e. measuring time-averaged quantities over a time much longer than the characteristic times of the phenomena under study), for spin glasses it is (like neutron scattering) more like a snapshot. Hence, like the other kinds of experiments we have discussed here, this technique does not allow us to conclude anything about the existence of a phase transition.

Mössbauer spectroscopy is limited as a tool for studying spin glass dynamics because it is not possible to vary the measurement frequency systematically to probe different ranges of characteristic relaxation rates. All that one can do is get a few isolated points by using different Mössbauer nuclei with different resonance linewidths. Nevertheless it has some merit because it provides a rather direct measurement of the internal field distribution.

One observes, at least in $\underline{Au}3at\%Fe$ and $\underline{Rh}5at\%Fe$, a fairly sharp onset of 'static' hyperfine fields or a nonvanishing spin expectation value $\langle S \rangle$

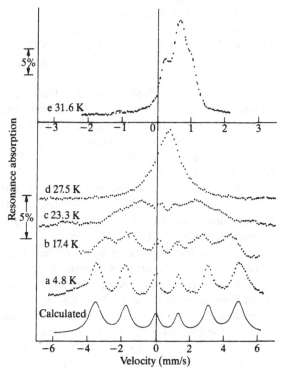

Figure 9.10: Mössbauer spectrum of A̲u̲Fe, showing the splitting in the spin glass internal field (Violet and Borg, 1965)

which can be fit fairly well below T_f by a Brillouin function. This would be another indication of spin glass freezing. However, as in all dynamic experiments, it is impossible to distinguish between true static and dynamic processes (Meyer et al, 1985, 1986). In addition, the Mössbauer effect is a local probe which is extremely sensitive to the formation of metallurgical clusters and hence to the thermal treatment of the system.

NMR and ESR

Magnetic resonance experiments are a rather indirect probe of spin glass dynamics, but they also fill in a little information in the frequency gap between the 'good' techniques ac susceptibility and neutron spin echo. In a magnetic resonance experiment, the internal fields (static or dynamic) that we are interested in here make themselves felt as little local perturbations of the main resonance at the Zeeman splitting of the spin states of a proton or electron: The resonance can be either shifted or broadened (or both) by the protons's or electron's interaction with the rest of the system. One generally has to make nontrivial calculations to connect the observed lineshifts and shapes with hypotheses about the spin glass dynamics that one wants to test.

The more or less straightforward regime is that in which the intrinsic spin dynamics are much faster than the experiment's characteristic time, the inverse of the Zeeman frequency. Then (Abragam, 1961) the linewidth is just proportional to the spin relaxation time that we would like to measure. This works well at high temperatures (see, e.g. Levitt and Walstedt, 1977 (Fig. 9.11a) for NMR and Salamon and Herman, 1978 (Fig. 9.11b), and Dahlberg et al, 1979 for ESR), but as the slowing-down of the spin dynamics starts to occur as we approach the neighbourhood of the spin glass transition, relaxation times become very long and we pass out of this regime just as the measurement was getting interesting.

In the other limit, where the technique is in its 'snapshot' regime, the line is broadened according to the distribution of internal fields and shifted by any net field (such as that produced by a remanent magnetization). This is useful for the study of low-temperature properties (Chapter 7).

Here we meet a similar problem to that encountered for the Mössbauer data. One measures a spin configuration which on the characteristic time-scale for NMR or ESR looks frozen, but cannot determine whether it is static or dynamic (Alloul, 1979). However, both ESR and NMR low-temperature experiments are important for the investigation of anisotropy, as discussed in Section 7.3.

Muon spin rotation and relaxation

These kinds of experiments exploit the fact that muons have an intrinsic correlation between their direction of motion and their spin (they are left-handed particles). The angular distribution of the electrons or positrons that result from the decay of a beam of muons is therefore asymmetric. Since the muon spin direction will precess in a magnetic field, the observed angular distribution of the decay spectrum will also be rotated and distorted, and analysis of these changes can be used to extract information about the internal field distribution in the material where the decay takes place. The frequency of this window is set by the muon lifetime, $2.2\mu s$.

The first sort of experiment on spin glasses exploiting this fact (Murnick et al, 1976) was done in a transverse (perpendicular to the beam direction) field. Then the angular distribution of the positrons rotates at the Larmor

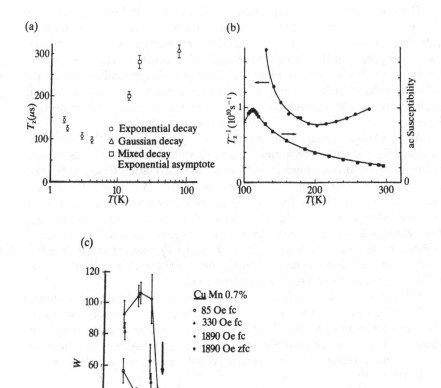

Figure 9.11: linewidths in C̲uMn:
(a) NMR (spin echo) (1% Mn) (Levitt and Walstedt, 1977)
(b) ESR (25% Mn) (Salamon and Herman, 1978)
(c) transverse μSR (0.7% Mn) (Murnick et al, 1976)

rate in the absence of the sample, and by measuring how it is damped
and smeared out one can extract a relaxation time which is analogous to
that measured in NMR. This is shown in Fig. 9.11c. The tendency toward
freezing of the spin fluctuations is quite evident, as is the strong effect of
the field.

These experiments can also be done in zero external field (Uemura et al, 1980), since the muons in the beam are nearly fully polarized along their velocity. Then it is simply the decay of the asymmetry between forward and backward positrons that is measured, and this relaxation time again has to be connected with the spin dynamics through a theory similar to that for NMR. Results of such measurements also indicate a strongly increasing spin correlation time in the region where there is a straightforward connection between it and the decay rate of the positron asymmetry.

As in magnetic resonance, the simple connection between the decay rate observed in the experiment and the correlation time of the spins we are trying to study breaks down in μSR when the correlation time reaches the order of the characteristic time of the experiment. Any spins slower than this appear frozen. The data then appear to be smeared over a distribution of static local fields. Emmerich et al (1985) extracted an effective q_{EA} (for this timescale) in this way from zero-field μSR data on $\underline{Cu}$Mn. They claimed to see some freezing even at $T \approx 1.6T_f$, but as the extraction of an order parameter from the data is necessarily rather indirect in this technique, the result is only a suggestive one.

A third way to do a μSR experiment is in a *longitudinal* field (Uemura et al, 1981). Here there is no precession, so the measurement is essentially the same as in the zero-field method. The novelty is that one can now study the field dependence of the relaxation times. However, again because of the indirect nature of the measurement, only qualitative conclusions can be drawn firmly.

The main difficult with this class of techniques is that, as in magnetic resonance, one needs a nontrivial level of theory to relate experimentally measured quantities to the theoretically interesting ones. This auxiliary theory is less developed here and more dependent on model assumptions than the theory of magnetic resonance experiments.

9.2   Ageing

As we noted above, ageing, i.e. the dependence of measured properties on the temperature and field history of the sample, is a particularly important phenomenon to understand because it is at the core of the nonequilibrium nature of the experimental situation one unavoidably faces in spin glasses. Taken together, the experiments we have described above have measured dynamical spin glass properties over 15 orders of magnitude of frequency, but any one experiment probed just one timescale at a time. In ageing, we probe the degrees of freedom acting on at least two timescales (including nontrivial interactions between them) simultaneously. Here we outline the theory of such effects (Koper and Hilhorst, 1988; Fisher and Huse, 1988a) along the lines of the scaling theory of Section 7.2.

Scaling theories are characterized by exponents. This one has one new exponent, λ, in addition to the two, y and ψ, that we had in the equilibrium theory. It describes how domains of a spin glass phase (remember that there are just two of them in this theory, global spin flips of each other) grow after a quench into the low-temperature phase.

One can study this kind of problem in a simpler system such as a ferromagnet, too. Here the canonical gedanken experiment is a sudden quench from a very high (formally infinite) temperature to a very low one. Then ferromagnetically ordered domains will grow larger and larger until eventually the whole system orders. However, in a formally infinite system, this equilibrium is never reached. The entire history of the system after $t = 0$ is one of a nonequilibrium process. This process is described by quantities like the correlation $C(t) \equiv \langle S_i(0)S_i(t)\rangle$, where $S_i(0)$ is the value of the spin i at the quench time and $S_i(t)$ is its value a time t later.

For a pure ferromagnet, the domain walls move diffusively, so the typical domain size $L(t)$ at a time t after the quench is proportional to $t^{\frac{1}{2}}$ (Lifshitz, 1962). This leads to power law decay of $C(t)$. The exponent is nontrivial, but we can put a bound on it by arguing that (ignoring correlations in the dynamics) the correlation function just varies like the inverse square root of the volume of the typical domain at time t, i.e.

$$C_{FM}(t) \propto L(t)^{-d/2} \propto t^{-d/4} \qquad (9.6)$$

The actual decay cannot be slower than this. Simulations (Fisher and Huse, 1988a) confirm this bound. In general, λ is the true exponent which occurs in (9.6) in place of the lower bound $d/2$.

In a spin glass, the corresponding information about the growth of the domains of the two pure spin glass phases can be obtained by considering a quench to a $T < T_f$ from a fully aligned state. (The fully aligned state is uncorrelated with the spin glass states just as the random infinite-T starting configuration in the ferromagnetic problem was uncorrelated with the ferromagnetic ground states.) In this case a measurement of the magnetization $M(t)$ remaining a time t after the quench just amounts to a measurement of the correlation function $C(t)$:

$$C(t) \equiv \frac{1}{N}\sum_i \langle S_i(0)S_i(t)\rangle = \frac{1}{N}\sum_i \langle S_i(t)\rangle \equiv M(t) \qquad (9.7)$$

using $S_i(0) = 1$.

Now, in contrast to the situation in the ferromagnet, domain growth requires excitation over barriers that grow with length scale, so we find, as in our equilibrium argument, that domains of the pure spin glass phases will grow *logarithmically* in time (cf. (8.51)):

$$L(t) \propto \left(\frac{T}{J} \ln t \right)^{1/\psi} \tag{9.8}$$

where ψ is the exponent that describes how the barrier heights change with length scale. That is, the nonequilibrium nature of the process we are considering here does not change the fundamental energetics: To relax the spins to equilibrium one has to surmount the same barriers that determine the equilibrium dynamics. As these domains grow, $C(t)$ (or the magnetization $M(t)$) will decay. As in the ferromagnet, we hypothesize that when the domains have grown to a size $L(t)$, $C(t)$ will be proportional to a (negative) power of L_t:

$$C(t) \propto \frac{1}{L(t)^\lambda} \propto \left(\frac{J}{T \ln t} \right)^{\lambda/\psi} \tag{9.9}$$

Comparing (9.9) with (8.51), we can see that λ plays a role for the nonequilibrium domain growth kinetics like that y does for equilibrium dynamics. We stress that there is no reason for λ to have the same value as y, since one describes equilibrium and the other a process far from equilibrium. Indeed, this equality would be inconsistent with our lower bound $\lambda > d/2$, given the small values of y we reported.

To illustrate how one can observe both nonequilibrium (domain growth) and effectively equilibrium relaxation experimentally we consider the following simple example (a thermo-remanent magnetization experiment): Suppose we cool a system rapidly, with a field on, to a temperature $T < T_f$, wait a time t_w, and then turn off the field and measure the decay of the magnetization. During the waiting time, the system will have evolved toward the equilibrium state at temperature T (in the field). Applying the kind of arguments we used in the equilibrium theory, equilibrium will have been established locally within domains of size up to

$$L_w \approx \left(\frac{J}{T \ln t_w} \right)^{1/\psi} \tag{9.10}$$

(cf. (8.50)). That is, the system at that timescale can be thought of as a collection of domains of typical size R_w. Within each domain there is an effective local equilibrium, but the system is of course not in equilibrium as a whole because of the domain structure still remaining on scales greater than L_w.

Now when the field is switched off, the equilibrium state toward which the system will evolve becomes a different one — the zero-field state rather than the equilibrium state in the cooling field. As we stressed in Chapter 8, these two equilibrium states are essentially the same on length scales up to a distance (equation (8.53))

$$L_h \approx \left(\frac{J}{h}\right)^{\frac{2}{d-2y}} \tag{9.11}$$

but uncorrelated with each other for longer distances.

What happens now depends on the ratio of L_w to L_h. We take first the case where $L_h \gg L_w$, i.e. the small field case. Then the domain growth in the waiting period will not have reached the point where the resulting state of the system is sensitive to the presence or absence of the cooling field. For very short times $t \ll t_w$ (actually the condition is $\ln t \ll \ln t_w$) the relaxation will occur by local spin rearrangements on lengthscales $\ll L_w$, i.e. within the effectively equilibrium domains. (Here t is the time between when the field is shut off and when the measurement is made.) Then the correlations (or the remanent magnetization) will decay like those in equilibrium:

$$\frac{M(t)}{h} \propto \left(\frac{J}{T \ln t}\right)^{y/\psi} \tag{9.12}$$

This will continue until the length scale of these processes grows to be of the same order as the domain sizes, i.e. when $t = O(t_w)$. Beyond this the dynamics is essentially that of nonequilibrium domain growth again (as it was before the field was turned off) and $M(t)$ will decay with the faster logarithmic power law

$$\frac{M(t)}{h} \propto \left(\frac{J}{T \ln t}\right)^{\lambda/\psi} \tag{9.13}$$

(Again, the strict condition for (9.13) is a logarithmic one: $\ln t \gg \ln t_w$.) The observed shift in the rate of relaxation around $t = t_w$ is thus a direct manifestation of the nonequilibrium nature of the state in which the system found itself at the end of the waiting time. The crossover time is just equal to the waiting or 'ageing' time; this is why this is know as an ageing effect (Lundgren et al, 1983). A crossover qualitatively consistent with this description is actually observed experimentally (Figs. 9.12, 9.13).

In the opposite case $L_w \gg L_h$, The difference between the two states is so great that the time spent waiting in the field will not play any role in the subsequent evolution after the field is turned off. The field might as well have been turned off right after the quench; the waiting time plays no role. One will observe effectively equilibrium behaviour, i.e. (9.12) $M \propto h/(\ln t)^{y/\psi}$, for $\ln t \ll \ln \tau_h \equiv \beta J L_h^\psi$, while for t much larger than τ_h

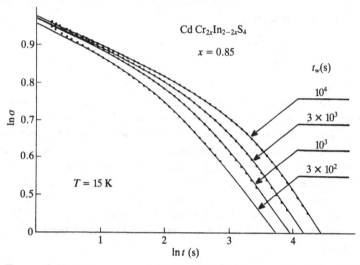

Figure 9.12: Thermoremanent magnetization versus time t for different waiting times t_w (Alba et al, 1987).

the dynamics will be dominated by domain growth, leading again to the faster decay (9.13) $M \propto h/(\ln t)^{\lambda/\psi}$.

Thus the two cases are similar in that each has a short-time regime with quasiequilibrium decay and a longer-time one with nonequilibrium decay of the remanent magnetization. The difference is in the crossover time, which is the smaller of the waiting time t_w and the field equilibration time τ_h. It is worth remarking that from (9.11) the longer the waiting time, the smaller the field necessary to see the waiting time effect:

$$h_{max}(t_w) \propto \frac{1}{(\ln t_w)^{\frac{d-2y}{2}}} \qquad (9.14)$$

Furthermore, it is clear that the qualitative nature of the effect does not depend on the particular form of experiment we have described. What is important is simply that after the waiting at some value of the field, the field has been shifted to another value by an amount Δh, which implies that the equilibrium state is rearranged for lengthscales greater than

$$L(\Delta h) \approx \left(\frac{J}{\Delta h}\right)^{\frac{2}{d-2y}} \qquad (9.15)$$

So the foregoing discussion applies to this more general field shift situation with h replaced in the formulas by Δh. (Another special case, of course, is

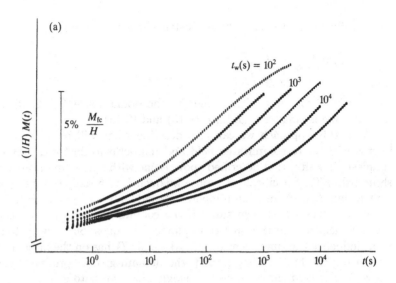

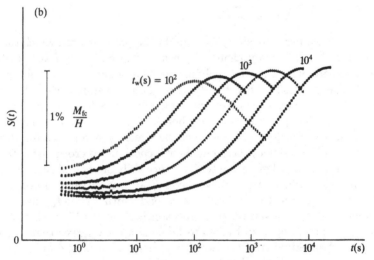

Figure 9.13: Zero-field-cooled susceptibility (a) and its derivative with rspect to $\ln t$ (b) for different waiting times (Granberg et al, 1988).

that of zero field cooling and the subsequent measurement of the magnetization after a field is turned on (Fig. 9.13).)

The same argument also applies to shifts in temperature. In this case we found in Chapter 8 that the characteristic length scale beyond which the equilibrium state has to be reconstructed is (cf. (8.45))

$$L(\Delta T) \approx \left(\frac{J}{\Delta T}\right)^{\frac{2}{d_s - 2y}} \tag{9.16}$$

Aging effects will be evident whenever the domain size L_w (9.10) at the waiting time is smaller than both (9.15) and (9.16).

With this approach we can also describe what will happen in experiments with more complicated field and temperature histories. The next simplest kind after that described above is one with two temperature or field shifts below T_f. Specifically, suppose we quench to a temperature $T_1 < T_f$, wait a time t_{w1}, change the temperature suddenly to $T_2 = T_1 \pm \Delta T$, wait a second interval t_{w2}, then turn off the cooling field and finally measure the remanent magnetization a time t later. The question we want to ask is what influence the time spent by the system at T_1 has on the measurement.

During the first waiting period, the domain growth process described above occurs, and the domains of a single phase grow to a size

$$L_{w1} \approx \left(\frac{J}{T_1 \ln t_{w1}}\right)^{1/\psi} \tag{9.17}$$

After the temperature shift, the equilibrium phase toward which the system will evolve is suddenly changed (on scales larger than $L_{\Delta T}$ (9.16)) from that toward which it was evolving before the shift. If ΔT is big enough that $L_{w1} \gg L(\Delta T)$, then the domains after the shift (referred to the new equilibrium state, of course) will suddenly be reduced to a size of order $L(\Delta T)$. If the opposite inequality holds, then is it only T_{w1} that limits the domain size; the domains remain at the size they had grown to. During the second waiting period, the domains will continue to grow, either from their previous size L_{w1} before the temperature shift or their shrunken size $L_{\Delta T}$, according to which of the above inequalities was satisfied. In order for this growth to amount to anything, the second waiting time t_{w2} has to be long enough for L_{w2} (given by an expression just like (9.17) with t_{w1} replaced by t_{w2} and T_1 replaced by T_2) to be significantly larger than the domain size at the beginning of this waiting period. That is, the final domain size at the end of the second waiting period is

$$L_w = \max(L_{w2}, \min(L_{\Delta T}, L_{w1})) \tag{9.18}$$

Now the field is shut off and the remanent magnetization measured at a time $t_{w1} + t_{w2} + t$ after the original quench, i.e. a time t after the field is turned off. The problem now is formally like our simpler one without the time spent at T_1, except that now the previous expression (9.10) for the domain size just before the field shutoff is replaced by (9.18), which can depend on the first waiting time and ΔT.

Again there is a question of the relative sizes of L_w and the field correlation length L_h (9.11). Taking the weak-field case, where there is waiting-time dependence, we get different possible behaviours according to the relative magnitudes of L_{w1}, L_{w2}, and $L_{\Delta T}$.

If ΔT is so small that $L_{\Delta T}$ is the largest of the three lengths, then L_w is just the larger of L_{w1} and L_{w2}. That is, it is essentially just the longer waiting time that counts. (This simplified account ignores the difference between T_1 and T_2 in the expressions for L_{w1} and L_{w2}. Also, we measured each of these T's relative to a single characteristic energy J, where we really should have used a temperature-dependent barrier $\Delta(T)$ which goes to zero as $T \to T_f^-$ with a characteristic critical exponent, so our analysis has to be modified near T_f; see Fisher and Huse (1988a) for details of this and other complications that we also ignore here.)

A second possibility is that ΔT is big enough (or t_{w2} long enough) for $L_{\Delta T}$ to be much smaller than L_{w2}. Then the first waiting time is irrelevant, and the resulting measurement will be essentially the same as if the system had been quenched directly to T_2. This is apparently the situation in the experiment reported by Refrigier et al (1987).

Finally, if $L_{w2} \ll L_{\Delta T} \ll L_{w1}$, (9.18) yields $L_w = L_{\Delta T}$, so both waiting times are irrelevant.

A way to analyze data, suggested by Fisher and Huse, is via the logarithmic derivative of the remanent magnetization with respect to $\ln t$:

$$\alpha_m(t) \equiv -\frac{d\ln m(t)}{d\ln t} \tag{9.19}$$

For the log-power decays of this theory, we would then have (from (9.12)

$$\alpha_m(t) \approx \frac{y/\psi}{\ln(t/\tau_0)} \tag{9.20}$$

(where we have restored explicitly the characteristic microscopic time $\tau_0 = O(10^{-12}\,\text{sec})$ in the short-time local equilibrium regime. For the simplest possibility (e.g. the McMillan assumption $\psi = y$), this gives an $\alpha_m \approx 0.03$ for observation times on the order of seconds, in at least rough agreement with experiment (Alba et al, 1987). In the long-time regime, where the dynamics are dominated by domain growth, we get the faster decay (9.13), leading to

$$\alpha_m(t) \approx \frac{\lambda/\psi}{\ln(t/\tau_0)} \tag{9.21}$$

Thus when t reaches t_w, the effective $\alpha_m(t)$ should get much bigger, in qualitative agreement with experiment (Refrigier et al, 1987; Alba et al,

1987; Nordblad et al, 1987). However, the largely qualitative nature of the theory makes it difficult to carry out quantitative experimental tests. This problem awaits further work, both theoretical and experimental.

The version of aging theory presented by Koper and Hilhorst is slightly different from that of Fisher and Huse, in that they assume a power law for the domain growth (as in the ferromagnetic example above) rather than the logarithmic power law (9.10) which emerged from the scaling theory of Chapter 8 and was also (therefore) adopted by Fisher and Huse. In our account here we have followed Fisher and Huse rather than Koper and Hilhorst, primarily because the power law growth assumption seems *ad hoc*. Nevertheless, the final test is experimental, and no decisive evidence in favour of one hypothesis or the other is available yet.

9.3   Phenomenological models

Because multi-timescale systems are not (at least until recently) very familiar to condensed-matter theorists, a number of people have constructed toy models which incorporate such dynamics in simple ways to gain some insight into the generic features of such systems and phenomena. In this section we explore one of the simplest of these models — a hierarchical (ultrametric) diffusion model due to Ogielski and Stein (1985) and, independently, to Schreckenberg (1985). We also mention some of the other models which have been found to have similar behaviour. Finally, we will show how models at this phenomenological level can be related to models at the level we usually deal with (such as the Ising EA model) in the context of the McMillan dynamical renormalization theory of Section 8.2.

These models are generally explicitly hierarchical. It is in this way that the multi-timescale nature of the problems is incorporated. It would of course be more satisfying to have models in which the hierarchical structure emerged naturally from more general conditions (such as randomness and frustration), but except in the case of the SK model this holy grail has generally eluded theorists. Nevertheless, it can still be very useful to study these models to see what consequences such hierarchical structure may have. They may also provide a framework for understanding experimental results. Indeed these models have nontrivial properties such as dynamical phase transitions which are not obviously put into them explicitly.

Most of these models are also formulated at a different level from conventional model descriptions of magnetic materials. That is, the 'sites' which occur in them do not generally represent the locations of spins, but rather points in the configuration or phase space of the system. The equation of motion is thus not a description of what happens to the spins; it is a (generally stochastic) description of how the point in configuration space describing the state of the system moves. Explicit assumptions are therefore necessary to tie such a description to measurable quantities. Fortunately, there are some simple assumptions about the connection with some quantities (like the spin autocorrelation function) which are natural for many systems.

Ultradiffusion: the Ogielski–Stein–Schreckenberg model

The simplest model in this category describes a 'particle' which can hop among a hierarchically grouped set of sites (Fig. 9.14). The probability per unit time to hop from one site to another depends on the ultrametric distance (how far up the family tree one has to go to find a common ancestor (cf. Section 3.4)) between them. Representing the (probabilistic) state of the system by a vector $\mathbf{P}$ with components P_i, the probability of being at site i, we then have the equation of motion

$$\frac{\partial \mathbf{P(t)}}{\partial t} = \epsilon \mathbf{P(t)} \tag{9.22}$$

where the transition rate matrix ϵ looks like

$$\epsilon = \begin{bmatrix} \epsilon_0 & \epsilon_1 & \epsilon_2 & \epsilon_2 & \epsilon_3 & \epsilon_3 & \epsilon_3 & \epsilon_3 \\ \epsilon_1 & \epsilon_0 & \epsilon_2 & \epsilon_2 & \epsilon_3 & \epsilon_3 & \epsilon_3 & \epsilon_3 \\ \epsilon_2 & \epsilon_2 & \epsilon_0 & \epsilon_1 & \epsilon_3 & \epsilon_3 & \epsilon_3 & \epsilon_3 \\ \epsilon_2 & \epsilon_2 & \epsilon_1 & \epsilon_0 & \epsilon_3 & \epsilon_3 & \epsilon_3 & \epsilon_3 \\ \epsilon_3 & \epsilon_3 & \epsilon_3 & \epsilon_3 & \epsilon_0 & \epsilon_1 & \epsilon_2 & \epsilon_2 \\ \epsilon_3 & \epsilon_3 & \epsilon_3 & \epsilon_3 & \epsilon_1 & \epsilon_0 & \epsilon_2 & \epsilon_2 \\ \epsilon_3 & \epsilon_3 & \epsilon_3 & \epsilon_3 & \epsilon_2 & \epsilon_2 & \epsilon_0 & \epsilon_1 \\ \epsilon_3 & \epsilon_3 & \epsilon_3 & \epsilon_3 & \epsilon_2 & \epsilon_2 & \epsilon_1 & \epsilon_0 \end{bmatrix} \tag{9.23}$$

for an 8-site model. We remark that (9.23) has the same kind of 'ultrametric' structure as the $q^{\alpha\beta}$ replica order parameter matrix (e.g. (3.67)) in the Parisi mean field theory of the SK model. That is, the rate to jump to a site an ultrametric distance j away is ϵ_j. In general, we denote the maximum ultrametric distance in the tree by n ($n = 3$ in (9.23)), so the number of sites is 2^n. By conservation of probability, the diagonal element $\epsilon_0 = -\sum_i^n 2^{i-1}\epsilon_i$.

The quantity we will be most interested in is the probability of remaining at the initial site 0: $P_0(t)$, given the initial condition $P_j(0) = \delta_{j,0}$. It is natural to identify this quantity with the spin autocorrelation function $\langle S_i(0)S_i(t) \rangle$ in a spin system we are trying to describe by this model.

A motivation for this kind of model might be something like this: We know that the states of the SK model really have an ultrametric structure.

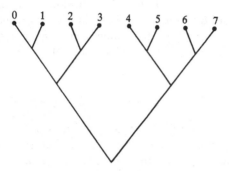

Figure 9.14: The ultrametric space of the Ogielski–Stein–Schrekenberg model. The distance between sites is just the number of steps toward the root of the tree before paths from the two sites converge. Thus for example sites 0 and 1 are separated by an ultrametric distance 1, sites 1 and 2 by a distance 2, and sites 3 and 4 by a distance 3. The 'ancestor sites' in the drawing are not part of the system; they are only drawn to indicate what the ultrametric distance measure is. The sites of the model are only those at the tips of the branches of the tree.

We cannot solve finite-range models exactly, but it is an interesting hypothesis that something like this ultrametric structure might carry over to more realistic models, albeit with finite barriers and hence finite relaxation times rather than the strict broken ergodicity of the SK model. Perhaps with the help of some further model assumptions, one can then see whether real spin glasses (or other complex systems we might suspect to be similar) behave according to the predictions of the model.

We note that no assumption is made here about the kind of ultrametricity we had in the SK model (of overlaps between configurations). Here one only assumes that the transition rates between states have this structure. It is not possible to make the connection to mean field theory more explicit without specific assumptions about a relationship between the mutual overlap between a pair of states and the transition rate between them.

We have to find the eigenvectors and eigenvalues of the general $2^n \times 2^n$ matrix ϵ. They are the following:

    1.   There is one eigenvalue $\eta_n = 0$, whose eigenvector has all elements equal. This is evident physically, since a uniform P_j will remain uniform.

    2.   There are 2^{n-1} degenerate eigenvectors which correspond to a

redistribution of the probability within pairs of ultrametric nearest neighbours; that is, they have value $\pm 1/\sqrt{2}$ for such pairs of sites and zero otherwise. Their common eigenvalue is easily computed to be $\eta_0 = \epsilon_0 - \epsilon_1$.

3. There are 2^{n-2} degenerate eigenvectors, with eigenvalue $\eta_1 = \epsilon_0 + \epsilon_1 - 2\epsilon_2$, each with nonzero components in a group of 4 sites which are ultrametric distance 1 or 2 from each other. They have the value $+\frac{1}{2}$ for two of the sites which are ultrametric nearest neighbours of each other and $-\frac{1}{2}$ for the other two. Thus these modes describe the redistribution of probability between one such pair and its closest relative.

4. Continuing more generally, there is a 2^{n-1-m}-fold degenerate eigenvalue

$$\eta_m = \epsilon_0 + \sum_{k=1}^{m} 2^{k-1}\epsilon_k - 2^m \epsilon_{m+1} \tag{9.24}$$

corresponding to redistribution of probability between two clusters of 2^m sites. The ultrametric distance between any pair of sites within each cluster is less than or equal to m and that between any site in one cluster and any site in the other is $m+1$. The last of these $(m = n - 1)$ corresponds to redistribution of probability between one half of the system and the other (the left and right halves, if we draw the system as in Fig. 9.14).

All these eigenvalues are negative except η_n, as required by the diffusive nature of the process. We can simplify things a little if we define the decay rates $\lambda_m \equiv -\eta_m > 0$ and $a_k \equiv 2^{k-1}\epsilon_k$, the total rate for hopping to *any* of the 2^{k-1} sites at a distance k. Then (9.24) becomes

$$\lambda_m = 2a_{m+1} + \sum_{k=m+2}^{n} a_k \quad (m < n - 1) \tag{9.25}$$

$$\lambda_{n-1} = 2a_n \tag{9.26}$$

From here it is just a little algebra to obtain the general solution

$$P_0(t) = 2^{-n} + \tfrac{1}{2}\sum_{m=0}^{n-1} 2^{-m}\exp(-2a_{m+1}t)\prod_{i=m+2}^{n}\exp(-a_i t) \tag{9.27}$$

Several special choices for the dependence of the transition rates on ultrametric distance are interesting. An interesting one is when the rates a_k decrease geometrically with k. If we write the a_k in terms of activation barriers as

$$a_k = \exp(-\beta\Delta_k) \tag{9.28}$$

this means that the barriers increase linearly with m:

$$\Delta_k = k\Delta \tag{9.29}$$

Then if we replace the sum (9.27) by an integral $P_0(t)$ can be evaluated exactly in terms of an incomplete gamma function. We are of course most interested in the leading behaviour for large t, which can be extracted very simply: On a timescale of the order of $a_k^{-1} = \exp(k\beta\Delta)$ it is the term with $m = k$ in (9.27) that dominates the sum. (All faster-decaying terms will have died away and all slower ones have such small amplitudes 2^{-m} as to be negligible.) Thus we find

$$P_0(t) \approx 2^{-T\ln t/\Delta} = t^{-T\ln 2/\Delta} \tag{9.30}$$

Remembering that if we have diffusion in a Euclidean space, $P_0(t) \propto t^{-d/2}$, we can describe this result by saying that there is a temperature-dependent effective dimensionality for this diffusion in an ultrametric space ('ultradiffusion' (Huberman and Kerszberg, 1985)) equal to $(2\ln 2)T/\Delta$.

If the barriers Δ_k grow too slowly with the ultrametric distance k, then the hopping becomes effectively infinite-ranged; $P_0(t)$ immediately spreads over all sites. 'Too slowly' means more slowly than $\ln k$.

The marginal case where $\Delta_k = \Delta \ln k$ (i.e. the rate $a_k \propto k^{-\beta\Delta}$) is quite interesting, because it is just here that one finds the famous Kohlrausch or 'stretched exponential' behaviour. Explicitly, evaluating (9.27), replaced by an integral, by steepest descents, we find (cf. (1.9) and (8.96))

$$P_0(t) \propto \exp(-t^{T/\Delta}) \tag{9.31}$$

In general, the faster the barriers grow with ultrametric distance, the slower the decay will be. For example, if Δ_k grows exponentially with k, $P_0(t)$ will be proportional to an inverse power of $\ln t$.

Other models for ultradiffusion

Ultradiffusion was first studied in a different model by Huberman and Kerszberg (1985), who considered the problem of diffusion in a *one-dimensional* array of sites between which barriers of different heights are arranged in a hierarchical fashion (Fig. 9.15). They considered the case where the barrier heights grew linearly with their hierachical order: $a_k \propto \exp(-\beta\Delta k)$.

We can see immediately that this model will have the same behaviour at low temperature as we found above for the corresponding kind of barrier growth in the Ogielski–Stein–Schreckenberg model. The reason is that at low T the rates for hopping over barriers of different heights will be so

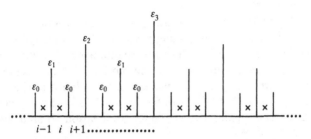

Figure 9.15: The Huberman–Kerszberg problem. A particle undergoes one-dimensional diffusion in a chain of sites separated by hierarchically arranged barriers.

different from each other that on any single timescale we can regard only one kind of hopping as active: Hopping over higher barriers will be so unlikely to have happened as to be negligible, while the probability will already have spread itself out evenly over all the sites which are only separated from each other by lower barriers. That is, on any timescale we can divide the system up into effectively noncommunicating cells. The size of these cells doubles at each succesive timescale as a new set of barriers can be climbed. But this is exactly the situation we had in the previous model at low lemperatures, leading to the power law (9.30) we found there.

This model can be solved exactly by real space renormalization (Teitel and Domany, 1985; Maritan and Stella, 1986). Doing this, one finds the decay

$$P_0(t) \propto t^{-\ln 2/(\ln 2 + \beta \Delta)} \tag{9.32}$$

and a *dynamical phase transition* at a critical temperature

$$T_c = \frac{\Delta}{\ln 2} \tag{9.33}$$

Above this temperature $P_0(t)$ goes over to the conventional result for one-dimensional diffusion $P_0(t) \propto t^{-\frac{1}{2}}$: The increase in barrier heights with level in the hierarchy is no longer strong enough to produce the more localized behaviour found at lower temperatures. (The exponent is continuous at T_c.)

Actually, the particular hierarchical arrangement of the barriers in this model is not essential to the model's properties. Well before this work it had been shown that the same kind of behaviour occurred in a chain with a *random* hopping rate distribution (Alexander et al, 1981). The important thing is just that the probability density of the hopping rate ϵ be proportional to $\epsilon^{-1+\ln 2/\beta\Delta}$ for small ϵ.

Kutasov et al (1986) have also studied a one-dimensional *Ising spin* model with bond strengths varying along the chain like the hopping rates in the Huberman–Kerszberg model. The low-temperature dynamical properties of an Ising chain are dominated by the motion of 'kinks' separating regions of up and down spins, and the motion of these kinks in this model is essentially the same as that of the diffusing particle in the Huberman–Kerszberg model. This model has no direct application to spin glasses, however.

Other people have studied another kind of ultradiffusion where (unlike the Ogielski–Stein–Schreckenberg model) the ancestor sites in Fig. 9.14 *are* sites onto which the particle can diffuse (Hoffmann et al, 1985, Grossmann et al, 1985; Sibani, 1986). This is a kind of random walk on a Cayley tree with an asymmetry in the rates toward and away from the root of the tree. Again, if one models the asymmetry of these rates proportional to $\exp(-\beta\Delta)$ in terms of an activation barrier Δ, the essential physics of the low-temperature limit is the same as in the other two models. The dynamics on a given timescale are dominated by processes on the lengthscale of regions within diffusive reach of each other on that timescale. Longer-range diffusion has not had a chance to get started, while well within these regions the probability has spread out essentially evenly. The same kind of power law in $P_0(t)$ is obtained.

A somewhat different model was introduced by Palmer et al (1984). It is different from all of the above ones in that the broad distribution of relaxation rates is not put in completely by hand. They stressed the idea that relaxation in complex materials involved serial as well as parallel processes: Slow or rare transitions in the relaxation process are rare because their occurrence depends on some particular configuration of the faster-relaxing degrees of freedom. In a (structural) glass, for example, we may imagine that there may be occasional processes involving coherent motion of clusters of atoms, but that they do not happen unless particular individual atoms get into the right configurations. We may suppose further that this kind of relationship repeats itself on successive levels involving bigger and bigger clusters. As we will see here, this kind of picture leads naturally to the sort of features and structure that we had in the models we considered above.

To be specific, let us assume a hierarchy of levels of conditional relaxation labelled by k. ($k = 0$ labels the shortest length scales and fastest processes, $k = 1$ the next fastest, and so on.) We think of each level as represented by a set of N_k (pseudo)spins with typical relaxation time τ_k.

We assume that a spin in the next level $k + 1$ can only flip if a set of μ_k of the level k spins assume a particular one of their 2^{μ_k} possible states. This sets the relaxation time for the level $k + 1$ spins:

$$\tau_{k+1} = 2^{\mu_k} \tau_k \tag{9.34}$$

so that

$$\tau_k = \tau_0 \exp(\sum_{j=0}^{k-1} \mu_j \ln 2) \tag{9.35}$$

In this way the kinds of hierarchies of relaxation times that we had above will occur naturally for suitable choices of the μ_k.

The spin autocorrelation function is then simply

$$C(t) \equiv \frac{1}{N} \sum_{i=1}^{N} \langle S_i(0) S_i(t) \rangle = \frac{1}{N} \sum_{k=0}^{\infty} N_k \exp(-t/\tau_k) \tag{9.36}$$

Further calculation depends on particular choices for the k-dependence of N_k and μ_k. Palmer et al considered a variety of possibilities, but for our purposes here the most interesting ones are those where N_k falls off exponentially with k

$$N_k \propto \exp(-Dk) \tag{9.37}$$

and μ_k either (a) is a constant μ_0, (b) falls off like an inverse power of k, or (c) grows exponentially with k.

Case (a) corresponds to linear barrier growth with order in the hierarchy, leading to power law decay like (9.30):

$$C(t) \propto t^{-d/\mu_0 \ln 2} \tag{9.38}$$

Case (b) is just the logarithmic growth of barriers which gave rise to the stretched exponential (9.31). Finally, case (c) gives decay like an inverse power of $\ln t$.

Connection with explicit models

Some insight into the relation between these sorts of models and more explicit ones can be gained by comparing expressions like (9.27) and (9.36) with expressions for the spin autocorrelation function in the McMillan renormalization theory of Sections 8.2 and 8.3. There we could evalute $C(t)$ as follows (McMillan, 1984): One must sum over all the steps (l) of renormalization (length scales) a product of the fraction of the degrees of freedom that remain on that lengthscale, the square of the effective moment of the block spins of this size, and the factor $\approx e^{-t/\tau_l}$ describing the relaxation of those block spins:

$$C(t) \propto \sum_l b^{-ld} b^{y_h} \exp(-t/\tau_l) \qquad (9.39)$$

Here b is the spatial rescaling factor and $y_h = \frac{1}{2}(d + 2 - \eta)$ is the exponent describing how the conjugate field (here, h^2) scales. The parallel with the expressions we have found for the models in this section is evident. So we have at least one theory at a more explicit level in terms of which we can identify the parameters of these models. For example, cases (a) and (c) for the model of Palmer et al above just correspond to the McMillan theory at and below T_f.

10

Specific heat, sound propagation, and transport properties

Specific heat, sound propagation, and transport properties have in common that they all are determined by the spin excitations, provided one subtracts contributions due to phonons and conduction electrons. For the specific heat, this is obvious since the heat capacity is a direct measure of the degrees of freedom of the system. The magnetic contribution to the resistivity and other transport properties of metallic spin glasses is due to the exchange coupling between the localized spins and the spins of the conduction electrons. This coupling leads to the scattering of conduction electrons by spin excitations and therefore to a resistivity which has some similarity with that due to electron–phonon scattering. The mechanism which leads to ultrasound attenuation and to a change of sound velocity is slightly more complicated. The sound waves modulate the exchange interactions J_{ij} between the spins and therefore 'feel' the spin excitations. In the specific heat and transport properties, the spin excitations enter in the form of integrals, which makes it plausible that no dramatic effects are seen in the freezing temperature. This is in contrast to sound propagation, which is directly determined by the dynamic susceptibility, as shown below.

10.1 Specific heat

We mentioned already in the Introduction that (at least at first sight) the magnetic contribution C_M to the specific heat shows no anomaly at the

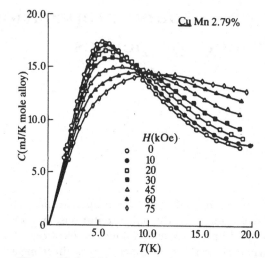

Figure 10.1: Magnetic contribution of the specific heat of CuMn 2.79% vs temperature in various fields (from Brodale et al, 1983).

freezing temperature T_f (see Fig. 1.8). This was taken as evidence against a possible phase transition. Rather, the specific heat has a broad maximum above T_f. For $T \ll T_f$, it shows a linear temperature dependence, as observed in CuMn, AuFe, EuSrS and other spin glasses (Wenger and Keesom 1976; Martin 1979, 1980a,b; Meschede et al, 1980). Experimentally, it is not easy to determine C_M accurately, since it is only a small fraction of the total specific heat. The latter consists of vibrational, electronic and magnetic parts, and the subtraction of the low-temperature electronic contribution (which is also proportional to T) is especially difficult.

The specific heat yields information about the excitations of a spin glass. It can be used to find the entropy difference up to a given temperature. One observes that a considerable amount of the total entropy (of the order of 70%) develops above T_f or is hidden in the broad maximum. Apparently, a considerable fraction of the spin degrees of freedom already freeze out in this temperature region. This holds for metallic and nonmetallic spin glasses and must be due to the formation of short range order, as mentioned in Section 8.4. A magnetic field reduces the maximum in $C_M(T)$ and enhances the specific heat at higher temperatures (Fig. 10.1): The spins now also have to overcome the field energy in order to rotate freely.

The absence of dramatic changes in $C_M(T)$ at T_f has several reasons.

The change of the internal energy might be too small to be observable, i.e. the magnetic anomalies such as the cusp in the susceptibility are produced by comparatively few degrees of freedom. A second reason is that the specific heat critical exponent is strongly negative ($\alpha \approx -2$) so that an anomaly shows up only in higher derivatives. The two reasons are actually related because the proximity to the lower critical dimensionality underlies them both. Very careful experiments on a 2.79% CuMn sample indeed show an anomaly in $\partial^2(C_M/T)/\partial T^2$ at T_f from which a peak in C_M/T can be extracted (Fig. 10.2). However, this peak is broad on a temperature scale on which the cusp in the susceptibility is rather sharp. Whereas the cusp in χ_{ac} is rounded off by extremely small fields (typically 10 to 100 Oe), the peak in $C_M(T)/T$ in a field of 75 kOe is only shifted to lower temperatures and becomes smaller. This shift obeys an h^2-law, as shown in the inset of Fig. 10.2d. Most likely, this field dependence agrees only accidentally with that of the GT line (5.30) (Fogle et al 1983).

A linear magnetic specific heat at low temperatures is easily explained in terms of modes μ with energy ϵ_μ which obey Bose statistics with the occupation number $\langle n(\epsilon_\mu) \rangle = [\exp(\beta\epsilon_\mu) - 1]^{-1}$ and with a density of states $\rho(\epsilon) \to$ const for $\epsilon \to 0$. At low temperature these modes lead to

$$E_M = \int_0^\infty d\epsilon \langle n(\epsilon) \rangle \epsilon \rho(\epsilon) \approx (k_B T)^2 \rho(0) \int_0^\infty dx\, x \langle n(x) \rangle,$$

$$x = \beta\epsilon \tag{10.1}$$

and thus to $C_M \propto T$ for $T \to 0$. The same linear relation is also obtained for 'two-level' systems in which a particle can tunnel between two states of roughly the same energy. One can again apply (10.1). However, because of the two levels, the Bose distribution now has to be replaced by the Fermi distribution (which leads to a different numerical constant). A phenomenological model with two-level systems 'explains' a considerable number of low-temperature experiments on real glasses, such as the specific heat, thermal conductivity, ultrasound propagation and damping, and possibly even a resistance anomaly (see von Löhneysen (1981) for details). For spin glasses, there seems to be little need to invoke such phenomenological two-level systems. In contrast to glasses, the basic interactions which lead to the spin glass structure and dynamics are well understood and tunneling processes are most likely irrelevant in spin glasses. We rather believe that in a single metastable state or 'valley', there is a fairly rich spectrum of spin excitations, as discussed below. The only possible mechanism which has some similarity with two-level systems would then be inter-valley transitions. The simplest example of that would be the change of helicity which we discussed in Section 7.1 for the XY model. However, this mechanism

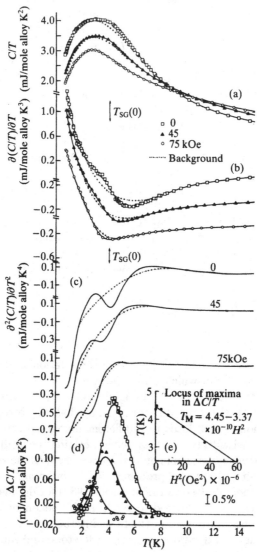

Figure 10.2: Specific heat $C(T)$ divided by T, and derived quantities for CuMn 0.279% vs temperature. The points represent data points in (a) and (d), and ratios of the differences between pairs of points separated by one other point to the corresponding temperature difference in (b). The solid curves represent spline fits to C/T in (a) and (d), spline fits to the ratios of differences in (b), and the derivative of the spline fit to the ratios of differences in (c). The dotted curves in (a)–(c) and the horizontal line through the origin in (d) represent the background heat capacity. The anomaly in C, ΔC, is shown in (d), and the locus of its maxima in (e) (from Fogle et al, 1983).

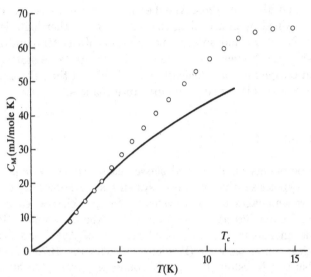

Figure 10.3: Specific heat vs temperature from computer simulations of dilute classical spins with RKKY interactions randomly embedded in an fcc lattice (solid line). The circles are data for CuMn 1.2% from Wenger and Keesom (1976). (From Walker and Walstedt, 1977.)

seems to be irrelevant for the specific heat and for all transport properties, though it is relevant for the decay of the remanent magnetization or other slow relaxation phenomena.

Evidence that the specific heat is due to intra-valley excitations comes from the computer simulations of dilute Heisenberg spin glasses with RKKY interactions mentioned in Section. 7.4 (Walker and Walstedt, 1977, 1980). In these simulations, the system goes into a metastable state and one investigates excitations within this state by diagonalization of the dynamic matrix. The corresponding density of states $\rho(\epsilon)$ turns out to be peaked at rather low energies ϵ and is possibly finite at $\epsilon = 0$. Fig. 10.3 shows the specific heat found when this density of states is inserted into (10.1). The agreement with the experimental data as shown in Fig. 1.8 (Wenger and Keesom, 1976) is rather good, except at higher temperatures where presumably the approximation of linear oscillations breaks down. In agreement with experimental data on CuMn and AuFe at lower temperatures (Martin, 1979, 1980a,b), the specific heat turns out not to be strictly linear in T.

Similar intra-valley processes presumably also lead to a T^2-law in the NMR spectrum and to a reversible contribution ΔM_R to the remanence $M_R(T)$ (Alloul and Mendels, 1985; Alloul et al, 1986; Mendels et al, 1987). The latter is obtained by first cooling the sample in a rather high field to a temperature $T_{fc} < T_f$ and subsequent variation of the temperature in zero field. Again, the system first is trapped in a metastable state (from which it would escape after a sufficiently long time) and then the amount of reversible excitations is varied by temperature changes.

10.2 Ultrasound

Ultrasound experiments on *structural glasses* have been very important in testing the hypothesis that two-level systems are responsible for the low-temperature anomalies of these systems. In spin glasses, so far no low-temperature anomalies have been observed in the ultrasound. However, one finds an anomaly in the sound velocity of AuCr and CuMn near the freezing temperature T_f (Hawkins and Thomas, 1978; Hawkins et al, 1979; Hsu and Marston, 1987). The sound propagation in metallic and nonmetallic spin glasses is different: In metallic spin glasses, no frequency dependence of the sound velocity anomaly or damping of the sound wave has been observed, in contrast to the semiconductors EuSrS, ZnMnSe and $(CoF_2)_{0.5}(BaF_2)_{0.2}(NaPO_3)_{0.3}$, which all show both effects (Baumann, 1981 (private communication); Mayanovic et al, 1988; Doussineau et al, 1988). This is similar to the ac susceptibility, where the dissipative part $\chi''(\omega)$ and the frequency dependence of the cusp temperature in $\chi'(\omega)$ are small in metallic systems and considerably larger in nonmetallic ones (as mentioned in Section 5.2). We will show that there is indeed a close relation between sound propagation and the local susceptibility $\chi(\omega)$. Both the cusp in $\chi'(\omega)$ and the change of sound velocity $\Delta v(\omega)$ are smeared out near T_f at high frequencies ($\omega \gtrsim 100$ MHz). For the susceptibility in $Eu_{0.43}Sr_{0.57}S$, this has been shown directly (Singh et al, 1983) and for CuMn 5 at% by neutron scattering (Mezei, 1982).

We now describe briefly how the change in sound propagation due to coupling to the spin system depends on the dynamic susceptibility (Bennett and Pytte, 1967; Hertz et al, 1981; Khurana, 1982; Fischer, 1983a; Beton and Moore 1983). In deriving this relation, we follow closely Fischer (1983a).

The sound waves 'see' the spin system due to the modulation of the exchange interaction $J_{ij} = J(\mathbf{R}_i - \mathbf{R}_j)$. To first order in the atomic displacements $\mathbf{u}_i(t) = \mathbf{R}_i(t) - \mathbf{R}_i^0$ out of the positions $\mathbf{R}_i^0$ and in the absence of sound one has

$$\delta J_{ij} = J_{ij} - J_{ij}^0 = (u_{i\mu} - u_{j\mu})\nabla_{i\mu}J_{ij} \tag{10.2}$$

with repeated indices $\mu = x, y, z$ summed over. The phonons are described in harmonic approximation by the Hamiltonian

$$H_{ph} = \sum_i \frac{p_i^2}{2M} + \tfrac{1}{2}\sum_{ij} \phi_{ij,\mu\nu} u_{i\mu}u_{j\nu} \tag{10.3}$$

with the coupling constants $\phi_{ij,\mu\nu} = \nabla_{i\mu}\nabla_{j\nu}\phi(\ldots \mathbf{R}_m \ldots)$. The total Hamiltonian

$$H = H_{ph} + H_s + H_{int} \tag{10.4}$$

consists of the phonon contribution H_{ph}, the spin contribution (2.5), and the interaction term

$$H_{int} = -\tfrac{1}{2}\sum_{ij}(u_{i\mu} - u_{j\mu})\nabla_{i\mu}J_{ij}\mathbf{S}_i \cdot \mathbf{S}_j \tag{10.5}$$

We treat the phonons quantum-mechanically and introduce the Green's function $D_{ij,\mu\nu}(\tau, \tau')$ for imaginary times $\tau = it$, $\tau' = it'$,

$$D_{ij,\mu\nu}(\tau, \tau') = -\langle T_\tau(u_{i\mu}(\tau)u_{j\nu}(\tau'))\rangle, \quad 0 \le \tau, \tau' \le \beta \tag{10.6}$$

with the time ordering operator T_τ and where $\langle \ldots \rangle$ is a trace taken with the full Hamiltonian H (see Abrikosov et al, 1963; Mahan, 1981). (The imaginary times $\tau = it$ or frequencies $i\omega_n$ have been introduced in order to do perturbation theory at finite temperatures.) The equation of motion for $D_{ij,\mu\nu}$ reads for $\tau' = 0$

$$M\frac{\partial^2}{\partial\tau^2}D_{ij,\mu\nu}(\tau) - \sum_l \phi_{il,\mu\sigma}D_{lj,\sigma\nu}(\tau) = \delta_{ij}\delta_{\mu\nu}\delta(\tau)$$

$$+ \sum_l \nabla_{i\mu}J_{il}\langle T_\tau(\mathbf{S}_i(\tau) \cdot \mathbf{S}_l(\tau)u_{j\nu}(0))\rangle \tag{10.7}$$

The interaction term on the right-hand side of (10.7) is treated in lowest order perturbation theory, which leads to

$$\frac{\partial^2}{\partial\tau^2}D_{ij,\mu\nu}(\tau) - \frac{1}{M}\sum_l \phi_{il,\mu\sigma}D_{lj,\sigma\nu}(\tau) = \frac{1}{M}\delta_{ij}\delta_{\mu,\nu}\delta(\tau)$$

$$- \int_0^\beta d\tau_1 \sum_{lmr} \Sigma_{ilmr,\mu\sigma}(\tau - \tau_1)(D_{mj,\sigma\nu}(\tau_1) - D_{rj,\sigma\nu}(\tau_1)) \tag{10.8}$$

with the self-energy

$$\Sigma_{ilmr,\mu\sigma}(\tau) =$$

$$(2M)^{-1}\nabla_{i\mu}J_{il}\nabla_{m\sigma}J_{mr}\langle T_\tau(\mathbf{S}_i(\tau)\cdot\mathbf{S}_l(\tau))(\mathbf{S}_m(0)\cdot\mathbf{S}_r(0))\rangle_0 \quad (10.9)$$

Here the average $\langle\ldots\rangle_0$ is taken for $H_{int}=0$. We learn from (10.9) that the sound couples to the spin system via a 4-spin correlation function. In the simplest approximation (the 'random phase approximation' (RPA)), one writes

$$\langle T_\tau(S_{i\mu}(\tau)S_{l\mu}(\tau)S_{m\nu}(0)S_{r\nu}(0)\rangle_0 =$$

$$\langle T_\tau(S_{i\mu}(\tau)S_{m\nu}(0))\rangle_0\langle T_\tau(S_{l\mu}(\tau)S_{r\nu}(0))\rangle_0$$

$$+\langle T_\tau(S_{i\mu}(\tau)S_{r\nu}(0))\rangle_0\langle T_\tau(S_{l\mu}(\tau)S_{m\nu}(0))\rangle_0 \quad (10.10)$$

where we dropped equal-time terms since they lead to static terms in the self-energy and do not contribute to a modification of the sound waves. The latter induce spin fluctuations $\delta\mathbf{S}_i = \mathbf{S}_i - \mathbf{S}_i^0$ out of the thermal equilibrium positions $\mathbf{S}_i^0 = \langle\mathbf{S}\rangle$. We learned already in Section 4.4 that the time dependent spin correlations $\langle\mathbf{S}_i(0)\mathbf{S}_j(t)\rangle - \langle\mathbf{S}_i\rangle\langle\mathbf{S}_j\rangle \equiv \langle\delta\mathbf{S}_i(0)\delta\mathbf{S}_j(t)\rangle$ are connected to the dissipative part χ''_{ij} of the susceptibility via the fluctuation-dissipation theorem (4.44). For imaginary times τ, one has

$$[\langle T_\tau(\delta S_{i\mu}(\tau)\delta S_{m\nu}(0))\rangle_0]_{av} = \chi_{im,\mu\nu}(\tau)$$

$$= T\sum_n e^{-i\omega_n\tau}\chi_{im,\mu\nu}(i\omega_n) \quad (10.11)$$

with the frequencies $\omega_n = 2\pi Tn$, $n = 0, \pm 1, \pm 2, \ldots$ (see Abrikosov et al, 1963). We can now express the self-energy (10.9) in terms of frequency-dependent susceptibilities. In order to do so, we have in principle to introduce susceptibilities which are not yet configuration-averaged. In the simplest approximation (which is in the spirit of the RPA (10.10)) we average all terms separately. We then consider the Fourier transform of the self-energy

$$\Sigma_{ilmr,\mu\nu}(i\omega_n) = \int_0^\beta d\tau\, e^{i\omega_n\tau}\Sigma_{ilmr,\mu\nu}(\tau) \quad (10.12)$$

This leads to

$$\Sigma_{ilmr,\mu\nu}(i\omega_n) = \frac{1}{M}[\nabla_{i\mu}J_{il}\nabla_{m\sigma}J_{mr}]_{av}(2[S_{i\alpha}^0 S_{m\beta}^0]_{av}\chi_{lr,\alpha\beta}(i\omega_n)$$

$$+T\sum_{n'}\chi_{im,\alpha\beta}(i\omega_{n'})\chi_{lr,\alpha\beta}(i\omega_n - i\omega_{n'})) \quad (10.13)$$

where we omitted terms proportional to $(\mathbf{S}_i^0 \cdot \mathbf{S}_l^0)(\mathbf{S}_m^0 \cdot \mathbf{S}_r^0)$ which again do not modify the sound waves. Now, we go back to real frequencies ω by means of the analytic continuation $i\omega_n \to z = \omega + i\eta$, $\eta = 0_+$, writing the susceptibility $\chi(z)$ in the spectral representation

$$\chi(z) = \int_{-\infty}^{\infty} \frac{d\omega'}{\pi} \frac{\chi''(\omega')}{\omega' - z} \tag{10.14}$$

and use the identity

$$T \sum_{\omega_n' = -\infty}^{\infty} \frac{1}{\omega - i\omega_n'} = \frac{1}{e^{\beta\omega} - 1} + \tfrac{1}{2} \tag{10.15}$$

to replace the remaining frequency sum by an integral with the Bose factor $n(\omega) = (\exp(\beta\omega) - 1)^{-1}$. This yields

$$T \sum_{n'} \chi(i\omega_{n'})\chi(z - i\omega_{n'}) =$$

$$\int_{-\infty}^{\infty} \frac{d\omega'}{\pi} \int_{-\infty}^{\infty} \frac{d\omega''}{\pi} \chi''(\omega')\chi''(\omega'') \frac{n(\omega') - n(\omega'')}{\omega'' - \omega' + z} \tag{10.16}$$

This expression simplifies considerably when we consider the classical limit $n(\omega) \approx T/\omega$ and apply the fluctuation–dissipation theorem, together with (10.14). Then (10.16) leads to

$$T \sum_{n'} \chi(i\omega_{n'})\chi(z - i\omega_{n'}) = \int_{-\infty}^{\infty} \frac{d\omega'}{\pi} C(\omega')\chi(\omega - \omega') - 2q\chi(\omega) \tag{10.17}$$

with the spin glass order parameter q. If we now restore all the site and spin component indices, the second term in (10.17) exactly cancels the first term in (10.13). Now we go back to (10.13) and (10.8). The Fourier transform of (10.8) with respect to space and time leads, after analytic continuation, to

$$(M D_\alpha(\mathbf{k}, \omega))^{-1} = \omega^2 - \omega_\alpha^2(\mathbf{k}) + \Sigma_\alpha(\mathbf{k}, \omega) \tag{10.18}$$

where $D_{\mu\nu}(\mathbf{k}, \omega) = \Sigma_\alpha D_\alpha(\mathbf{k}, \omega) e_\alpha^\mu(\mathbf{k}) e_\alpha^\nu(\mathbf{k})$ with the polarization vectors $e_\alpha^\mu(\mathbf{k})$ and the eigenfrequencies $\omega_\alpha(\mathbf{k}) = v_\alpha^0 k$ for the unperturbed sound wave. The spin system leads to a shift Δv_α of the velocity v_α^0 and to the damping γ_α of the sound waves

$$\Delta v_\alpha(\omega, T) = \frac{(v_\alpha^2 - v_\alpha^0)}{2v_\alpha^0} = -\frac{1}{2k^2 v_\alpha^0} Re\Sigma_\alpha(\mathbf{k}, \omega) \tag{10.19}$$

$$\gamma_\alpha(\omega, T) = \frac{1}{k^2\omega} Im\Sigma_\alpha(\mathbf{k}, \omega) \tag{10.20}$$

with the self-energy

$$\Sigma_\alpha(\mathbf{k}, \omega) =$$

$$\frac{e_\alpha^\mu e_\alpha^\nu}{2N} \sum_{ilmr} \Sigma_{ilmr,\mu\nu}(\omega)(e^{-i\mathbf{k}\cdot\mathbf{R}_i} - e^{-i\mathbf{k}\cdot\mathbf{R}_l})(e^{i\mathbf{k}\cdot\mathbf{R}_m} - e^{i\mathbf{k}\cdot\mathbf{R}_r}) \tag{10.21}$$

(We now write $\mathbf{R}$ instead of $\mathbf{R}^0$ for the unperturbed lattice sites.) For sound waves $|\mathbf{k}\cdot\mathbf{R}| \ll 1$ holds, and the last factor in (10.21) simplifies to $(\mathbf{k}\cdot(\mathbf{R}_i - \mathbf{R}_l))(\mathbf{k}\cdot(\mathbf{R}_m - \mathbf{R}_r)) \approx \frac{1}{3}k^2(\mathbf{R}_i - \mathbf{R}_l)\cdot(\mathbf{R}_m - \mathbf{R}_r)$. In a spin glass model with a symmetric bond distribution, the configuration-averaged susceptibility $\chi_{lr,\alpha\beta}(\omega)$ is site-independent and can be chosen to be diagonal in the coordinates α, β. For isotropic spin glasses, we have $\chi_{lr,\alpha\beta}(\omega) = \chi(\omega)\delta_{lr}\delta_{\alpha\beta}$. With

$$\tilde{J}^2 \equiv \sum_h R_h^2(\nabla J_h)^2, \quad \mathbf{R}_h = \mathbf{R}_i - \mathbf{R}_l \tag{10.22}$$

this leads to our final result

$$\Sigma_\alpha(\mathbf{k}, \omega) = \frac{k^2\tilde{J}^2}{3M} \int_{-\infty}^{\infty} d\omega'\pi C(\omega')\chi(\omega - \omega')$$

$$= \frac{k^2\tilde{J}^2}{3M}\left[2q\chi(\omega) + \int_{-\infty}^{\infty} \frac{d\omega'}{\pi} \tilde{C}(\omega')\chi(\omega - \omega')\right] \tag{10.23}$$

where the correlation function $\tilde{C}(\omega)$ is defined by (5.21).

It tells us that both the velocity shift (10.19) and the damping (10.20) of the sound wave depend on the susceptibility $\chi(\omega)$, where the correlation function $C(\omega)$ can be expressed in terms of $\chi''(\omega)$ and the spin glass parameter q by means of the fluctuation–dissipation theorem (4.45). In order to estimate the integral in (10.23), we go back to the MFT and expand $\chi(\omega) = \chi_{loc}(\omega)$ in eigenstates of the Hessian $A_{ij} = \partial^2\beta F/\partial m_i\partial m_j$ (3.105). For relaxational dynamics (such as Glauber or Langevin dynamics) (3.110) then has to be generalized to

$$\chi(\omega) = \frac{\beta}{N} \sum_\lambda \frac{1}{r_\lambda - i\omega\tau_0} = \int_0^\infty dr \frac{\rho(r)}{r - i\omega\tau_0} \tag{10.24}$$

(A proof of (10.24) can be found in Beton and Moore (1983).) We are particularly interested in the temperature and frequency dependence of $\chi(\omega)$ near T_f where the density $\rho(r)$ varies for small r as $\rho(r) \approx \rho_0 r^{\frac{1}{2}}$. With the fluctuation–dissipation theorem, this leads to the self-energy (10.23)

$$\Sigma_\alpha(\mathbf{k}, \omega) = \frac{\tilde{J}^2 k^2}{3M} [2q(\chi_{loc}(0) - \rho_0\pi(-i\omega)^{\frac{1}{2}})$$

$$+\chi_{loc}^2 T - (-2i)^{\frac{1}{2}}\pi T\rho_0^2\omega|\ln\omega|] \tag{10.25}$$

with $\chi_{loc} = \beta(1 - q)$ from (3.109). (The first two terms on the right-hand side of (10.25) are proportional to $\chi_{loc}(\omega)$ to order $\omega^{\frac{1}{2}}$.) The shift Δv_α (10.19) turns out to be frequency independent to lowest order, whereas the damping factor γ_α varies proportionally to $|\ln\omega|$ for $T > T_f$ and diverges like $\omega^{-\frac{1}{2}}$ at and just below T_f. In MFT, $\rho(r) \approx \rho_0 r^{\frac{1}{2}}$ for all $T \leq T_f$, and this divergence should persist in this temperature range. However, the MFT predicts a cusp in $\Delta v(T)$ at T_f which is not observed.

The result (10.23) is not restricted to MFT. We can obtain a more realistic estimate of the frequency dependence of Δv_α and γ_α by using susceptibilities and correlation functions from the scaling theory presented in Chapter 8. For $T < T_f$, the term proportional to $q\chi_{loc}(\omega)$ yields the strongest frequency dependence. From the spin autocorrelation function $\tilde{C}(\omega)$ (8.52) and the fluctuation–dissipation theorem $\chi_{loc}''(\omega) \propto |\ln\omega|^{-(1+y/\psi)}$. The real part $\chi_{loc}'(\omega)$ can then be obtained through the Kramers–Kronig relation. For slowly varying functions, one has approximately $\chi_{loc}' \approx \chi_{loc}(\omega = 0) - \beta\tilde{C}(t = 1/\omega)$ or, with (8.51), $\chi_{loc}'(\omega) = \chi_{loc}(\omega = 0) - \text{const}|\ln\omega|^{-y/\psi}$ (Fisher and Huse, 1988b). This leads to a logarithmic frequency dependence of the sound velocity shift (10.19) and to a damping γ_α of the sound wave proportional to $\omega^{-1}|\ln\omega|^{-(1+y/\psi)}$. (Fisher and Huse (1988b) estimate $y/\psi \approx 1$.) Unfortunately, the experimental data are still too inaccurate to allow for a detailed comparison with these theoretical predictions.

10.3 Electrical resistivity and thermopower

The temperature dependence of the electrical resistivity of metallic spin glasses has some similarity to that of the specific heat. There seems to be no anomaly at the freezing temperature T_f. All data show a broad maximum at a temperature T_{max} well above T_f which, however, does not coincide with the maximum of the specific heat. At low temperatures, one observed a T^2 or $T^{3/2}$ behaviour. The $T^{3/2}$ behaviour holds approximately for $0.1T_f \leq T \leq T_f$. The T^2 law is observed at very low temperatures, at least in $\underline{Cu}Mn$ and $\underline{Rh}Fe$ (Mydosh et al, 1974; Ford and Mydosh, 1976; Laborde and Radhakrishna, 1973; Campbell et al, 1982). This temperature dependence holds when the contribution to the resistivity from electron–phonon scattering ρ_{ph} has been subtracted. Unfortunately, this subtraction procedure does not yield directly the magnetic contribution ρ_M: One has deviations from Matthiessen's rule $\Delta\rho = \rho - \rho_{ph} - \rho_M$ which are strongest if ρ_M and ρ_{ph} are of the same order of magnitude and are difficult to assess at high temperatures. For this reason, the significance of a $\ln T$-dependence observed above the resistivity-maximum temperature T_{max} is hard to evaluate. Such a decrease of resistivity with increasing temperature is characteristic of the Kondo effect. We will show below that the thermopower yields even more convincing evidence that the Kondo effect exists, though in modified form, in spin glasses.

The Kondo effect is normally observed in very dilute magnetic alloys with negative exchange J_{sd} between the host conduction electrons and the impurity spins. In its ideal form, it is a single-impurity effect and leads to a resistivity maximum at $T = 0$. It can be described by an effective interaction $J_{sd}^{eff}(T)$ which increases with decreasing temperature and finally leads to a singlet state formed by a single impurity spin and the spins of the surrounding conduction electrons (Kondo, 1969; Nozières, 1974; Wilson, 1975). At larger concentrations, the impurity–spin interactions become important and lead to a partial destruction of this singlet state. The impurities again become 'magnetic'. Conventional spin glasses have typically a few percent magnetic impurities, and the Kondo effect seems to be irrelevant for the formation of the spin glass state since $T_f \gg T_K$. Here, the 'Kondo' temperature T_K is characteristic for the crossover from a magnetic to a nonmagnetic single-impurity state. The impurity–spin interactions lead to a resistivity maximum at a finite temperature T_{max} even in the Kondo regime. With increasing concentration, T_{max} moves to larger temperatures and exceeds T_f in spin glasses. This picture has been confirmed in $\underline{Cu}Mn$ with 0.1 to 1 at% Mn and $30 mK \leq T \leq 4K$ by Laborde and Radhakrishna (1973). Thus the resistivity maximum in spin glasses is indeed a remnant of the Kondo effect.

Below T_{max}, the spin glass resistivity $\rho_m(T)$ is determined by the spin excitation spectrum $\chi_{ij}''(\omega)$ or dynamic susceptibility, and the static spin disorder. In order to show this, we have to investigate the scattering of the conduction electrons by the magnetic impurities. Each impurity leads to spin-independent scattering by a potential V and to spin-dependent scattering by the exchange interaction. We assume that the reader is familiar with standard Boltzmann transport theory and with the formalism of 'second quantization' (see, for instance, Mahan 1981). The total Hamiltonian

$$H = H_0 + H_s + H_{sd} \qquad (10.26)$$

of the electron-impurity system consists of three parts. We have the energy of the conduction electrons for which we assume a single band

$$H_0 = \sum_{\mathbf{k},s} \epsilon_{\mathbf{k}} c_{ks}^+ c_{ks} \tag{10.27}$$

where c_{ks}^+ creates a Bloch electron in the state $\mathbf{k}, s$ with energy ϵ_k and spin index $s = \pm 1$. The spin-interaction term H_s has been defined in (2.4) for a model with random sites, and the interaction between the conduction electrons and impurities at the sites $\mathbf{R}_i$ is written as

$$H_{sd} = \frac{1}{N} \sum_{\substack{\mathbf{kk'} \\ \sigma\sigma'}} \sum_i c_i e^{i(\mathbf{k'}-\mathbf{k})\cdot\mathbf{R}_i} c_{\mathbf{k}'s'}^+ c_{\mathbf{k}s} (V\delta_{ss'} - \tfrac{1}{2} J_{sd} \mathbf{S}_i \cdot \vec{\sigma}_{ss'}) \tag{10.28}$$

where N is the number of electrons and $\vec{\sigma}$ are the Pauli matrices. We assume J_{sd} and V to be small and treat the scattering in Born approximation. The transition probability for a scattering event $\mathbf{k}s \to \mathbf{k}'s'$ is calculated by means of the 'Golden Rule'

$$\Gamma_{\mathbf{k}s \to \mathbf{k}'s'} =$$

$$f_{\mathbf{k}s}(1 - f_{\mathbf{k}'s'}) \frac{1}{N^2} \sum_{ij} c_i c_j e^{i(\mathbf{k'}-\mathbf{k})\cdot(\mathbf{R}_i - \mathbf{R}_j)} \int d\omega \delta(\omega - \epsilon_{\mathbf{k}s} + \epsilon_{\mathbf{k}'s'})$$

$$\times \int_{-\infty}^{\infty} dt\, e^{i\omega t} [\tfrac{1}{4} J_{sd}^2 \langle \mathbf{S}_i(t) \cdot \mathbf{S}_j(0) \rangle + V^2 + s J_{sd} V \langle S_{iz} \rangle] \tag{10.29}$$

(Mills and Lederer, 1966). Here, $s = \pm 1$, $f_{\mathbf{k}s}$ is the Fermi function and the time dependence of $\mathbf{S}_i(t)$ is defined by the spin interaction term H_s

$$\mathbf{S}_i(t) = c^{iH_s t} \mathbf{S}_i c^{-iH_s t} \tag{10.30}$$

The quantization axis z is taken to be the same for all spins and conduction electrons. The transport relaxation time $\tau_{\mathbf{k}s}$ is obtained by means of the Boltzmann equation for the conduction electrons

$$-\frac{1}{\tau_{ks}} \frac{\partial f_{ks}}{\partial \epsilon_{ks}} = \frac{1}{T} \sum_{k's} [\Gamma_{\mathbf{k}s \to \mathbf{k}'s'}]_{av} (1 - \cos\theta_{\mathbf{k}\mathbf{k}'}) \tag{10.31}$$

where $\theta_{\mathbf{k}\mathbf{k}'}$ is the scattering angle and where we averaged the transition probability over the site occupation numbers $c_i = (0,1)$ (Ziman, 1960; Mahan, 1981). The fluctuating part of the spin correlations in (10.29) can be expressed in terms of the dissipative susceptibility $\chi_{ij}''(\omega)$. In order to do this, we have to generalize the fluctuation–dissipation theorem (4.44) to quantum spins with the components $\mu, \nu = x, y, z$

$$\frac{\chi_{ij,\mu\nu}(\omega) - \chi_{ij,\mu\nu}(-\omega)}{i(1 - e^{-\beta\omega})} =$$

$$\int_{-\infty}^{\infty} dt\, e^{i\omega t} [\langle S_{i\mu}(0)S_{j\nu}(t)\rangle - \langle S_{i\mu}\rangle\langle S_{j\nu}\rangle] \tag{10.32}$$

(Kadanoff and Martin, 1962). For not too concentrated spin glasses, one needs only calculate the averaged transition probability in (10.31) to first order in the concentration c, apart from the concentration dependence of $\chi_{ij}''(\omega)$. One has then only to consider terms $i = j$ and the second term under the integral in (10.32) reduces to the spin glass order parameter (2.71). In this approximation, the transition probability (10.29) becomes independent of $\mathbf{k}$ and $\mathbf{k}'$, and one finds from (10.31) with $f_{ks}(1 - f_{ks}) = -T\partial f_{ks}/\partial\epsilon_{ks}$ and the density of states $N(\epsilon_k) = N^{-1}\sum_{\mathbf{k}'s}\delta(\epsilon_{\mathbf{k}s} - \epsilon_{\mathbf{k}'s})$ the elastic inverse relaxation time

$$\tau_{el,s}^{-1}(\epsilon) = c\pi N(\epsilon)\left(\frac{1}{4}J_{sd}^2 q(T) + V^2 + sJ_{sd}VN^{-1}\sum_i\langle S_{iz}\rangle\right)$$

$$\equiv \tau_{el}^{-1} + s(\tau_{el}')^{-1} \tag{10.33}$$

with $q = \sum_\mu q_{\mu\mu}$. The last term on the right-hand side of (10.33) vanishes in zero field but yields an important contribution to the magnetoresistivity. The inelastic relaxation time is independent of the spin direction s and (10.29)–(10.32) leads to

$$-\frac{\partial f}{\partial\epsilon}\tau_{inel}^{-1}(\epsilon) =$$

$$\frac{1}{4T}J_{sd}^2 N(\epsilon)\int_{-\infty}^{\infty} d\omega\, \frac{f(\epsilon - \omega) - f(\epsilon)}{(e^{\beta\omega} - 1)(1 - e^{-\beta\omega})}\frac{1}{N}\sum_{i,\mu}\chi_{ii,\mu}''(\omega) \tag{10.34}$$

where we used the identity

$$f(\epsilon)(1 - f(\epsilon - \omega)) = \frac{f(\epsilon) - f(\epsilon - \omega)}{1 - e^{\beta\omega}} \tag{10.35}$$

The electrical resistivity $\rho = (e^2 K_0)^{-1}$ is obtained from the transport integrals

$$K_n = \frac{n_0}{2m^*}\sum_{s=\pm 1}\int d\epsilon\,\epsilon^n \tau_s(\epsilon)\left(-\frac{df}{d\epsilon}\right) \tag{10.36}$$

for $n = 0$ with $\tau_s^{-1} = \tau_{el}^{-1} + \tau_{inel}^{-1}$; n_0 is the electron density, and m^* the effective electron mass. Equation (10.36) holds if the electron density of states $N(\epsilon)$ is constant near the Fermi energy. This expression simplifies considerably for $J_{sd}^2 \ll V^2$. In this case one has

$$\rho(T) = \frac{m^*}{n_0 e^2}(\tau_{el}^{-1} + \langle\tau_{inel}^{-1}\rangle_\epsilon - \tau_{el}(\tau_{el}')^{-2}) \tag{10.37}$$

with

$$\langle\tau_{inel}^{-1}\rangle_\epsilon = \int d\epsilon \tau_{inel}^{-1}(\epsilon)\left(-\frac{df}{d\epsilon}\right)$$

$$= \frac{\pi}{4}J_{sd}^2 N(0)c \int_{-\infty}^{\infty} \frac{d\omega}{\pi} \frac{\beta\omega N^{-1}\sum_{i,\mu}\chi_{ii,\mu\mu}''(\omega)}{(e^{\beta\omega}-1)(1-e^{-\beta\omega})} \tag{10.38}$$

This justifies our claim that the inelastic contribution to the resistivity is determined by the excitation spectrum of the spins, i.e. in this case by the spin glass dynamics. As a simple test, one can insert into (10.38) the spectrum calculated for diluted Heisenberg spins with RKKY interactions which led to the specific heat shown in Fig. 10.3 (Walker and Walstedt 1977). Below T_f, the resistivity $\rho_m(T)$ calculated in this way indeed agrees fairly well with the experimental data for AuFe, CuMn, AgMn, AuMn, and AuCr (Campbell, 1981; Campbell et al, 1982). This shows that $\rho_m(T)$ has a fairly universal behaviour below T_f and that there is a close relation between the specific heat and the resistivity. At $T = 0$, (10.37) leads with $q = S^2$ to a resistivity proportional to $V^2 + \frac{1}{4}J_{sd}^2 S^2$, a result which holds also for dilute antiferromagnets (Yosida, 1957). The opposite limit $T \to \infty$ leads with the Kramers–Kronig relations and $\chi'(\omega = 0) = \chi(T) = \beta S(S+1)/3$ to the 'Yosida limit'

$$\rho_m = \frac{m^*}{n_0 e^2}c\pi N(0)[V^2 + \frac{1}{4}J_{sd}^2 S(S+1)] \tag{10.39}$$

and the resistivity saturates. This result can be improved if the term proportional to V^2 is replaced by the exact expression for non–magnetic scattering in terms of phase shifts (Ziman, 1960; Kittel, 1964). The observed T^2 and $T^{3/2}$ dependences of the low–temperature resistivity can be explained in various ways. Oscillating modes with a density of states $N(\omega) \to$ const for $\omega \to 0$ lead with (10.1) to a linear specific heat and with (10.38) to $\rho_m \propto T^2$, but the same result is also obtained from diffusive modes. The mean field result $\chi''(\omega) \propto \omega^{1/2}$ (4.21) leads to $\rho_m \propto T^{3/2}$, but the same $T^{3/2}$ law is also obtained from spin waves with ferromagnetic dispersion $\omega_k \propto k^2$. Hence, the observed temperature dependence of the resistivity gives no unique information about the spin glass dynamics (Fischer, 1979, 1980).

So far we considered the resistivity in Born approximation in the interaction J_{sd} (10.28). Higher order terms in J_{sd} lead to resistivity contributions proportional to $\ln T$ which are characterisitic of the Kondo effect and which increase with decreasing temperature. However, sufficiently strong spin interactions destroy the Kondo effect and most likely a $\ln T$ term survives only in the high-temperature resistivity. The combination of the resistivity (10.37) with this $\ln T$ term then leads to the observed resistivity maximum. The calculation of the Kondo effect of interacting spins is fairly complicated and so far only crude estimates have been performed (Fischer, 1981a).

The calculation of the magneto-resistivity is considerably more complicated and we do not go into details. In elastic scattering, one now has to distinguish between the longitudinal and transverse spin glass parameters q_L and q_T, respectively. In the spin dynamics (10.30), one has in addition to the interaction term H_s the field term $-h \sum_i S_{iz}$, which modifies the spectrum $\chi''_{ij,\mu\mu}(\omega)$. In contrast to the case $h = 0$, nonlocal terms $i \neq j$ now can no longer be ignored in the susceptibility $\chi_{ij}(\omega)$ and the interference term proportional to $J_{sd}V$ in (10.23) becomes important. So far, there is no theory which takes into account all these effects. Experiments on AuFe, AuMn, CuMn, and AgMn indicate a negative magnetoresistivity $\Delta\rho = -A(T,c)h^2$ with $A > 0$ for small fields (Nigam and Majumdar, 1983).

The thermopower of metallic spin glasses typically shows a huge minimum (or sometimes maximum), followed by a small maximum (minimum) at lower temperatures. The minimum appears in systems like AuFe, CuFe, AuCo and CuCo, in which the 3d shell is more than half-filled, the maximum in AuV, CuV, CuCr, etc. with less than half-filled d-shells. In all cases, the thermopower is fairly concentration independent and, apart from the small maximum (minimum), is fairly similar to that of Kondo systems with extreme dilution. The temperature at which it changes sign is in most cases near the freezing temperature T_f (Ford et al, 1973; Foiles, 1978; Cooper et al, 1980).

In general, one has a 'diffusive' and a 'phonon-drag' contribution to the thermopower. In magnetic alloys, the latter can most likely be ignored, since the diffusive part S_d is 'giant'. From the Boltzmann equation one finds

$$S_d = \frac{1}{eT}\frac{K_1}{K_0} \tag{10.40}$$

with the transport integrals (10.36) and $e < 0$ (Ziman, 1960). From (10.36), one sees immediately that only the antisymmetric part of the relaxation time $\tau_s(\epsilon)$ contributes to S_d (assuming a constant $N(\epsilon)$ near the Fermi level). However, the additional factor ϵ in (10.36) leads to an additional factor ω in the integral of (10.38) and the elastic and inelastic relaxation times (10.33) and (10.34) are both symmetric. Hence we have to go beyond the Born approximation in order to explain the data.

The lowest order terms which yield a giant thermopower are proportional to J_{sd}^2 and J_{sd}^3. This is known in the theory of the Kondo effect (see Fischer, 1970). For spin glasses, the calculation is more complicated and we give only the result. One obtains

$$S_d = S_d^{(1)} + S_d^{(2)}$$

$$= \frac{k_B \rho_0}{e \rho_{tot}} (\pi N(0))^3 V J_{sd}^2 [\pi J_{sd} N(0) S_{eff}^2(T) + (6\pi)^{-1} \tilde{S}_d] \qquad (10.41)$$

with $\rho_0 = m^* c / n_0 e^2 \pi N(0)$ and the total resistivity $\rho_{tot}(T)$. The effective spin $S_{eff}^2(T)$ is defined by the integral in (10.38) and reduces to $S(S+1)$ for $T \to \infty$. This 'Yosida limit' leads to a giant temperature-independent contribution to $S_d(T)$, apart from the factor ρ_0/ρ_{tot} which represents the 'Nordheim–Gorter' rule. For independent spins, one has $S_{eff}^2 = S(S+1)$ for all temperatures. A summation of similar 'Kondo' terms to infinite order in J_{sd} and V then leads, depending on the sign of V, to the giant minimum or maximum observed in Kondo alloys.

The second term in (10.41) with

$$\tilde{S}_d = 2 \int_0^\infty d\omega p(\beta \omega) \frac{1}{N} \sum_{i,\mu} \chi_{ii,\mu\mu}''(\omega) \qquad (10.42)$$

$$p(x) = x \left[1 + \frac{x}{2\pi} Im \psi^{(1)} \left(\frac{ix}{2\pi} \right) \right] \qquad (10.43)$$

with the trigamma function $\psi^{(1)}$. It vanishes for elastic scattering (and hence for independent spins) and for $J_{sd} < 0$ has the opposite sign from the first term. Both terms together produce a change of sign of the thermopower at a temperature T_0 which is of the same order of magnitude as T_f. For oscillating modes such as the Walker–Walstedt modes discussed above, $S_d^{(1)}$ should vary like T^2 and $S_{sd}^{(2)}$ as T for low temperatures. Details can be found in the original paper (Fischer, 1981b).

The terms proportional to $V J_{sd}^2$ and $V J_{sd}^3$ considered in (10.41) are the lowest order terms in which the Kondo anomaly shows up in the thermopower. In spin glasses the Kondo effect is partly suppressed owing to the spin interactions. This should be true in particular at low temperatures, where the interaction energy exceeds the thermal energy. Thus one can hope that the terms considered are sufficient for a qualitative discussion of the thermopower in spin glasses, at least if the freezing temperature is large compared to the Kondo temperature T_K. For most spin glasses, this seems to be the case.

11

Competition between spin glass and ferromagnetic or antiferromagnetic order

The spin-glass properties discussed so far are observed only in a restricted concentration range of the magnetic atoms. Very dilute metallic magnetic alloys exhibit the Kondo effect (see Section 10.3) and very dilute insulating systems with short-range exchange interactions like $Eu_xSr_{1-x}S$ with $x \leq 0.13$ remain superparamagnetic at all temperatures. In the opposite limit of large concentrations, one has magnetic order. However, there is an interesting concentration range in which one observes a competition between spin-glass and ferro- or antiferromagnetic long range order. This is the subject of this chapter. We will be particularly interested in whether a system which has made a paramagnetic to ferro- or antiferromagnetic transition will go into a 'reentrant' spin glass state at some lower temperature or whether there is a coexistence of spin-glass and conventional magnetic order.

11.1 Mean field theory

The first hint of an answer to this question comes from MFT. The SK model for Ising spins with the Gaussian bond distribution (3.29), as discussed in Section 3.3, predicts the phase diagram of Fig. 3.7. If the width J of the bond distribution is smaller than the variance J_0, one has a spin glass state without spontaneous magnetization M_s. For $J_0 > J$, the system goes with decreasing temperature first into a ferromagnetic state with finite order parameters M_s and q and at a lower temperature into a spin glass state with replica symmetry breaking. In this state, the order parameter q has to be replaced by the Parisi order parameter function $q(x)$ (Section 3.4), whereas the spontaneous magnetization M_s remains finite. Below the 'reentrant' transition, one has typical spin glass properties with soft modes and a large number of metastable states or solutions of the TAP equations (3.86), together with ferromagnetic order. This spin glass state with a finite spontaneous magnetization is sometimes called a 'mixed' phase or 'magnetized spin glass'.

It is interesting to know whether there are other mean field theories which also predict such a mixed phase. This is indeed the case for a 'Bethe lattice' with a random distribution of $\pm J$ nearest neighbour interactions and suitable boundary conditions. In a Bethe lattice, there is a multifurcation of each 'branch' into $Z - 1$ branches, where Z is the coordination number. An example is indicated in Fig. 11.1 for the branching ratio $Z - 1 = 2$. Such a model indeed shows spin glass properties similar to those predicted by the SK model. This includes an 'AT line', a spin glass phase and a mixed phase, all of which are defined by the divergency of the spin glass susceptibility (2.64). Spin glass properties on Bethe lattices have been investigated by Chayes et al (1986), Carlson et al (1988), and others.

In many cases, real systems show a transition of a different type. Mössbauer data to be discussed below indicate 'spin canting' at a temperature well below the Curie temperature T_c. The system develops a local *transverse* magnetization as produced by the freezing of the transverse spin components analogous to the Gabay–Toulouse (GT) line (6.33) for vector spins. In order to describe such an effect in reentrant systems, we have to develop a MFT for vector spins which includes the magnetization $\mathbf{M(h}, T)$ as an additional order parameter. It turns out that the SK theory for vector spins with $J_0 \neq 0$ just leads to a replacement of the field h by $h + M J_0/J$. The GT line then reads

$$1 - \frac{T_{c_1}}{T_f} = A(h + M J_0/J)^2 \tag{11.1}$$

with $A = (m^2 + 4m + 2)/4J^2(m + 2)^2$, where m is the number of spin

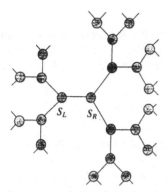

Figure 11.1: Isotropic Bethe lattice with branching ratio $Z - 1 = 2$ or coordination number $Z = 3$. The magnetization $M_L = \langle S_L \rangle$ of the spin S_L on this lattice can be derived from the half-space magnetizations M_L and M_R, where the latter are computed assuming that the bond between S_L and S_R is removed (see Chayes et al, 1986).

components and $k_B T_f = J$. The generalized GT line (11.1) is defined by the vanishing of the transverse spin glass parameter $q_T = 0$, where q_L, q_T, and M have to be determined self-consistently. It turns out that the spontaneous magnetization M_s remains finite below the line (11.1). With $M = M_s + \delta M$ and the field-induced magnetization $\delta M = \chi h$, (11.1) then leads immediately to a *linear* field dependence of the transition temperature T_{c_1}, in contrast to the h^2 dependence of the GT line and the $h^{2/3}$-law of the AT line (3.28). (The same linear field dependence also holds at the triple point $J = J_0$, $T_f = T_c$ where $M_s = 0$.) Near the Curie temperature T_c, the spontaneous magnetization M_s is proportional to $(T_c - T)^{1/2}$ as in the MFT of an ideal ferromagnet. Near the transition line (11.1) the longitudinal spin glass parameter q_L becomes important. From the self-consistent equation for $M_s \ll 1$ one finds

$$M_s = \beta J_0 M_s [1 - (\beta J)^2 q_L] - \tfrac{1}{5}(\beta J_0 M_s)^3 \qquad (11.2)$$

With $q_L \propto M_s$, this leads to

$$M_s = \frac{2^{\frac{1}{2}}(T_c - T)T_f}{T_c^2} \qquad (11.3)$$

Compared to the ideal ferromagnet with $J = 0$, M_s is strongly reduced and varies *linearly* with $T_c - T$. Inserting (11.3) back into (11.1) and using $T_c = J_0/k_B$ with the result of Gabay and Toulouse (1981), we have

$$1 - \frac{T_{c_1}}{T_f} = \frac{2A(J_0 - J)^2}{J^2} \qquad (11.4)$$

(Dubiel et al, 1987).

A slightly modified version of the Ising SK model describes a transition from an antiferromagnetic state into a spin glass. One introduces two types of spins S_{1i}, S_{2i} on the lattice sites. The Hamiltonian (3.1) is then replaced by

$$H = -\tfrac{1}{2} \sum_{ij} J_{ij} S_{1i} S_{2j} - h \sum_i (S_{1i} + S_{2i}) \qquad (11.5)$$

where one assumes exchange interactions only between different types of spins. With the bond distribution (3.29) and $J_0 < 0$, the formalism presented in Section 3.3 again applies and, for $h = 0$, leads to a symmetric extension of the phase diagram of Fig. 3.7 to the region $J_0 < 0$, with an antiferromagnetic state for $-J_0 > J$. If one adds a homogeneous field, the situation becomes more complicated than in the ferromagnetic case because one now has to solve self-consistent equations for the magnetization $M = \tfrac{1}{2}(M_1 + M_2)$, for the staggered magnetization $M_{stagg} = \tfrac{1}{2}(M_1 - M_2)$ and for q (Korenblit and Shender, 1985; Fyodorov et al, 1987).

11.2   Reentrant transitions

There is a large class of systems which with decreasing temperature exhibit first a transition from a paramagnetic state to a ferromagnetic one and then a second transition from the ferromagnetic state to a spin glass. Examples of these 'reentrant' systems are (in particular concentration ranges) AuFe, (Eu,Sr)S, FeCr, NiMn, AlFe, $(Pd,Fe)_{1-x}Mn_x$, (Eu,Sr)Te, (Eu,Sr)As, amorphous (a-) FeNi, a-FeMn, a-FeCr, and a-ZrFe. In most cases, the amorphous substances contain an additional glass-stabilizing metalloid such as $P_{16}B_6Al_3$. The phase diagrams of some typical reentrant systems have been shown already in Figs. 2.10a–d. Reentrant transitions from an antiferromagnetic state have been observed in $FeMgCl_2$ and possibly from more complicated modulated spin structures in YTb, GdSc, and TbSc. A typical example of a system with a paramagnetic to ferromagnetic and a ferromagnetic to spin glass transition is $Eu_xSr_{1-x}S$ with $0.51 \leq x \leq 0.65$ (Fig. 2.10b). However, we will see that in this and other systems the transitions are not well defined.

The ferromagnetic state is conveniently investigated by means of an Arrott plot: One determines the spontaneous magnetization $M_s(T)$ by plotting the total magnetization M^2 versus h/M. In a mean field ferromagnet, one has straight lines near T_c and the extrapolation to $h = 0$ yields a finite value for M_s for $T < T_c$. In reentrant EuSrS, these lines have a strong curvature for small fields and no straightforward extrapolation to $h = 0$. Often one plots directly the reversible magnetization $M(T)$ versus temperature, as shown in Fig. 11.2 for $(Fe_{0.70}Mn_{0.30})_{75}$ $P_{16}B_6$ Al_3. The shape of $M(T)$ changes even for the smallest measured field (1 Oe) and makes an exact determination of T_c extremely difficult. Below T_c, a ferromagnetic state appears in the measured magnetization as a demagnetization-limited plateau. At low temperatures, one observes a decrease of magnetization, which is taken as evidence for a transition to a spin glass state. The determination of the transition temperatures from the ac and dc magnetizations of AuFe, a-FeMn, a-FeNi, and a-NiMn leads to similar difficulties. (We remember that the cusp in the ac susceptibility of a good spin glass like CuMn is also extremely sensitive to a magnetic field.) In the case of reentrant AuFe, one also has difficulties due to chemical clustering: Most likely, this system has a phase transition from a superparamagnet to a spin glass phase for iron concentrations between 8 and 15.5 at%, as indicated in Fig. 2.10a.

In order to distinguish clearly between a superparamagnet and a ferromagnet, a zero-field experiment is of advantage because it avoids errors due to an induced magnetization. In zero-field neutron scattering, one observes a ferromagnetic state in the form of Bragg peaks or in the small angle scattering as the correlation length diverges below T_c. Both types of experiments have been performed on $Eu_{0.52}Sr_{0.48}S$ (Shapiro et al, 1986). Within the experimental resolution, no magnetic Bragg scattering could be observed at any temperature whereas the inverse correlation length κ becomes indistinguishable from zero in a certain temperature region (Fig. 11.3). (In determining the Bragg scattering, great care has to be taken to separate off diffusive contributions. The κ-values are obtained by fitting the diffusive scattering to a Lorentzian line.) So, even in the 'ferromagnetic' region one has only an extremely small spontaneous magnetization, if any. The temperature at which κ first seems to vanish lies considerably below the T_c value obtained from magnetization measurements indicated in Fig. 2.10b. We have to conclude that even in this 'classical' reentrant system there is no clear evidence for a true ferromagnetic state.

The transition from ferromagnetism to a spin glass is even more difficult to determine. In most cases, one defines it by the onset of deviations from the plateau in the magnetization (see Fig. 11.2). Unfortunately, this onset again depends on the field and possibly on remanence effects. Other criteria are the onset of viscosity (or of time effects in the magnetization) and remanence. The neutron scattering data for EuSrS (Fig. 11.3) show a fairly well defined temperature below which the inverse correlation length κ becomes nonzero. However, elastic light scattering on $Eu_{0.54}Sr_{0.46}S$ shows

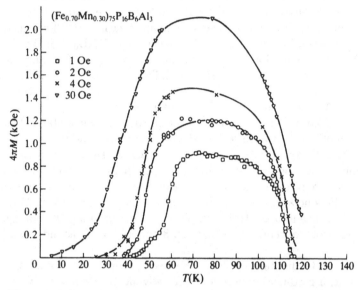

Figure 11.2: Reversible magnetization vs temperature of amorphous $(Fe_{0.70}Mn_{0.30})_{75}P_{16}B_6Al_3$ at various applied fields. For elliptic samples, one has the demagnetization field in direction $\mu \parallel H_{d\mu} = -N_\mu M_\mu$ with the demagnetization tensor component N_μ. With $M_\mu \chi^{-1} = H_\mu + H_{d\mu}$, this leads to the effective susceptibility $\chi_{eff} = \chi(1 + \chi N_\mu)^{-1}$ or for $\chi \to \infty$ to $\chi_{eff} = N_\mu^{-1}$ (from Manheimer et al, 1983).

only a very modest change of the correlation length as one goes from the ferromagnetic to the spin glass region (Geschwind et al, 1987).

So far, we did not specify this 'spin glass' state very well, though the neutron scattering and magnetization data seem to indicate that it has zero magnetization and is not a 'mixed' state with spin glass and ferromagnetic properties. Evidence that in some systems one has fairly long-ranged ferromagnetic correlations even in this 'reentrant' region comes from inelastic neutron scattering data. In AuFe, NiMn, and a-FeMn one observes spin waves with ferromagnetic dispersion which persist in the spin-glass region down to very low temperatures and which coexist with strong quasi-elastic scattering. The corresponding spin correlation length should be at least of the order of $2\pi/Q$, where $Q \approx 0.04$ Å^{-1} is the scattering vector (Murani, 1983; Hennion et al, 1983, 1984, 1986). However, these results seem to be in contrast to the spin waves observed in a-FeNi, AlFe, and FeCr, which appear to be strongly damped and seem to loose their stiffness below the

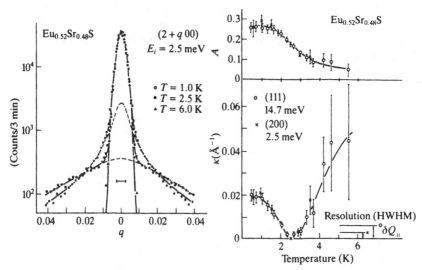

Figure 11.3: Neutron diffraction measurements on $Eu_{0.52}Sr_{0.48}S$: Temperature dependence of Lorentz amplitude A and half–width κ if the scattering cross–section is fitted by $S(\mathbf{Q}) = A/\{\mathbf{Q} - \mathbf{G})^2 + \kappa^2\} + B\delta(\mathbf{Q} - \mathbf{G})$. ($\mathbf{Q}$ is the momentum transfer, $\mathbf{G}$ a reciprocal lattice vector, and κ an inverse correlation length.) The Lorentzian and δ–function terms represent the diffusive and Bragg scattering, respectively (from Shapiro et al, 1986).

reentrant transition (Motoya et al, 1983; Erwin et al, 1985; Shapiro et al, 1981).

Mössbauer data on AuFe and CrFe below the reentrant transition indicate spin canting without significant change of the longitudinal magnetization component. Again, this is a strong indication of ferromagnetic spin correlations which persist into the spin glass region. Unfortunately, the Mössbauer effect gives information only about local fields and cannot answer the question whether or not there is true coexistence of spin glass and ferromagnetic order in some temperature region (Lauer and Keune, 1982; Campbell et al, 1983; Abd-Elmeguid et al, 1986; Dubiel et al, 1987).

In antiferromagnets, such coexistence seems to be possible. Pure $FeCl_2$ is an antiferromagnet with strong uniaxial anisotropy. Dilution with about 50% Mg gives both antiferromagnetic and spin glass properties. In $Fe_{0.55}Mg_{0.45}Cl_2$, one observes a Néel temperature $T_N = 7.5K$, both from a cusp in the susceptibility and from the onset of Bragg peaks. These

Bragg peaks persist at all temperatures below T_N. However, below a second temperature $T_f \approx 3K$ the zero-field-cooled and field-cooled susceptibilities differ in their temperature dependence, and the ac susceptibility becomes frequency dependent, as in ordinary spin glasses (Bertrand et al, 1982; Wong et al, 1985a,b). Similar effects have been observed in $Fe_x Mn_{1-x} TiO_3$ (Yoshizawa et al, 1987).

12

One-dimensional models

As in many problems, one-dimensional models can give some insight into the physics of spin glasses because they are easier to solve than those in higher dimensionality, but this advantage can be limited because the physics of the one-dimensional problem may be essentially different. Typically, for example, the higher-dimensional problem we really want to study has a broken symmetry phase at low temperature, while one-dimensional systems (with short-range interactions) do not exhibit broken symmetry at finite temperature.

Nevertheless, short-range one-dimensional systems can have broken symmetry at zero temperature, so one can study this relatively simple kind of order and the approach to it as the temperature is lowered to zero. This *zero-temperature phase transition* can be compared with finite-temperature transitions in higher dimensionality. This is the motivation for Section 12.1, where we will see that the critical point at $T = h = 0$ in a one-dimensional Ising spin glass with nearest-neighbour interactions has interesting features in common with finite-temperature spin glass transitions, and that they can be calculated analytically, sometimes rather simply.

Further insight can be gained by studying one-dimensional systems with long-range (typically power law) interactions. Then if the exponent describing the falloff of the interaction strength with distance is small enough, finite-temperature phase transitions are possible. Though such problems are harder to solve than nearest-neighbour ones, they can still be easier than the full higher-dimensional ones. Furthermore, by varying this exponent one can produce a range of different models which correspond, at least crudely, to short-range models in a range of dimensionalities. This is the reason for studying the model described in Section 12.2.

12.1 The nearest-neighbour Ising spin glass chain

The thermodynamics of Ising chains with nearest-neighbour interactions can be calculated systematically in a renormalization group approach by decimation (e.g. carrying out the trace over every other spin and expressing the result in terms of new effective interactions between the remaining spins (Grinstein et al, 1976)). Here, however, we follow Chen and Ma (1982) in deriving the interesting low-temperature properties, including time-dependent ones, from a domain-wall picture.

The main idea in this picture is to focus not on the spin variables themselves, but rather on the domain wall or 'kink' variables

$$\tau_i = \text{sgn}[J_{i,i+1}S_iS_{i+1}] \tag{12.1}$$

associated with the bonds: A $\tau_i = -1$ describes a broken bond between sites i and $i + 1$. The energy of such a broken bond is $2|J_{i,i+1}|$, i.e. the energy of the whole system is

$$H = +\sum_i |J_{i,i+1}|\tau_i \tag{12.2}$$

(Of course, one can always write the energy of a nearest-neighbour Ising system in this form, but in higher dimensionality the domain wall variables on different sites are not independent and calculations are more difficult.)

We will be most interested in the case of a bond distribution $p(J)$ (e.g. a Gaussian) which is continuous at $J = 0$, but it will also be interesting to consider distributions where

$$p(J) \propto J^n \tag{12.3}$$

at small J, for $n > -1$. We will restrict our attention to symmetric distributions.

A useful way to picture the state of the system (Fig. 12.1) is in terms of the effective potential $|J_{i,i+1}|$ for kinks: It is favourable for a kink to move (flipping a spin at each site it moves through) to local minima (i.e. bonds weaker than neighbouring ones) of this random potential (though at finite temperature uphill moves will also occur with some probability).

At temperature $T = \beta^{-1}$, the equilibrium probability of a kink on bond $i, i + 1$ is just

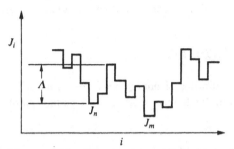

Figure 12.1: Effective potential for kinks in a one-dimensional random magnet (from Chen and Ma, 1982).

$$f(J_{i,i+1}) = \frac{1}{1 + \exp(2\beta|J_{i,i+1}|)} \qquad (12.4)$$

Thus it is simple to compute equilibrium quantities like the specific heat

$$C = \frac{1}{N}\frac{\partial E}{\partial T} = \int dJ p(J) \frac{\partial f(J)}{\partial T} \propto T^{n+1} \qquad (12.5)$$

The susceptibility

$$\chi = \beta \sum_{ij} \langle S_i S_j \rangle \qquad (12.6)$$

is trivially a Curie law in the limit of zero field because the $i \neq j$ terms in (12.6) vanish because of the symmetry of the bond distribution (independent of n).

We now look at dynamics. If we start the system in a configuration with many kinks (e.g. a fully magnetized state, which breaks half the bonds), they will sometimes collide as they try to move to weaker bonds. If two kinks collide, they will annihilate each other. After this fast initial period of local equilibration is over, there will be either one or zero kinks left in each local minimum.

If we are exactly at $T = 0$, nothing more will happen, since any spin flip would involve moving a kink uphill. Since exactly one bond in three is a local minimum and there are two possibilities (kink or no kink) at each minimum, there are (Li, 1981) $2^{N/3}$ distinct metastable states (cf. the corresponding result (3.128) for the SK model). Another immediate implication of this picture, based on the argument that the remanent magnetization comes from those pairs of ferromagnetically coupled spins through which no kinks pass, is a zero-temperature remanent magnetization per spin of 1/3 (Fernandez and Medina, 1979).

At finite T, uphill kink motion can occur, and the magnetization will relax further. At time of order t, barriers of height $T \ln t$ will have been climbed and will no longer contribute to the remanent magnetization. Thus we have

$$\frac{M(t)}{N} = \int_0^\infty p(J)D(J, T \ln t)dJ \tag{12.7}$$

where $p(J)D(J, \Lambda)dJ$ is the number of energy wells per spin lying in the interval $[J, J + dJ]$ with a barrier height of at least Λ. This quantity can be written

$$D(J, \Lambda) = \sum_{n=1}^\infty nP^2(J + \Lambda, \infty)P^{n-1}(J, J + \Lambda) \tag{12.8}$$

where

$$P(a, b) = \int_a^b p(J)dJ \tag{12.9}$$

Each term in the sum in (12.8) represents the contribution from a well n bonds wide. The $n - 1$ factors of $P(J, J + \Lambda)$ are the probability of $n - 1$ bonds between the minimum J and the maximum $J + \Lambda$, the factor n is because the weakest bond (the bottom of the well) could be in any of the n positions, and the $P^2(J + \Lambda, \infty)$ is the probability of finding the *two* bonds to the left and right of the well larger than Λ (so that the region is a well of the requisite depth).

For an exponential bond distribution $p(J) = \exp(-J)$, the quantity $D(J, J + \Lambda)$ and the integral (12.7) can be evaluated exactly. At short times ($T \ln t \ll 1$) one finds

$$\frac{M(t)}{N} = \frac{1}{3}\left(1 - \tfrac{1}{2}T \ln t + \cdots\right) \tag{12.10}$$

and at long times the simple power law

$$\frac{M(t)}{N} \propto t^{-T} \tag{12.11}$$

This power-law decay is qualitatively similar to what is seen in the ultradiffusive models of Section 9.3 and at the finite-temperature spin glass transition in 3 dimensions.

A similar calculation gives the decay of the energy density

$$\frac{E(t)}{N} = \frac{1}{18}\left(1 - \frac{7}{8}T \ln t + \cdots\right) \tag{12.12}$$

for short times, and

$$\frac{E(t)}{N} \approx \frac{T \ln t}{2t^{2T}} \tag{12.13}$$

at long times.

So far we considered the system in zero external field. We now calculate the response to an external field, initially at zero temperature. In the presence of the field, it will now become favourable to break some (weak) bonds in the ground state. The argument about how to do this is essentially like the Imry–Ma argument about the effect of a random field on a ferromagnet. Flipping a region of length l spins can gain a field energy of order $hl^{\frac{1}{2}}$ and can be accomplished at the cost of breaking two bonds. For a given h, the distance between a pair of bonds both less than a value J_c is $l \approx J_c^{-(n+1)}$. Setting J_c equal to $hl^{\frac{1}{2}}$ and eliminating J_c from this pair of equations gives a typical domain size

$$l(h) \propto h^{-\frac{2(n+1)}{n+3}} \tag{12.14}$$

and thus a magnetization per site

$$M(h) \propto [l(h)]^{-\frac{1}{2}} \propto h^{\frac{n+1}{n+3}} \tag{12.15}$$

Thus, just as at finite-dimensional spin glass transitions, there is a non-analytic dependence of magnetization on field at the critical temperature $T_f = 0$. However, the analogy is limited: If we express the result (12.15) in terms of the critical exponent δ as (cf. (8.6))

$$\chi_{nl}(h) \propto \frac{\partial^3 M}{\partial h^3} \propto (h^2)^{\frac{1}{\delta}-1} \tag{12.16}$$

we obtain a negative $\delta = -(n+3)$. A positive δ would require an exponent greater than 1 in (12.15).

Another comparison one can make is with the nonanalytic term in $M(h)$ proportional to $h^{d/(d-2y)}$ (8.54) predicted by the scaling theory of Section 8.2. Again, there is an important difference if there is a finite-temperature transition ($y > 0$), because then the scaling theory exponent will always be bigger than 1, while that in (12.15) is less than 1. The correspondence does make sense, on the other hand, if one does the scaling theory in one dimension directly (Bray and Moore, 1986), obtaining $y = -d = -1$ exactly (for $n = 0$): Then (8.54) is reassuringly the same as the $n = 0$ case of (12.15).

12.2 A long-range model

As indicated above, a way to get a finite transition temperature in a one-dimensional model is to take interactions which fall off like a sufficiently slow power of distance. If this power is small enough, a phase transition will be possible; indeed, one can go all the way to the infinite-range limit by taking a limit where the power goes to zero.

So suppose we have a one-dimensional spin glass where the variance of the interactions falls off like $|i - j|^{-2\sigma}$:

$$[J_{ij}^2]_{av} = \frac{J^2}{a^2|i - j|^{2\sigma}} \tag{12.17}$$

(Kotliar, Anderson, and Stein, 1983). In order that there be a well-defined thermodynamic limit, the sum $\sum_j [J_{ij}^2]_{av}$ must be finite, so here we have to have $\sigma > \frac{1}{2}$. (Otherwise, we have to make $[J_{ij}^2]_{av}$ N-dependent, as, for example, in the SK model.)

A Landau–Ginzburg expansion in the replica formalism for this model would look just like the one (8.77) we made for the short-range model in Section 8.3, except for the presence of a nonanalytic nonlocal term like (8.89) (but with a different definition of σ)

$$\delta H_{eff} = \tfrac{1}{2}c' \sum_k k^{2\sigma-1} tr(\mathsf{q}_k \mathsf{q}_{-k}) \tag{12.18}$$

If $\sigma < 3/2$, this new term will dominate the usual k^2 term, and critical behaviour will be different from the short-range case. The value of σ which corresponds to the upper critical dimensionality of the short-range problem is determined as in the corresponding short-range problem by when the loop diagram with three critical propagators,

$$\int dk \left(\frac{1}{k^{2\sigma-1}}\right)^3 \tag{12.19}$$

diverges (cf (8.62)), i.e. at $\sigma = 2/3$.

For $\frac{1}{2} < \sigma < 2/3$, then, the spin glass transition in this model (in zero field) has mean field critical behaviour. For σ just below 2/3, one can expand in $\epsilon = 3\sigma - 2$ in the same way we did in Section 8.3 in $6 - d$ in the short-range system. The difference between that calculation and this one comes about because there are *no* corrections to the coefficient c' of the new nonlocal term (12.18) from the perturbative elimination of Fourier components q_k with k in the outer shell (since all such perturbative contributions to the new effective Hamiltonian are analytic in k). So, just as in the argument leading to (8.90) we get only the contributions from the rescaling of the fields (a factor $b^{\frac{1}{2}(2-\eta)}$ for each q) and of the momenta (a factor $b^{1-2\sigma}$), leading to

$$\frac{dc'}{dl} = (2 - \eta + 1 - 2\sigma)c' \tag{12.20}$$

which fixes $\eta = 3 - 2\sigma$. For $\sigma > 3/2$, we then find a fixed point (r^*, w^*) of the renormalization group equations (8.63–64) of order ϵ and extract non-classical exponents to first order in ϵ in the standard way. The result to this order for the correlation length exponent ν is

$$\nu = \frac{3}{1 - 4\epsilon} \tag{12.21}$$

This result is for the Ising case but the m-vector can be done the same way.

In the above treatment we have not used the one-dimensionality of the model; it is straightfwardy generalized to arbitrary d. The one-dimensionality is useful, however, near another (larger) value of σ which corresponds to the *lower* critical dimensionality, beyond which long range order disappears. The present model has the advantage that this critical value of σ is very easy to find analytically, in contrast to the short–range higher-dimensional spin glass, where it is not known exactly.

To see what this critical value is and to get some insight into behaviour at this point we begin with the remark (already implicit in the discussion above) that the effective interaction in the replica formalism is $\Delta_{ij} \equiv [J_{ij}^2]_{av}$ (see, e.g. (2.33)), which in the present model (12.17) is proportional to $|i - j|^{-2\sigma}$. Now we know from other one-dimensional problems (notably the corresponding ferromagnet) that when interactions fall off like the inverse square of the separation between spins there is a very special situation with a characteristic interesting kind of physics, and that indeed this value of the exponent governing the falloff of interactions between spins is the critical value separating a region where a phase transition can happen from one where it cannot.

To understand just why an inverse square law is special, we resort, as we did in the previous section, to a domain wall or 'kink' picture of the configurations of the system. To get the physical ideas clear, we first consider the ferromagnetic chain with $1/r^2$-interactions. Now at low temperatures we expect that there will at least be a lot of strong local order, if not a long-range ordered state. So the typical configuration will have occasional kinks or antikinks separating ordered up or down regions. (In fact the kinks and antikinks will alternate.) Just as in the short-range case of the preceding section, it is useful to express the energy of the system in terms of kink variables instead of the original spins. The kink field is just the derivative of the spin one, i.e.

$$\psi(x) = \frac{d}{dx} S(x) \tag{12.22}$$

where a suitable continuum limit has been taken. In the same limit, the energy can be written (with a couple of integrations by parts) as

$$H = -\tfrac{1}{2} \int dxdy S(x)\frac{1}{|x-y|^2}S(y)$$

$$\rightarrow -\tfrac{1}{2}\int dxdy\psi(x)\ln|x-y|\psi(y) + \mu\int dx\psi^2(x) \qquad (12.23)$$

(up to an additive constant), where μ is the energy of a single kink or antikink.

So, in contrast to the kinks of the short-range problem, which do not interact with each other except when they meet and annihilate each other, the kinks of this problem interact with a very long-range logarithmic potential. Kinks attract antikinks and repel other kinks. The problem is equivalent to that of a neutral gas of logarithmically interacting charges, with μ their chemical potential.

Logarithmic interactions are very special because they are a critical borderline case between ones which fall off to zero at long distances and ones which grow like a power of the distance in that limit. They occur in other important problems as well: In the two-dimensional XY model, vortex excitations interact logarithmically (Kosterlitz and Thouless, 1973), while the long-time dynamics of the Kondo problem (a magnetic impurity in a Fermi sea) can be mapped (with time replacing space) into exactly the $1/r^2$ ferromagnet we are discussing here (Anderson, Yuval and Hamann, 1970; Anderson and Yuval, 1971).

In this special logarithmic case, there turns out to be a critical temperature below which kinks are bound to neighbouring antikinks in 'molecules', but above which they are unbound and form a free plasma. If we look at the problem in the low-temperature phase on a length scale longer than the typical binding radius of the kink–antikink molecules, we will essentially see no kinks or antikinks. That is, the magnetization will have long-range order. In the high-temperature phase, on the other hand, there will be free kinks all over, and the phase will be magnetically disordered.

If we change the power law governing the interaction just a little, however, so the interactions fall off more rapidly, then sufficiently distant kinks will not interact. On large enough length scales the kink gas will *always* be in the unbound (plasma) phase, so the system will be magnetically disordered. (We have not presented these arguments carefully here; our aim is just to give their flavour. The interested reader is referred to the original articles (particularly Kosterlitz and Thouless) for details.)

So by going over to kink variables we are able to establish the maximum value of the exponent describing the falloff of interactions with distance if there is to be magnetic order: $1/r^2$-interactions are critical because they correspond to logarithmic interactions between kinks.

What now about the spin glass problem? Now (see (2.33) again) the variables that interact are products $S_i^\alpha S_i^\beta$, where α and β are replica indices, or. equivalently, $Q_i^{\alpha\beta}$-matrices. But again we expect that there will at least be strong short-range order at low temperatures, and a picture in terms of kinks should be valid. Here the kinks are more complicated objects than in the ferromagnet because of the replica indices, but the argument about the distance-dependence of the interactions between kinks is the same as before. Thus $\sigma = 1$ gives logarithmic interactions between kinks and is the largest value of σ for which there is spin glass order in this model.

Kotliar et al studied this gas by means of the renormalization group, extending the treatment to σ just below 1 by making a formal expansion in $\epsilon = 2(1 - \sigma)$. (This is analogous to expansions for critical properties of short-range m-vector ferromagnets just above two dimensions in powers of $d - 2$.) They find a correlation length exponent $\nu \approx 0.9\epsilon^{-\frac{1}{2}}$. Using the scaling law (8.9) $d\nu = 2 - \alpha$, this indicates a large and negative specific heat exponent, in agreement with experimental and numerical results on three-dimensional short-range systems discussed earlier. One cannot make any quantitative comparisons, of course, but in the absence of any short-range model for which one can make a corresponding calculation, it gives some insight into the physics of spin glasses near a lower critical dimension, which is presumably the situation we have in three-dimensional short-range Ising systems.

13

Random fields and random anisotropy

In most of this book the important randomness we were concerned with was random exchange. In this chapter we will examine systems characterized by two other kinds of randomness: random external fields and random anisotropy. They both share some general features with random exchange, but they have different properties as well. They also differ significantly from each other.

We studied random anisotropy, especially of the Dzyaloshinskii–Moriya sort, in Chapter 6, and again, briefly, in Chapter 7. However, there we were interested in it as a small perturbation on a system whose physics was determined primarily by random exchange. Its importance lay in the fact that it changed the overall symmetry of the system, and this had qualitative consequences at long lengthscales.

A similar remark applies to random fields: Whenever we have been interested in the effects of uniform external fields (as in the AT line, for example) they might as well have been random fields; for symmetrically distributed exchange interactions, the only condition on the external field was that it be uncorrelated with the interactions. Thus we could say that we have implicitly been studying the effects of random fields in spin glasses, but, as with the random anisotropies we studied, we were mainly interested in their effects as perturbations on the spin glass state produced by the random exchange.

Both these kinds of randomness are *local*, so they only produce trivial behaviour by themselves. They are only interesting in interaction or competition with other kinds of order. In the situations mentioned above, the other order was spin glass order. In this chapter we will be interested in what happens when they compete with ferromagnetism or other ordered magnetism.

13.1   Random external fields

The most important question about the competition between an external random field and ferromagnetic order is 'Who wins?' That is, we want to know how strong the random field has to be to destroy the ferromagnetically ordered state. This is not merely a quantitative question: In low enough dimensionality, *any* random field, no matter how small, will turn out to destroy the ordered state.

Let us see how this comes about. To be specific, consider a ferromagnet with nearest–neighbour couplings of strength J and a random field of zero mean and variance

$$[h_i h_j]_{av} = h^2 \delta_{ij} \tag{13.1}$$

We note in passing that the case of *strong* random fields ($h > 4J$ for a '$\pm h$' model) is trivial. Then the energy can be minimized simply by aligning the spins with the random field site by site; even if one could gain exchange energy on all four bonds connected to a single site by flipping the spin at that site, the energy gain ($8J$ in the $\pm h$ model) would be outweighed by the cost ($2h$) in field energy.

Thus the nontrivial question is whether a *very small* random field can break up the order. Although it obviously cannot beat the stronger exchange site by site, there is the possibility that it may make it favourable to break the system up into fairly large domains of different (up or down) ferromagnetic phases.

How this happens for Ising magnets was first shown by Imry and Ma (1975) in the argument we already gave in Chapter 8 in connection with the effect of a field on the spin glass state. Briefly, the argument was that in domains of size L, the field energy was of order

$$E_{field} \approx -hL^{d/2}, \tag{13.2}$$

while the cost in exchange energy of the wall around the domain (assuming it compact) was

$$E_{exch} \approx 2JL^{d-1}, \tag{13.3}$$

so domain formation of size

$$\xi \approx \left(\frac{J}{h}\right)^{\frac{2}{2-d}} \tag{13.4}$$

is favourable for $d < 2$ (cf (8.53)). There is long-range ferromagnetic order for $d > 2$ for weak enough random fields, but none is possible below it. Notice that this argument deals only with energy competition; there are no thermal fluctuations such as go into the old lower-critical-dimensionality arguments for pure systems. So it is about whether long-range order breaks down in the presence of small random fields *even at* $T = 0$; temperature is irrelevant to the argument.

For d exactly equal to 2, more subtle arguments (Binder, 1983) indicate no long range order, but a correlation length that diverges like $\exp[c(J/h)^2]$ in the limit of small randomness.

This was for Ising systems. For systems with continuous symmetry, our domains are somewhat fuzzier objects because the walls are not sharp. Because the spins have angular degrees of freedom, they can rotate slowly as one passes from one domain to the neighbouring one, giving a lower wall energy than in the Ising case. This means that the wall thickness is of the order of L itself, and, therefore, that the wall energy is

$$E_{exch} \approx 2JL^{d-2} \tag{13.5}$$

(This comes from multiplying the wall area L^{d-1} by the average exchange energy $2J/L^2$ by the wall thickness L.) Thus we now have to compare $d/2$ with $d - 2$, and the lower critical dimensionality becomes 4.

Comparing these lower critical dimensionalities with the corresponding ones found in Chapter 2 for stability against thermal fluctuations ($d_l = 1$ for Ising and $d_l = 2$ for Heisenberg spins), we can see that random fields have a much stronger effect. The critical dimensionalities are doubled in both cases.

There is another way to get this result for Heisenberg spins which is quite analogous to the Peierls argument presented in Chapter 2 for the destruction of long range order by thermal fluctuations for $m > 2$ in pure systems. We argue that the random fields, since they point in all directions, will in general tip the magnetization away from its equilibrium direction. Let us compute the mean square of these magnetization fluctuations. We have

$$\langle \mathbf{S}(\mathbf{x}) \rangle = \int d^d y \, G_\perp(\mathbf{x} - \mathbf{y}) \mathbf{h}_\perp(\mathbf{y}) \tag{13.6}$$

Since we work only to lowest order in the random fields we take $G_\perp$ to be that for the pure system. The notation should be evident: the $\perp$ means that we are looking only at the fluctuations perpendicular to the equilibrium magnetization. Squaring and averaging, we find

$$[\langle \mathbf{S}(\mathbf{x}) \rangle^2]_{av} = \int d^d y \, G_\perp^2(\mathbf{x} - \mathbf{y}) \cdot h^2$$

$$= \int \frac{d^d k}{(2\pi)^d} G_\perp^2(\mathbf{k}) h^2 \left(\frac{m-1}{m}\right) \tag{13.7}$$

where h^2 is the variance of the random field:

$$[\mathbf{h}(\mathbf{x}) \cdot \mathbf{h}(\mathbf{y})]_{av} = h^2 \delta(\mathbf{x} - \mathbf{y}) \tag{13.8}$$

But recall from Chapter 2 that $G(\mathbf{k}) \propto k^{-2}$. Therefore (13.7) diverges at or below $d = 4$, in agreement with the Imry–Ma argument.

This argument can be extended by doing a calculation of the magnetization $\mathbf{M}$ in the presence of an external uniform field $\mathbf{H}$ in addition to the random field. That is, we will find the equation of state relating M and H. We do this for the Landau–Ginzburg model (7.16) with a field term

$$- \int d^d x [\mathbf{h}(\mathbf{x}) + \mathbf{H}] \cdot \mathbf{S}(\mathbf{x}) \tag{13.9}$$

added. To lowest nontrivial order we obtain the diagrams of Fig. 13.1, which give (Aharony and Pytte, 1980)

$$M = \left(\frac{1}{r}\right) H - \left(\frac{1}{r}\right) \frac{u}{2} M^3$$

$$- \left(\frac{1}{r}\right) \frac{u}{2} M h^2 \left(\frac{m-1}{m}\right) \int \frac{d^d k}{(2\pi)^d} G_\perp^2(\mathbf{k}) \tag{13.10}$$

There are contributions to the h^2 term from fluctuations both parallel and perpendicular to $\mathbf{M}$, but we keep only the perpendicular ones because they are divergent as k and $H \to 0$.

We now need to know $G_\perp(\mathbf{k})$ in the presence of the field H. But this can be found exactly, as at $k = 0$, from rotational invariance alone. The argument is simply that adding an infinitesimal field $\delta \mathbf{H}$ perpendicular to $\mathbf{H}$ just amounts to rotating the field through an angle $\delta H/H$. And this just rotates $\mathbf{M}$ by the same angle i.e.

$$\frac{\delta H}{H} = \frac{\delta M}{M} \tag{13.11}$$

But the perpendicular susceptibility is just $\delta M/\delta H$, so we have

$$G_\perp^{-1}(\mathbf{k}) = \frac{H}{M} + O(k^2) \tag{13.12}$$

Proceeding, we evaluate the integral in (13.10) using (13.12) and keeping only the most divergent term in M/H:

$$\frac{H}{M} = r + \frac{u}{2} M^2 + C \frac{u}{2} h^2 \left(\frac{M}{H}\right)^{\frac{1}{2}(4-d)} \tag{13.13}$$

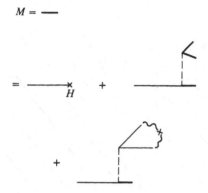

$M = \text{———}$

Figure 13.1: Diagrams for the equation of state of am m-vector Landau–Ginzburg model in a random field. The thick solid line stands for M, the thin solid line for G, the dotted line for u and the wavy line for the random field. The averaging of a pair of factors of the random field is indicated by tying the two wavy lines together with a cross.

where the constant is $O(1)$.

Following the experimental custom, we make an 'Arrott plot' of M^2 against H/M (Fig. 13.2). In an ordered state, there is a nonzero intercept at $H/M = 0$; for mean field theory without the random field we just get a set of parallel straight lines for different values of T (i.e. of r). We see that in the presence of the random field we never get any magnetization for zero field, though for low T (large negative r) the susceptibility is very large.

In the preceding arguments, the important thing for the divergent integrals was the presence of the factors

$$G^2(\mathbf{k})h^2 = \qquad\qquad \tag{13.14}$$

It turns out on further inspection (Grinstein, 1976; Parisi and Sourlas, 1979) that the most divergent diagrams in perturbation theory in general have this factor where the theory without the random field would just have G itself. This means, crudely speaking, that the effective dimensionality is shifted by 2 (every diagram diverges in a dimension 2 higher than in the corresponding pure system). Thus corrections to mean field exponents begin at $d = 6$ instead of $d = 4$. This is also consistent with the shift of the lower critical dimensionality from 2 up to 4 in the case of the random field Heisenberg model. However, there is a clear discrepancy between this argument, which would predict a lower critical dimensionality of 3 for the random field Ising model, and the Imry–Ma argument, which gave the lower

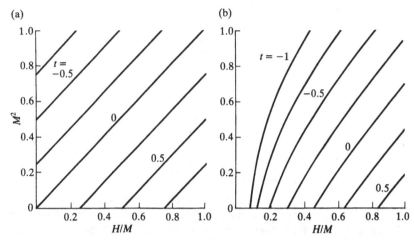

Figure 13.2: M^2 as a function of H/M ('Arrot plots') for (a) no random field and (b) finite random field (from Aharony and Pytte, 1980).

critical dimensionality 2.

An exact proof that the lower critical dimensionality is 2 has been given by Imbrie (1984,1985). The initially appealing argument for an effective dimensionality shift of 2 must therefore be wrong. One simply cannot trust an argument based on perturbation theory so far from the upper critical dimensionality. (That the upper critical dimensionality is 6, however, is not in question.)

Now a few words about experiments. It is hard to make a random field in the laboratory. However, Fishman and Aharony (1979) showed that a weakly diluted antiferromagnet in a uniform external field acts like a ferromagnet in a random field, so one can actually do experiments. To see this, consider the unit cell of a simple antiferromagnet (Fig. 13.3) with spins S_1 and S_2. The order parameter for the cell is $S_1 - S_2$ and the total magnetization is $S_1 + S_2$. Now in a cell where one of the spins is missing because of the dilution, the external field couples to one (or the other) spin instead of the sum. Since this single spin can be expressed as either the even or odd linear combination of the staggered and total magnetization in the cell, this means that in these cells, the external field couples to both the total magnetization and the order parameter. The sign of the coupling is random because either spin could be absent with equal probability.

After several years of controversy the experimental situation resolved itself in favour of the Imry–Ma prediction (Belanger et al, 1983; Birgeneau

Sublattices

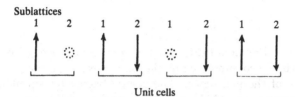

Unit cells

Figure 13.3: One-dimensional dilute antiferromagnet (see text for explanation).

et al, 1984). That is, there appears to be no long range order in $d = 2$ but there is in $d = 3$. The controversy arose because, as is spin glasses, there are two possible ways to get from a high-T, zero-field state to the low-T state in the field. One is to turn on the field already at high T, and the other is to cool in zero field and then turn on the field at low T. As in spin glasses, experiments done in these two ways give different results. The former appear to have no long range order and were first thought to support the hypothesis that the lower critical dimensionality was 3, while the latter showed showed long range order in three dimensions (Hagen et al, 1983; Wong et al, 1984). It now appears that the field-cooled state is not an equilibrium one, however, but relaxes very slowly (logarithmically) — so slowly that we could not expect to see the equilibrium ordered state in any resonable experimental time. Thus, as far as equilibrium is concerned, the Imry–Ma result is correct, but the dynamics of the slow decay of the field-cooled state is an interesting problem in itself. It has something in common with the slow decays we have discussed in spin glasses.

One cannot use this Fishman–Aharony trick except in the Ising case because the effective random field generated by the dilution points in only one direction. However, charge-density-wave systems allow a realization of the XY ($m = 2$) case. Here the order parameter is a periodic variation of the charge density, so ordinary impurities act like a random field would in the magnetic case. Studies of three dimensional charge-density-wave systems such as NbSe$_3$ are consistent with the existence of a characteristic domain size (the so-called Lee–Rice length) obtained from the Imry–Ma argument (Efetov and Larkin, 1977; Lee and Rice, 1979).

We can get a rough understanding of the dynamics of the relaxation of the field-cooled state (which presumably has domains of the wrong orientation) from the following argument (Villain, 1984; Grinstein, 1985). We consider first $d = 2$. Suppose that we have such a domain, of size L. In order to shrink, it must get activated over an energy barrier which we estimate as follows. For a shrinkage of size δL the energy change is given roughly by

$$\delta E \approx h(L \cdot \delta L)^{\frac{1}{2}} - J\delta L \qquad (13.15)$$

(where h is the typical random field). That is, it costs field energy to shrink the domain, but that is counteracted by the gain in exchange energy caused by the shortening of the perimeter. The energy barrier for shrinking the domain is the maximum of (13.15), which occurs for a shrinkage

$$\delta L \approx \left(\frac{h}{J}\right)^2 L \qquad (13.16)$$

This gives a barrier of order

$$\Delta \approx \frac{h^2 L}{J} \qquad (13.17)$$

(which can be very large for a large domain, even for small h). The relaxation time associated with this barrier is

$$\tau(L) = \tau_0 \exp\left(\frac{h^2 L}{JT}\right) \qquad (13.18)$$

Inverting this, we have that at time t domains of size

$$L(t) = \frac{JT}{h^2} \ln(t/\tau_0) \qquad (13.19)$$

will have equilibrated. This gives an estimate of the observed correlation length at time t after a quench from the high-T state. (As in other such estimates we have made, we take $\tau_0 = \mathrm{O}(10^{-12}\ \mathrm{sec})$, a typical microscopic time for magnetic systems.)

In d dimensions, δE becomes

$$\delta E \approx h(L^{d-1}\delta L)^{\frac{1}{2}} - JL^{d-2}\delta L \qquad (13.20)$$

and repeating the previous calculation gives

$$\delta L \approx \frac{h^2}{J^2} L^{3-d} \qquad (13.21)$$

which leads to the same barrier estimate (13.17) as for $d = 2$. We remark that the resulting domain growth law (13.19) is very similar to that (8.50) for spin glasses. The underlying common feature, of course, is that the dynamics in both systems are barrier-dominated. The difference is that while in the random-field system domains of the wrong orientation shrink

in size by typical jumps $\delta L \ll L$ (13.16), in spin glasses the shrinkage is of
the order of L itself (Fisher and Huse, 1988a).

For realistic experimental T we get $\xi \leq 5000$ lattice spacings, which is
consistent with experiment. Also consistent with experiment are that $d = 2$
and $d = 3$ look qualitatively the same and that $\xi \propto 1/h^2$ (Birgeneau et al,
1984).

So far we have only talked about the random-field system well below
its critical temperature, if it has a positive one. The transition in three-
dimensional random-field magnets is also very interesting, though far from
fully understood. We do not even know for certain whether the transition
is first or second order (Belanger et al, 1985; Birgeneau et al, 1986).

On the hypothesis that it is actually second order, we can describe the
transition in renormalization-group terms by the flow diagram of Fig. 13.4.
It shows three fixed points (the trivial infinite-temperature fixed point lies
off the graph):

1. The point $T = h = 0$, which is the sink of the RG flow in the low-
   temperature phase.

2. The fixed point at $h = 0$, $T = T_{c0}$ which describes the critical point
   of the pure system.

3. The fixed point at $T = 0$, $h = h_c$ which describes the critical behaviour for any nonzero h.

It is the last of these which is of interest to us here because it controls
the critical behaviour at the transition. Because it is not the same as
the pure system fixed point, the critical exponents in finite random field
are different from those with no randomness. For example, the exponent
β which describes the temperature-dependence of the magnetization just
below T_c is found to be very near zero (see, e.g. Ogielski, 1986). (The
smallness of β makes it very hard to determine experimentally whether the
transition is first or second order.)

The fact that this fixed point lies at $T = 0$, i.e. that temperature
is an irrelevant variable, makes the critical behaviour different from the
usual sort. Although the temperature goes to zero on large lengthscales
like $T(L) \propto L^{-\theta}$ with a characteristic exponent θ, physical quantities can
depend on T in a singular way, and their critical behaviour is changed as
a result. We met an example of such a 'dangerously irrelevant' variable
in Section 8.3 in connection with the AT line: The fourth-order term in
the Landau–Ginzburg expansion (8.77) was formally irrelevant above six

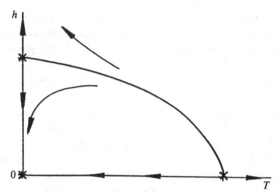

Figure 13.4: Renormalization group flows for the random-field magnet above its lower critical dimensionality.

dimensions, but the scaling of the external field depended on its coefficient, and this led to a violation of the usual scaling laws in the exponent giving the shape of the AT line. There are similar violations of scaling laws in the present problem. (For a more complete discussion see Nattermann and Villain (1988).)

Of particular interest is the violation of conventional dynamical scaling (Fisher, 1986). The argument goes like this: To determine the relaxation time, we renormalize to a length scale L equal to the correlation length $\xi \propto (T - T_c)^{-\nu}$. On this scale, the characteristic time varies as $\exp[C/T(\xi)]$, so we obtain

$$\tau(T) \propto \exp\left(\frac{C}{|T - T_c|^{\nu\theta}}\right) \tag{13.22}$$

Thus we find a critical dependence of the characteristic time of the system quite different from what we would have had with conventional dynamical scaling ($\tau(T) \propto |T - T_c|^{-\nu z}$). This happens because the relaxation time depends exponentially on the inverse temperature, which is singular because of its 'dangerous irrelevance'. Spin glasses, in contrast, obey conventional dynamical scaling (8.70, 8.100), because the fixed point describing the spin glass transition is a finite-temperature one.

13.2 Random anisotropy

We mentioned already in Sections 2.6 and 11.2 that spin glasses can also be amorphous. Amorphous $Gd_x Al_{1-x}$ with concentrations $0.30 \leq x \leq 0.40$

shows all typical spin glass properties, and a-FeNi, a-FeMn, a-FeCr, and a-ZrFe are reentrant systems in a certain concentration range. All these alloys have in common that their orbital moments are very small: In Gd systems, one has S-state ions, in transition metals like Fe and Mn the orbital moment is quenched, and Ni has no local moment at all.

The situation becomes more complicated in amorphous rare earth alloys such as TbAu, DyCu, TbFe, or $Dy_x Gd_{1-x}Ni$. The Tb and Dy ions possess an orbital moment which is influenced by the electrostatic field of the surrounding ions. In a crystal, this 'crystal field' leads to single-ion anisotropy which depends on the symmetry of the crystal. Owing to the strong spin–orbit interaction of the rare earth the spin of a single ion is forced together with the orbital moment into the easy direction of the lattice. In hexagonal crystals, there is a single easy axis, and the anisotropy energy takes the form $-D \sum_i S_{iz}^2$ as in (6.51). (We tacitly replaced the total angular momentum $\mathbf{J}$ by $\mathbf{S}$.) In cubic crystals, the anisotropy is of higher order in the spin variables. In amorphous materials, the lattice symmetry is broken and the leading anisotropy term again is of the order $S_{i\mu}^2$ ($\mu = x, y, z$). This suggests the model (Harris et al, 1973)

$$H = -\tfrac{1}{2}\sum_{ij} J_{ij} \mathbf{S}_i \cdot \mathbf{S}_j - \tfrac{1}{2}\sum_i D_i(\hat{\mathbf{n}}_i \cdot \mathbf{S}_i)^2 - \mathbf{h} \cdot \sum_i \mathbf{S}_i \qquad (13.23)$$

In what follows, we will simplify (13.23) somewhat, replacing the anisotropy strengths D_i by a constant D. This seems to be a good approximation in most random anisotropy systems (Moorjani and Coey, 1984). The anisotropy axes $\hat{\mathbf{n}}_i$ are assumed to be distributed randomly in all space directions with a probability $P(\hat{\mathbf{n}}_i) = \text{const}$. However, there might be correlations between two axes on the sites i and j. Fig. 13.5 shows the ground state of a spin chain with ferromagnetic exchange between nearest neighbours and random anisotropy: Owing to the exchange, the angles of adjacent spins differ only little. However, the random anisotropy destroys the long-range ferromagnetic order.

There are now four possibilities for randomness, and all exist in amorphous materials: If the exchange J_{ij} is mostly ferromagnetic and the anisotropy weak, one has a good ferromagnet. This seems to be the case for a-GdAu since all Gd alloys have extremely small anisotropy. However, a-GdAg seems to be near to a spin glass state and a-GdAl has sufficently strong negative exchange interactions that it becomes a spin glass in a certain concentration range. In a-$Fe_{91}Zr_9$ and a-$Fe_{98}Tb_2$, the ferromagnetic state is also destroyed by negative exchange, and in a-$Tb_{75}Fe_{25}$ and a-TbFe by random anisotropy and negative exchange. A standard random-axis system with strong anisotropy but ferromagnetic exchange is DyCu. It shows all typical spin glass properties, such as a linear low-temperature specific

Figure 13.5: Ground state of a spin chain with ferromagnetic exchange between nearest neighbours and random anisotropy (from Dieny, 1985).

heat, different zero-field-cooled and field-cooled magnetizations and slowly decaying remanence (von Molnar et al, 1982a,b). In a-DyNi, one observes scaling properties of the nonlinear susceptibility, but the critical exponents are apparently different from those of the 'canonical' spin glasses CuMn or a-GdAl (Dieny and Barbara, 1985, 1986). All amorphous magnets have in common with spin glasses the fact that their saturation field is extremely high.

In the simplest theoretical approach to amorphous magnets, we consider the Hamiltonian (13.23) in the framework of the SK model and assume for the exchange the Gaussian distribution (3.29) between all pairs J_{ij}. The anisotropy in (13.23) is a single-site interaction. Hence all manipulations of the SK model (Section 3.3) leading to the free energy (3.50) with (3.51) and to the MF equations in the replica-symmetric approach (3.52) and (3.53) go through as before, apart from two exceptions: First, we now have vector spins with the order parameters $q_{\mu\nu}$ and $p_{\mu\nu}$ (6.11–12). Second, the trace over the spins can no longer be performed. In the absence of internal strains or for completely random anisotropy axes, we can choose a coordinate system in which $q_{\mu\nu}$ and $p_{\mu\nu}$ are diagonal. The free energy and self-consistency equations then read (Fischer and Zippelius, 1986)

$$-\beta f = (\tfrac{1}{2}\beta J)^2 \sum_\mu (q_\mu^2 - p_\mu^2) - \tfrac{1}{2}\beta J_0 M^2$$

$$+ \int \prod_\mu \frac{dz_\mu}{(2\pi)^{\frac{1}{2}}} e^{-\frac{1}{2}z^2} [\ln \bar{Z}]_D \qquad (13.24)$$

with the partition function

$$\bar{Z} = tr \exp[\tfrac{1}{2}(\beta J)^2 \sum_\mu (p_\mu - q_\mu)S_\mu^2 + \bar{\eta}] \qquad (13.25)$$

$$\bar{\eta} = \beta \sum_\mu (Jq^{\frac{1}{2}}z_\mu + J_0 M_\mu + h_\mu)S_\mu + \beta D(\hat{\mathbf{n}} \cdot \mathbf{S})^2 \qquad (13.26)$$

where $z^2 = \sum_\mu z_\mu^2$ and

$$q_\mu = \left[\int \prod_\mu \frac{dz_\mu}{(2\pi)^{\frac{1}{2}}} e^{-\frac{1}{2}z^2} \langle S_\mu \rangle^2 \right]_{av} \tag{13.27}$$

$$\langle S_\mu \rangle = \bar{Z}^{-1} tr S_\mu \exp[\tfrac{1}{2}(\beta J)^2 \sum_\mu (p_\mu - q_\mu) S_\mu^2 + \bar{\eta}] \tag{13.28}$$

Similar expressions hold for the magnetization M_μ and for the order parameter $p_\mu = [\langle S_{i,\mu}^2 \rangle]_D$ (see (6.21)), where $[...]_D$ means averaging over the axes $\hat{n}_i$.

The result (13.24–13.28) has some interesting properties: The single-site anisotropy has the tendency to transform the vector spins into Ising spins. In the strong-anisotropy limit $D \to \infty$ all spins are confined to the directions $\hat{n}_i$. For $S_i^2 = 1$, one has

$$\mathbf{S}_i = \hat{n}_i \sigma_i \quad (D \to \infty) \tag{13.29}$$

with the Ising spin variables $\sigma_i = \pm 1$. The anisotropy term $D(\hat{n}_i \cdot \mathbf{S}_i)^2$ in the Hamiltonian (13.23) becomes a constant and for $J > J_0$ one finds the usual spin glass with the AT instability (3.27), apart from a numerical constant.

Here we are more interested in the opposite limit of strictly ferromagnetic exchange ($J = 0$, $J_0 \neq 0$). The equations (13.24) to (13.26) and (13.28) then no longer contain the spin glass parameters q_μ and p_μ. The only order parameter left is the magnetization $\mathbf{M}$, and no replicas are needed to derive the free energy. Hence one has no replica symmetry breaking, either, and all interesting spin glass properties disappear. One has a ferromagnet with reduced remanent magnetization for all values of D since the spins even at $T = 0$ are not parallel but lie in one hemisphere. In the limit $D \to \infty$, one derives a linear low-temperature specific heat but no other spin glass effects (Derrida and Vannimenus 1980). This is surprising: In the limit $D \to \infty$, the Hamiltonian (13.23) transforms for $J_{ij} = J_0/N$ with (13.29) into

$$H = -\tfrac{1}{2} \sum_{ij} \tilde{J}_{ij} \sigma_i \sigma_j - \sum_i h_i \sigma_i + \text{const} \tag{13.30}$$

with

$$\tilde{J}_{ij} = (J_0/N)\hat{n}_i \cdot \hat{n}_j, \quad \mathbf{h}_i = h\hat{n}_i \tag{13.31}$$

For random axes the exchange $\tilde{J}_{ij}$ is also random but with the restriction $\sum_\mu \hat{n}_{i\mu}^2 = 1$. Apparently, this restriction is strong enough to produce a ferromagnetic state for infinite-range interactions. This is in contrast to a short-range model with nearest neighbour interactions which does have spin glass properties: Computer simulations in three dimensions and renormalization group calculations in two dimensions indicate critical exponents similar to those in 3d and 2d short-range spin glasses (Chakrabarti and Dasgupta, 1988; Bray and Moore, 1985b).

We suspect that most amorphous rare earth alloys have partly ferromagnetic and partly spin glass properties. A convenient method for measuring the magnetic structure of a system is neutron scattering. The frequency-integrated magnetic scattering form factor $S_{Q,\mu}$ is proportional to the equal time spin correlation function $[\langle S_{i\mu} S_{j\mu} \rangle]_D$ which can be expressed in terms of the susceptibility $\chi_{ij,\mu}$ and the magnetization correlation function $q_{ij,\mu} = [m_{i\mu} m_{j\mu}]_D$ (which generalizes the spin glass parameter $q = [m_{i\mu}^2]_D$). From linear response we have

$$T\chi_{ij,\alpha} = [\langle S_{i\mu}, S_{j\mu} \rangle]_D - q_{ij,\mu} \tag{13.32}$$

which generalizes (1.2). The form factor $S_{Q,\mu}$ is just the Fourier transform of the spin correlation function:

$$S_{Q,\mu} = \sum_{ij} e^{i\mathbf{Q}\cdot(\mathbf{x}_i - \mathbf{x}_j)}(T\chi_{ij,\mu} + q_{ij,\mu}) \equiv T\chi_{Q,\mu} + q_{Q,\mu} \tag{13.33}$$

where $q_{Q,\mu}$ reduces for a homogeneous ferromagnet with $\mathbf{m}_i = \mathbf{M}$ to the Bragg term $M_\mu^2 \delta_{\mathbf{Q},\mathbf{K}}$ with the reciprocal lattice vectors $\mathbf{K}$.

The small-angle form factor $S_{Q,\mu}$ with $Q \to 0$ can be analysed in terms of a Lorentzian and the square of a Lorentzian. A single Lorentzian $S_Q = A(Q^2 + \xi^{-2})^{-1}$ is characteristic of an ideal ferromagnet. In amorphous materials, one observes in addition a term $B(Q^2 + \xi^{-2})^{-2}$, which vanishes above the characteristic temperature T_f in zero field. Fig. 13.6 shows the temperature dependence of the coefficients A and B for a-TbFe$_2$ where clearly $B = 0$ for $T > T_f = 409\,K$. The correlation length $\xi(T)$ has a more or less sharp maximum at T_f but remains finite below T_f. Typical examples are shown in Fig. 13.7: a-Tb$_{75}$Fe$_{25}$ and a-Tb$_2$Fe$_{98}$. We will show that the Lorentzian and Lorentzian squared terms can be identified with the two terms in (13.33).

A crude estimate of a correlation length ξ_D which depends on the exchange J of nearest neighbours and on the anisotropy energy D is obtained from the domain argument of Imry and Ma (1975) that we applied to random–field ferromagnets in the preceding section. For vector spins we argue that the formation of a domain with misaligned spins of size L in an otherwise ferromagnetic medium of d dimensions costs the exchange energy $E_J \sim JL^{d-2}$ since for slowly rotating spins at a distance b

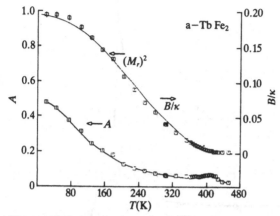

Figure 13.6: Temperature dependence of the Lorentzian coefficient A and the Lorentzian squared coefficient B (divided by $\kappa = \xi^{-1}$) for a-TbFe$_2$. The solid line in the upper curve is the temperature dependence of the field-induced moment extrapolated to $H = 0$ (from Rhyne, 1986).

$$\mathbf{S}(x) \cdot \mathbf{S}(x + b) = \cos(\pi b/2L) \approx 1 - \tfrac{1}{2}(\pi b/2L)^2 \qquad (13.34)$$

Since the spins would like to align also along the random axes, they gain in direction of net anisotropy the anisotropy energy $E_D \sim -DL^{d/2}$. The total energy is minimized for

$$L = \xi_D = (J/D)^{2/(4-d)} \qquad (13.35)$$

which leads to a critical dimension $d = 4$. For $d > 4$, the formation of domains is unfavourable and the ferromagnetic state stable. For $d < 4$, the system decays into domains. For $d = 3$ the typical size of these domains is $L = \xi_D \sim (J/D)^2$.

This estimate agrees for $d > 4$ with the result of the SK theory with infinite-range ferromagnetic exchange. For $d = 3$, we consider an improved site dependent MFT with short-range ferromagnetic exchange (Chudnovsky et al, 1986; Fischer, 1987). The conventional site-independent MFT of a ferromagnet leads to independent modes with momentum $\mathbf{Q}$ and violates the constraint $\sum_i S_i^2 = \sum S_Q^2 = 1$ (Brout, 1965). In the treatment presented in Section 4.1, this deficiency could be removed by including the Onsager term. A convenient way to remove it in a general manner is the spherical approximation (Section 3.6a) in which one adds the constraint $\sum_i S_i^2 = 1$ by means of a Lagrange parameter to the Hamiltonian. In our

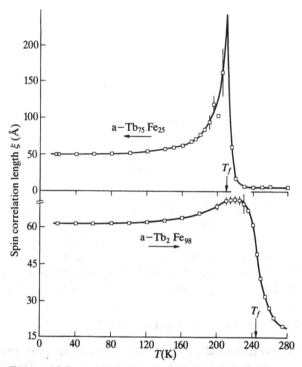

Figure 13.7: Spin correlation lengths ξ derived from a fit of the neutron scattering cross-section by a Lorentzian and the square of a Lorentzian. Note the differences in ξ near the characteristic temperature T_f, with that for a-Tb$_{75}$Fe$_{25}$ reaching a resolution-limited maximum value (from Spano and Rhyne, 1985).

case, we add the term $\frac{1}{2}\sum_i \lambda_i(S_i^2 - 1)$ to the Hamiltonian (13.23) and determine the parameters λ_i self-consistently. These parameters λ_i contain a term $\lambda\sum_l J_l$ of the pure ferromagnet (with $D = 0$) which renormalizes the Curie temperature and will be omitted. There remain contributions $\bar{\lambda}$ and $\delta\lambda_i$ which depend on the anisotropy parameter D. The parameter $\bar{\lambda}$ is determined by the normalization condition $q_{ij} \rightarrow q$ for $i = j$ (assuming a macroscopically isotropic state below T_f), and the parameters $\delta\lambda_i$ from the condition that the sponaneous magnetization $[m_{i,\mu}]_D$ should vanish in zero field since we look for a nonferromagnetic solution. We find with $J_Q \approx J_0(1 - b^2 Q^2)$ and $A = b^2 J_0$ the form factor for small momentum transfer $\mathbf{Q}$ and $D \rightarrow 0$

$$S_{Q,\mu} = \frac{T}{A}\frac{1}{Q^2 + \xi_\mu^{-2}} + \frac{4\pi\xi_D^{-1}(q - q_\mu)}{(Q^2 + \xi_\mu^{-2})^2} + M_\mu^2\delta_{Q,0} \qquad (13.36)$$

which indeed consists of a Lorentzian and a Lorentzian squared term. The correlation length ξ_μ with $\xi_\mu^{-2} = \xi_{FM,\mu}^{-2} + \xi_D^{-2}$ consists of a ferromagnetic contribution ξ_{FM} which vanishes in zero field below the characteristic temperature T_f, and a term $\xi_d \sim (A/D)^2$ which agrees with the Imry–Ma estimate (13.35). In zero field the form factor (13.36) is isotropic and the magnetization correlations q_{ij} decay exponentially in all directions for $T < T_f$ with the same correlation length ξ_D. In a finite magnetic field $\xi_{FM,\mu}$ remains finite and leads to a constant contribution (the last term in (13.36)) in the longitudinal correlation and to a reduction of the transverse correlation length. For $\xi_{FM} \gg \xi_D$ with $\xi_{FM} \approx h/AM \approx h/AS$ the latter is proportional to $h^{-\frac{1}{2}}$.

This calculation, together with the experimental data, suggests a spin glass state with a large but finite isotropic susceptibility and considerable ferromagnetic short-range order. Like spin glasses, this state is very sensitive to small fields. There is a local freezing of the spins at T_f, leading to finite magnetizations $\mathbf{m}_i = \langle \mathbf{S}_i \rangle$ which vanish on the average: $[m_{i\mu}]_D = 0$. For small anisotropy D, the directions of these moments fluctuate only little from site to site, but destroy the ferromagnetic order over a distance ξ which decreases with increasing anisotropy. So far, there is no analytic theory for random anisotropy materials which explains the observed spin glass properties such as remanence effects, a singular nonlinear susceptibility and relaxation which extends over long times.

14

The physics of complexity

Spin glass theory has had a rather large and unexpected impact on some problems far removed from spin glasses themselves. It turns out that a number of problems in fields outside physics share some of the essential features — randomness and frustration — that characterize spin glasses, and insights and techniques can be borrowed from spin glass theory and brought usefully to bear on them. This is especially true of the special novel concepts of mean field theory: broken ergodicity is a fundamental concept and broken replica symmetry may be a basic tool for analysing complex systems. While these methods probably do not apply to real spin glasses, many of these other problems have effectively infinite-ranged interactions, and mean field theory (sometimes with replica symmetry breaking) is applicable to them.

In the long run, these problems, rather than spin glasses themselves, may be the reason for reading the first part of this book. The hope that understanding spin glasses could be a key that unlocks the secrets of many other complex systems in and out of physics has been an important factor in making spin glass theory such an active field in the last decade.

In this chapter we describe a few representative examples of such problems. We will examine first some *combinatorial optimization problems*, in particular, the weighted matching, 'travelling salesman', and graph bipartitioning problems. Then we will describe some of the recent progress in understanding collective computational networks. Our account is necessarily sketchy, since a full treatment would fill another book this size. Our aim is just to illustrate how these problems can be formulated 'in physics language' and to show how they have some spin-glass-like features. We will have some descriptive things to say about their solutions, but we will not solve them here. Readers interested in going deeper into these and other such problems from a spin glass perspective should consult the excellent book by Mézard, Parisi and Virasoro (1987).

14.1 Combinatorial optimization problems

The SK spin glass may be viewed as an optimization problem — one wants to find the configuration which minimizes the energy. There are many other such problems that arise in general optimization theory, with applications in technological and economic systems. The problems which have been most intensively studied mathematically are, like the SK spin glass, extreme idealizations of the situation one wants to model, but, as in the SK case, one hopes to learn generic things from simple models amenable to mathematical analysis. Spin-glass-like theoretical methods have recently been applied to these problems, a few of which we now examine. Our purpose here is just to exhibit how one formulates them as statistical–mechanical problems, not to solve them.

The weighted matching problem

The first problem is called the weighted matching problem (WMP)(Orland, 1985; Mézard and Parisi, 1985). One has a set of N points, with d_{ij} the distance betwen the ith and the jth points. The points may be taken as randomly distributed in some space and the d_{ij} calculated from that, or we may consider a version of the model in which the d_{ij} are independent random variables with some distribution $P(d_{ij})$. The problem is then to link the points together in pairs so that the total length of these links is the mimimum of all its possible values (Fig. 14.1). That is, taking $n_{ij} = n_{ji} = 1$ if there is a link between point i and point j, and $n_{ij} = 0$ otherwise, the task is to minimize

$$E = \sum_{ij} d_{ij} n_{ij} \tag{14.1}$$

subject to the constraint that each point is only linked to one other. This constraint can be expressed as

$$\sum_j n_{ij} = 1 \quad \text{(all } i) \tag{14.2}$$

We can make this into an Ising-like model in the $N(N-1)/2$ variables n_{ij}. We first soften the constraint, writing a partition function

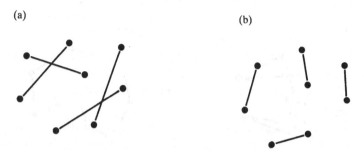

Figure 14.1: The weighted matching problem. (a) shows a poor solution; (b) shows a good solution.

$$Z = Tr_n \exp\left(-\beta \sum_{\langle ij \rangle} n_{ij} d_{ij}\right) \prod_i \exp\left[-\gamma\left(1 - \sum_j n_{ij}\right)^2\right] \quad (14.3)$$

In the limit $\gamma \to \infty$, the last factor will vanish unless (14.2) is satisfied for all i. In the limit $\beta \to \infty$, $-lnZ/\beta$ is then the minimal length of the connecting links. Thus we have an effective Hamiltonian or *cost function*

$$\beta H_{eff} = \sum_{\langle ij \rangle}(\beta d_{ij} - 4\gamma)n_{ij} + \gamma \sum_{ijk} n_{ij}n_{ik} \quad (14.4)$$

In spin language, there is a (random) field acting on the link variables, plus an antiferromagnetic interaction between them. The d_{ij}'s are random, so one has to resort to replicas or some other way of averaging over their distribution.

The formulations one finds in the literature are slightly different, but the present one exhibits nicely the analogy with a random spin system. Notice that here the randomness is all in the first (random field) term, while the frustration is contained in the antiferromagnetic interactions between all pairs of links, in contrast to the spin glass, where the randomness and frustration were both in the exchange term.

(a)   (b)

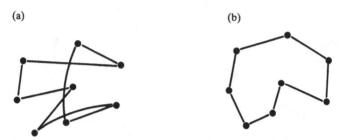

Figure 14.2: The travelling salesman problem. (a) shows a poor solution; (b) shows a good solution.

The travelling salesman

The second problem, called the travelling salesman problem (TSP) (Kirkpatrick and Toulouse, 1985; Hopfield and Tank, 1985; Mézard and Parisi, 1986) is rather similar. We start with the same sort of distribution of points as before, but here the problem is to find the shortest path passing through each point (city) exactly once (Fig. 14.2). Let us define a variable $n_{ja} = 1$ if the jth city is the ath stop in the tour, otherwise $n_{ja} = 0$. The total length of the tour is

$$E = \sum_{\langle ij \rangle, a} d_{ij} n_{ia}(n_{j,a+1} + n_{j,a-1}) \tag{14.5}$$

The constraints on the n_{ja} are (1): each city appears exactly once in the tour i.e.

$$\sum_a n_{ja} = 1 \quad \text{(all } j) \tag{14.6}$$

and (2) each stop on the route contains just one city:

$$\sum_j n_{ja} = 1 \quad \text{(all } j) \tag{14.7}$$

Thus by analogy with the WMP, the partition function is

$$Z = Tr_n \exp\left[-\beta \sum_{\langle ij \rangle, a} d_{ij} n_{ia}(n_{j,a+1} + n_{j,a-1}) \right]$$

$$\times \prod_a \exp\left[-\gamma \left(\sum_i n_{ia} - 1 \right)^2 \right] \prod_i \exp\left[-\gamma \left(\sum_a n_{ia} - 1 \right)^2 \right] \tag{14.8}$$

and the effective Hamiltonian or cost function is given by

$$\beta H_{eff} = \beta \sum_{\langle ij \rangle, a} d_{ij} n_{ia}(n_{j,a+1} + n_{j,a-1}) + 2\gamma \sum_{\langle ij \rangle, a} n_{ia} n_{ja}$$

$$+ \gamma \sum_{iab} n_{ia} n_{ib} - 4\gamma \sum_{ia} n_{ia} \qquad (14.9)$$

(The solution is of course $2n$–fold degenerate, since the tour can start at any city and go in either direction.) Again, randomness and frustration are both present, and one expects many 'metastable tours'. The thermal fluctuations present in a finite-temperature simulation allow one to escape locally stable tours which are not of the shortest length. 'Simulated annealing' is a strategy for solving such problems in which one tries to find the best (or a very good) solution by gradually lowering the temperature (Kirkpatrick et al, 1983).

These problems are technically harder than the spin glass, because one needs a whole family of q matrices, one for each cumulant of the field, instead of just the Edwards–Anderson one for the second cumulant. For this reason, they have only been solved in the replica-symmetric ansatz. The stability of the solutions has not been tested, so the possibility of replica symmetry breaking or broken ergodicity remains unsettled. There is some evidence (Kirkpatrick and Toulouse, 1985) for ultrametricity in the TSP; this would be hard to reconcile with a stable replica-symmetric solution if it is true of equilibrium, i.e. of the shortest tours.

Graph bipartitioning

Since we have emphasized the importance of replica symmetry breaking as a basic tool for characterizing complex structure, it would be nice to have an example of a combinatorial optimization problem where we know it happens. Graph bipartitioning is such a problem. It may be described as follows (Fu and Anderson, 1986).

We begin with a general *graph*, i.e. a set of points i and a set of *edges* which connect pairs ij of the points. A simple example would be a set of points which form a Euclidean lattice with edges connecting only nearest-neighbour points. Here we will be concerned instead with random graphs in which there is a fixed probability p of a given point being connected with any other point. The average number pN of other points to which a given one is connected is called the *valency* of the graph. We can consider both *extensive* (i.e. $p = O(1)$) and *intensive* valency ($pN = O(1)$). Here we will mostly be concerned with the extensive case. Given such a graph, the task is to divide the points into two sets of equal size in such a way that

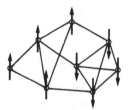

Figure 14.3: A solution of a graph bipartitioning problem

the number of edges connecting points in different sets is minimized (Fig. 14.3).

Thus, defining $J_{ij} = 1$ if points i and j are connected and $J_{ij} = 0$ if they are not, and associating a variable S_i with each site which is +1 if the site is in one set and −1 if it is in the other, we want to minimize

$$L = -\sum_{\langle ij \rangle} J_{ij} S_i S_j \tag{14.10}$$

subject to the constraint

$$\sum_i S_i = 0 \tag{14.11}$$

In magnetic terms this is a randomly diluted ferromagnet with the constraint of zero total magnetization.

A way to enforce the constraint softly (in analogy to what we did in (14.3) and (14.8) is to add a term in the effective Hamiltonian which strongly penalizes total magnetizations different from zero. Thus we have a cost function

$$H = -\sum_{\langle ij \rangle} J_{ij} S_i S_j + \mu \left(\sum_i S_i \right)^2 \tag{14.12}$$

The second term adds a constant antiferromagnetic interaction between all pairs of 'spins', so now some of the interactions will be positive and some negative.

So essentially we have an infinite-range spin glass! (Even though we only recover the exact SK model this way with a symmetric bond distribution for $p = \frac{1}{2}$ and a special value of μ, one can make an exact mapping to an SK model (Fu and Anderson, 1986) for any $p = O(1)$). Thus the solution of the graph bipartitioning problem indeed has the full richness of nontrivial broken ergodicity (broken replica symmetry) that we studied in the mean field theory of spin glasses.

The bipartitioning problem with *intensive* valency corresponds to the kind of infinite-range spin glass where each spin only interacts with a finite ($O(1)$) number of other spins (but these other spins could be anywhere in the system). As we remarked in Chapter 3, the simple replica-symmetric solution of such a model is unstable, and no stable solution has been found yet.

There are obvious generalizations of this bipartitioning problem to multipartitioning (dividing the points up into three or more sets instead of two). These would be similarly related to infinite-range Potts glasses (Section 3.7).

14.2 Collective computation and neural networks

The other area we will mention where ideas from spin glass theory have played an important role is that of 'neural networks'. This kind of theory has two aims: (1) to understand the nature of collective computation in artificial networks composed of many simple processors, and (2) to model the operation of the brain. While these problem areas are large and complex, there are some general questions where relatively simple models may be able to provide insight.

Underlying the potential relevance to real neural systems is the following (very simplified) picture of how nerve cells (neurons) in the brain operate (Fig. 14.4). (For more background on the nervous system, see Kuffler et al, 1984; and Kandel and Schwartz, 1985.) Electrical signals in the nerves cause special chemicals, called transmitter substances, to be released at the synaptic junctions where the nerves almost touch. This leads via some complicated process at the membrane of the receiving cell to a local flow of ions in or out of the cell, which raises or lowers the electrical potential inside it. Now the internal dynamics of the cell are such that if the potential exceeds a certain threshold, a soliton-like wave propagates from the cell body down the axon (we say that the cell 'fires'). This then leads to the release of transmitters at the synapses to other cells, which react in the same way. The nervous system is a huge ($\sim 10^{11}$ cells) highly interconnected assembly of such cells.

An important point is that, to a good approximation in many kinds of neurons, every firing of a given cell is identical with every other firing. That is, a cell has effectively just two meaningful states, firing and not firing. It

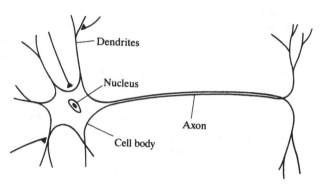

Figure 14.4: A crude picture of a nerve cell with synapses from and to other cells.

can therefore be described by a binary variable. We adopt a spin analogy, calling the variable characterizing cell i $S_i(= \pm 1)$, $(i = 1, \ldots N)$. We also know that cells can act either to inhibit or to excite each other. That is, writing the voltage ('postsynaptic potential') in cell i as

$$h_i = \sum_j J_{ij} S_j \tag{14.13}$$

the parameters J_{ij} describing the influence of cell j on cell i can be either positive (ferromagnetic) or negative (antiferromagnetic, in magnetic language). The idealized dynamics we have described may be formulated (in a discrete-time picture) (McCullough and Pitts, 1943; Caianiello, 1961; Little, 1974) as

$$S_i(t+1) = \text{sgn}\left[\sum_j J_{ij} S_j(t) - \theta_i\right] \tag{14.14}$$

where θ_i is the threshold of the ith cell. If the cells are updated one at a time in random sequence, this corresponds exactly to the zero-temperature limit of Monte Carlo dynamics. J_{ij} is an exchange interaction and $-\theta_i$ is an external field on spin i. The fact that both positive and negative J_{ij} are known to occur even hints at a possible analogy with spin glasses, with many metastable states, etc.

According to this picture, we should view the brain in computational terms as a multiprocessor system with billions of processors each executing very simple programs (14.14) in parallel. This is essentially the opposite of the way a conventional serial computer works, where one very complex processor (or maybe a few of them) executes a very complex program.

Of course, much has been left out in this picture, notably dynamical aspects such as the dependence of the firing thresholds on the recent firing history of the cell, which makes the dynamics of the system much more complex. Furthermore, there is a huge variety of cells, differing both in morphology and mode of computation (some do not fire in the above discrete fashion, transmitting graded potentials instead). Serious computational neuroscience has to take such variety into account. On the other hand, even this oversimplified picture is very rich as a source of inspiration for design of artificial computing networks, so we restrict our discussion here to it.

One difference between standard spin models and the present kind of system lies in the fact that the systems we generally meet in physics have symmetric interactions: $J_{ij} = J_{ji}$. This is not the case for neural systems, either natural or artificial, at all. This makes them much more difficult, in principle, and in fact hindered progress in understanding such models for a long time. However, we will follow what has turned out to be a rather successful strategy, and study first the symmetric case. We will see later how the features we discover are affected by the introduction of asymmetry.

The analogy with spin systems can also be extended to nonzero temperature. This has a biological basis in real neural systems: The amount of transmitter substance released at a synapse when a cell fires can vary somewhat, so it is possible that, for example, a cell does not fire, even though $\sum_j J_{ij}S_j$ exceeds θ_i. We can therefore introduce a stochastic dynamics in which $S_i(t+1) = +1$ with probability

$$P_+(h_i) = \frac{1}{1 + \exp[-2\beta(h_i - \theta_i)]} \tag{14.15}$$

Any sigmoidal function of h_i with limiting values of 0 and 1 at $-\infty$ and $+\infty$ will lead to similar behaviour, but this choice is particularly convenient because it corresponds exactly to finite-T Glauber dynamics. For symmetric J_{ij}, then, the system is guaranteed to obey equilibrium statistical mechanics: It has a stationary distribution $\exp(-\beta H)$ with the Hamiltonian (Hopfield, 1982, 1984; Amit et al, 1985a)

$$H = -\tfrac{1}{2} \sum_{ij} J_{ij}S_iS_j + \sum_i \theta_i S_i \tag{14.16}$$

The equilibrium phases or pure states (labelled by superscript a) are characterized by (possibly zero) expectation values $\langle S_i \rangle^a \equiv m_i^a$. In terms of cells firing or not firing, then $(m_i^a + 1)/2$ is the mean firing rate of cell i in state a.

In the context of artificial computing networks it can actually be advantageous to introduce this kind of noise, as we will see below.

So far we have not addressed the question of how the J_{ij}'s get their values and what these have to do with the functioning of the system. Here we follow the suggestion of Hebb (1949) who hypothesized that the connection between two cells was strengthened when the firing of one succeeded in causing the other to fire, while if it failed to do so the synapse would be weakened. In this way, he argued, information about the history (experience) of the system would be stored in the values of the synaptic connection strengths, and this could be the neural basis of memory.

In the language of our spin system the Hebb hypothesis amounts to something like (Caianiello, 1961)

$$\frac{\partial J_{ij}}{\partial t} \propto S_i(t)S_j(t) \tag{14.17}$$

(Other, slightly different forms are also consistent with and maybe closer to both Hebb's original formulation and the real biological situation, but we will use (14.17) in the model calculations we discuss here.) While the real details have still not been sorted out, there is experimental evidence for Hebb's hypothesis, and most neuroscientists seem to accept as the only plausible working hypothesis that (long-term) memory is stored in some such way in the synaptic connection strengths.

We are especially interested in questions such as how a neural network can recall stored items on the basis of incomplete or noisy input (like recognizing a badly drawn number or letter, or a face from a degraded picture), even though the individual neurons do not operate reliably or precisely. This property is called *associative memory*. It is obviously characteristic of real brains, but not of ordinary computers. The hope is that an artificial computing network based on (14.14) and (14.17) would also have such desirable features.

To get an idea how this can happen, we consider the simplest possible situation, where the $S_i(t)$'s in (14.17) have been forced into particular patterns ξ_i^μ by a very strong external field. (If we are imagining this to be happening in the brain, the external field comes from cells in the sensory cortex which transmit signals from the external world to the associative areas of the cortex. These signals are of course highly transformed representations of the original physical stimuli, but this transformation is not part of the problem we are discussing here.) For the sake of formulating a simple model amenable to analysis we suppose further that all the patterns imposed in this way are binary ($\xi_i^\mu \pm 1$) and statistically independent, and that different patterns receive equal exposure. Then we get (Hopfield, 1982, 1984; Amit et al, 1985a)

$$J_{ij} = \frac{1}{N} \sum_{\mu=1}^{p} \xi_i^\mu \xi_j^\mu \qquad (14.18)$$

(The factor of $1/N$ is just a convenient choice for setting the scale of the couplings.) We call the ξ_i^μ *training patterns* and (14.18) the *Hebb rule*. In what follows, we set the thresholds $\theta_i = 0$, though the theory can also be worked out without this simplification. We also treat the J_{ij}'s as frozen, so the influence of the recall process on the memory is ignored. We are thus making something like a Born–Oppenheimer approximation (Caianiello, 1961). This is apparently reasonable in practice; indeed, it would be hard to imagine how the system would function without this separation of timescales.

How can such a system function as an associative memory? One can get a hint by considering the trivial case $p = 1$ where there is just one uniform training pattern $\xi_i^0 = 1$. Then (14.18) is just an infinite-range ferromagnet. Its stable states are uniform (all up or all down), like the training pattern. More generally, for $p = 1$ with any training pattern ξ_i^0, the states $S_i = \pm\xi_i^0$ will be the ground states. If the system is put in any other configuration, it will eventually evolve to one of these two. This is an elementary example of association: In the uniform example, the system will go to the all-up state if the majority of spins in the initial state are up. In general, it finds the stable state which most resembles the starting configuration. This shows how the memory acts to correct errors in the input (starting) configuration, just by following its dynamics to its nearest stable state. Since the evolution of the system is driven by the collective dynamics of the spins, we also use the term 'collective computation' to describe what the system is doing.

None of this should be very surprising. What is less obvious but also easy to show is that even when there are many different patterns imprinted in the synaptic strengths via the Hebb rule (14.18), every one of them is a locally stable state, as long as they are few in comparison to the size N of the system. To see this, suppose the system is in state ν, and compute the field acting on spin i. We split the sum on training patterns μ in the synaptic strength J_{ij} (14.18) into the term with $\mu = \nu$ and all the rest, finding

$$h_i^\nu \equiv \sum_j J_{ij}\xi_i^\nu = \xi_i^\nu + \frac{1}{N} \sum_{\mu\neq\nu,j} \xi_i^\mu \xi_j^\mu \xi_j^\nu \qquad (14.19)$$

If the patterns are uncorrelated, the second term is random, with typical size $[(p-1)/N]^{\frac{1}{2}}$, so if $N \to \infty$ this 'crosstalk' is negligible in comparison with the first term. Thus each spin is lined up along its local field; the state is stable *for every* ν.

Incidentally, this shows that if the ξ_i^μ are orthogonal to each other, there is no crosstalk at all; the system can thus store up to the maximum number N of possible mutually orthogonal patterns. The noise in this model comes about only because the ξ_i^μ are random and therefore have some random mutual overlap. By the same reasoning, if the training patterns are correlated more than randomly, the Hebb rule (14.18) will not lead to reliable recall.

Returning to the case of independent binary random patterns, we now consider the effects of finite temperature. The system can still recall all p patterns (Amit et al, 1985a) as long as $T < T_c = 1$ (though no longer perfectly) for any finite number of patterns p, in the limit $N \to \infty$. To see this, we simply make the ansatz that for each training pattern ν there is a stable pure state ('memory state') in which the local spontaneous magnetization is proportional to ξ_i^ν:

$$\langle S_i \rangle = m\xi_i^\nu \qquad (14.20)$$

We now proceed in mean field theory, replacing S_i in (14.13) by (14.20) and computing

$$m\xi_i^\nu = \tanh(\beta N^{-1} \sum_{j,\mu} \xi_i^\mu \xi_j^\mu \cdot m\xi_j^\nu) \qquad (14.21)$$

Making a decomposition of the internal field like (14.19), we can again throw the crosstalk term away and obtain simply

$$m = \tanh(\beta m) \qquad (14.22)$$

which is just the mean field equation (2.13) for the magnetization of a ferromagnet. So up to $T_c = 1$ the equilibrium state will have a finite overlap with the training pattern. The shape of $m(T)$ is such that the average recall error $\frac{1}{2}(1 - m)$ in each bit of the pattern is still quite small until T gets rather close to 1.

As we remarked above, finite T may even be useful: One finds 'spurious states' other than the training patterns (states where $\langle S_i \rangle$ is proportional to a sum of several different ξ_i^μ's) which are stable for low T. But it turns out that none of these spurious states has a transition temperature above 0.46, so by choosing the temperature just above this one can avoid them while still keeping the memory states stable and their average recall error very small (a few percent).

It is important to recognize the difference between this model and the SK spin glass, though the system described by the Hamiltonian (14.16) with the interactions given by the Hebb rule (14.18) can be thought of as a random-bond system because the patterns ξ_i^μ are random. In the limit

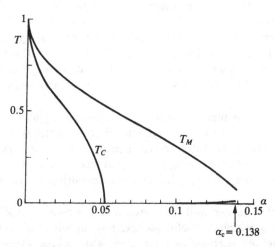

Figure 14.5: The phase diagram of the Hopfield neural network in $\alpha - T$ space (Amit et al, 1985b).

$p \ll N$ which we have been studying so far, this system has nothing like the hierarchically correlated states that the SK glass had. For $0.46 < T < 1$, it has just $2p$ uncorrelated states, for example. At low temperatures, there are very many spurious states in addition to the memory ones, but they have no ultrametric organization.

It is of great interest to study the capacity of the memory. That is, what happens when the parameter $\alpha \equiv p/N$ becomes finite? Then the crosstalk term in (14.19) can no longer be ignored, even for uncorrelated training patterns.

This case was also studied by Amit et al (1985b, 1987a). We do not reproduce their calculations here because of their length, but they are rather straightforward applications of the replica method like those of Chapter 3. Basically, they find that memory states with $S_i \propto \xi_i^\mu$ remain locally stable (with a small amount of error) for small enough α, but beyond a critical value $\alpha_c(T)$ they suddenly become unstable and one goes over to a spin glass phase. At $T = 0$, α_c attains its maximum value ≈ 0.14. There is therefore a roughly triangular region of the $\alpha - T$ plane where the memory states are locally stable (see Fig. 14.5). The order parameter m jumps discontinuously from a finite value to zero at the stability limit. The jump is quite large at low T; for $T = 0$ one goes from $m > 0.97$ (a recall error of less than 2%) to $m = 0$ at α_c.

The spin glass phase above $\alpha_c(T)$ has the ultrametric structure and the

other characteristic properties of the SK spin glass. Its many pure states
are uncorrelated with the ξ_i^μ. At sufficiently high temperature, the spin
glass phase melts to a paramagnetic one just the way the SK model does.

Actually, the spin glass phase still has a lower free energy than the
memory states in a region $\alpha_1(T) < \alpha < \alpha_c(T)$. Below α_1 the memory
states have a lower free energy, so the line $\alpha_1(T)$ is a standard first-order
phase transition line. The spin glass phase remains locally stable for all
$\alpha < \alpha_1(T)$.

At low enough T, the memory states also undergo replica symmetry
breaking, indicating the growth of an ultrametric tree of states from each
memory state, but the quantitative consequences of this are very small.
This is also shown in the figure, though the region where these phases occur
is hard to see except just below α_c because the transition temperature is
so small.

Thus this model is rather well understood. More recent work has fo-
cused on the degree to which its useful features remain robust when various
aspects of the model are changed or restrictive assumptions relaxed: One
can also see how it works when the up–down symmetry is broken by an
external field (Amit et al, 1987b) and how to modify it when the patterns
are correlated. Of particular interest is the case of hierarchically correlated
patterns (Parga and Virasoro, 1986, Cortes et al, 1987, Ioffe and Feigelman, 1987, Krogh and Hertz, 1988). One can also change the dynamics,
for example, to one involving synchronous updating of all the spins (Amit
et al, 1985a).

Another modification, called 'clipping', involves rounding the synaptic
strengths in (14.18) to the nearest of a small discrete set of values; the
extreme case would be (van Hemmen, 1986)

$$J_{ij} = \frac{p^{\frac{1}{2}}}{N}\mathrm{sgn}\left(\sum_{\mu=1}^{p}\xi_i^\mu\xi_j^\mu\right) \tag{14.23}$$

The system also seems to exhibit a degree of robustness against this clipping: The capacity of the system at $T = 0$ is just reduced from the value
$\alpha_c \approx 0.14$ given above to $\alpha_c \approx 0.10$. This is very important for the design
of memory chips based on these principles, since the J_{ij}'s are represented
by resistors, and one cannot control their resistances that precisely.

One can also study what happens when asymmetry is introduced into
the synapses, so that $J_{ij} \neq J_{ji}$ (Hertz et al, 1986a,b). Specifically, suppose
we multiply (14.18) by a random variable $w_{ij} = 0$ or 1, taking w_{ij} and w_{ji}
as independent variables, otherwise using the same dynamical prescription
as before. The asymmetry of J_{ij} means that the existence of a stationary
Gibbs distribution no longer follows, so we cannot use equilibrium statis-

tical mechanical methods. However, we can write formal kinetic equations describing the spin evolution as in Chapters 4 and 5 and average them over the stochastic 'thermal' noise and over the random training patterns ξ_i^μ. We find the following: (1) At $\alpha = 0$ (i.e. p finite, while $N \to \infty$) the memory states remain stable for low T; the only difference is that the transition temperature T_c is lowered to $\frac{1}{2}$. This is easy to understand in terms of simple dilution. (2) The spin glass state found at finite α in the symmetric case is suppressed. This may actually be helpful to the functioning of the system as a memory, since it cannot any longer be trapped in spin glass configurations. On the other hand, numerical work (Kinzel, 1986b) indicates the existence of persistent cycles at low T, so these dynamical traps may be replacing the static traps of the spin glass phase as the price one pays for beginning to overload the capacity of the system.

It is not intended to take this kind of model very literally, i.e. in terms of specific neurons in the brains of advanced animals. On the other hand, it serves to illustrate the nature of the computation which the system must carry out. Establishing just how real neural hardware carries out such a computation is then a further problem in neuroscience. This is an area where interaction between theoretical formulations like this and experimental neuroanatomy and electrophysiology can be very fruitful.

Despite its virtues, this model is still incomplete and deficient as a basis for practical artificial computational networks. The capacity found for the Hebb rule $p_{max} \approx 0.14N$ (and that only for uncorrelated patterns) is rather small compared to the potential capacity of $2N$ independent patterns (Gardner, 1987). An approach that improves dramatically on the limited capacity is to forget trying to find an explicit formula like (14.18) for the requisite J_{ij}, concentrating instead on finding an iterative algorithm for improving the J_{ij} until satisfactory performance is achieved. Such an algorithm can be found for the kind of model we have discussed here, including the generalization to the case where the patterns ξ_i^μ are only defined on some of the cells i of the network (Hinton and Sejnowski, 1986). There is at present a great deal of activity concentrated on this and other algorithms and their applications to practical problems.

At the time of writing, the actual construction of artificial neural networks is at a rather primitive stage, though real hardware implementations are expected to begin to be common within a few years. Nevertheless, simulations of these networks have proved to be very valuable in themselves. The reason is that the close connection with a kind of physical system over which we already have some intuitive command makes computations based on them more transparent and easier to program than traditional algorithms, even though they are still implemented on conventional serial computers.

15

A short history of spin glasses

The first indication that well-known alloys such as A̲u̲Fe or C̲u̲Mn might have unusual properties came from measurements of the low temperature specific heat $C(T)$. De Nobel and Chantenier (1959) and Zimmermann and Hoare (1960) found in A̲g̲Mn and C̲u̲Mn a linear term in $C(T)$ which could not be attributed to the conduction electrons and turned out to be independent of the Mn concentration. Marshall (1960) and Klein and Brout (1963) assumed that these alloys could be described by an Ising model with RKKY interactions. Simple assumptions lead either to a Lorentzian or a Gaussian distribution $P(h_{eff})$ of effective fields h_{eff}, with $P(0) \propto c^{-1}$, where c is the impurity concentration. The free energy

$$F = -\beta[\ln Z]_{av} = -\beta \int_{-\infty}^{\infty} dh_{eff} P(h_{eff}) \ln 2 \cosh \beta h_{eff} \qquad (15.1)$$

then immediately leads to a linear specific heat as $T \to 0$. Ising spins, of course, represent a special case of two-level systems. Such systems (albeit of different physical origin) exist in true glasses, which also exhibit a linear specific heat. Anderson et al (1972) attempted to explain both types of systems by a distribution of two-level systems with different energies and tunnelling probabilities.

Originally, characteristic spin glass properties such as the linear specific heat, a time-dependent remanent magnetization (Kouvel, 1960, 1961; Beck, 1972), and anomalies in ESR spectra (Owen et al, 1956) were known only in metallic systems with RKKY interactions between the spins. The RKKY interactions decay asymptotically as a function of distance R between two magnetic impurities as R^{-3}. Blandin (1961) predicted that for small concentrations this dependence should lead for the field dependent magnetic specific heat $C_m(T, h, c)$ and magnetization $M(T, h, c)$ to the scaling properties

$$C_m(T, h, c) = cf_1\left(\frac{T}{c}, \frac{h}{c}\right), \quad M(T, h, c) = cf_2\left(\frac{T}{c}, \frac{h}{c}\right) \tag{15.2}$$

which have indeed been observed by Souletie and Tournier (1969).

Spin glasses are interacting spin systems, and these interactions lead to magnetic order in periodic systems. This raises two questions: (1) Is there also a sharp phase transition in spin glasses, and if yes (2) what type of magnetic order develops below the characteristic temperature? These central questions remained unanswered for more than a quarter of century.

Owen et al (1956) interpreted their susceptibility and ESR data on dilute CuMn as evidence for a phase transition with antiferromagnetic order, whereas the Grenoble group explained their magnetization data on CuFe and AuFe by a theory of Néel (1961) for a collection of fine magnetic particles (Tholence and Tournier, 1974). This theory predicts a continuous freezing of the spins without any sharp phase transition and seemed to agree with the broad maximum in the susceptibility $\chi(T)$ as observed in fairly high fields. A similar description of a spin glass as a collection of fairly independent spin clusters was also advocated by Kouvel and Beck, who found a displacement of the magnetization hysteresis loop after cooling in a field (Kouvel, 1960) and fit the magnetization $M(T, h)$ by a superposition of Brillouin functions (Beck, 1972). Beck called these metallic spin glasses 'mictomagnets'.

Mössbauer data on AuFe in turn yielded evidence for a sharp phase transition instead of a continuous freezing of the spins (Violet and Borg, 1965, 1967). This interpretation found strong support in the famous experiments of Cannella et al (1971) and Cannnella and Mydosh (1972), who observed a sharp cusp in the ac susceptibility of CuMn, AgMn, AuMn, and AuCr. In contrast to earlier dc susceptibility experiments, these measurements were performed with very small fields of the order 5 to 20 Oe, and the cusp turned out to be extremely sensitive to the field strength. The cusp temperature turned out to be frequency-dependent (Tholence, 1980). A simple demonstration of this frequency dependence was the different freezing temperatures observed in Mössbauer and ac susceptibility measurements. One could wonder whether the cusp temperature really remained finite for $\omega \to 0$. However, an indication that $T_f(\omega = 0)$ was finite came from the earlier work of Nagata et al (1979), who had observed a similar cusp in the dc susceptibility in fields of the order of 5 Oe. This experiment also showed different field-cooled and zero-field-cooled magnetizations below $T_f(\omega = 0)$. This difference had been demonstrated already earlier for larger fields (Tholence and Tournier, 1974).

The cusp in $\chi(T)$ suggested a sharp phase transition to a so far unknown new magnetic state. Neutron scattering data indicated no Bragg reflections, even at the lowest temperatures (Arrott, 1965). Hence the spins did not order periodically and the staggered or homogeneous magnetizations could not be the proper order parameter.

This led Edwards and Anderson [EA] (1975) to their famous theory in which they introduced the quantity $q_{EA} = [\langle S_i \rangle^2]_{av}$, equation (2.37), as spin glass order parameter. The parameter q_{EA} also enters into the local susceptibility (2.44), and the vanishing of $q_{EA}(T)$ for $T \geq T_f$ led to a simple explanation of the cusp in the susceptibility. Unfortunately, this theory had two flaws: It was a mean field theory, and mean field theories sometimes predict phase transitions in systems in which there is none, and the solution had some unphysical properties, as discussed below.

The EA theory was an enormous step forward: It was based on the assumption that a spin glass can be described by spins connected by a random distribution of bonds and did not rely on the assumption of RKKY interactions. Hence it should hold for a much larger class of systems. The justification for this assumption came only a little later when Holzberg et al (1977), Maletta and Convert (1979), and Maletta and Felsch (1979, 1980) discovered that the insulator $Eu_xSr_{1-x}S$ with $0.13 \leq x \leq 0.51$ also showed all important spin glass properties, such as the cusp in the susceptibility, no Bragg reflections, and a slowly decaying remanent magnetization which differs in field-cooled and zero-field-cooled states. In $Eu_xSr_{1-x}S$, one has essentially only nearest and next-nearest neighbour spin interactions and the system has a well-defined percolation threshold at the concentration $x_p = 0.13$, below which the spin glass goes over into a superparamagnet. This superparamagnet has distinctly different properties from those of a spin glass in its ac and dc susceptibilities (Eiselt et al, 1979). This difference is perhaps the strongest evidence against a model in which a spin glass is described by independent clusters of spins.

Extremely short-range spin glasses such as $Eu_xSr_{1-x}S$ are also the best examples of the concept of frustration, introduced by Toulouse (1977) for Ising spin glasses. This concept permitted a distinction between trivial and nontrivial disorder and has also been applied to XY (Villain, 1977, 1978) and Heisenberg spin glasses (Dzyaloshinskii and Volovik, 1978).

However, subsequent experiments brought evidence against a sharp phase transition: A very careful investigation showed that the cusp in $\chi(T)$ is rounded even in the smallest available fields (Mulder et al, 1981, 1982). At low temperatures, part of the relaxation processes turned out to be so slow that one could doubt that thermal equilibrium could ever be reached;

and scaling plots seemed to work nearly equally well for $T_f = 0$ and $T_f \neq 0$ (Bontemps et al, 1984). Moreover, no anomaly could be detected in the zero field specific heat at or near T_f (Wenger and Keesom, 1975, 1976). The situation became even more complicated when one observed that the magnetization depended not only on the order in which the sample is cooled and the field applied but also on the 'waiting' time, i.e. the time between a change in temperature and the application of a field. The sample 'ages' (Lundgren et al, 1981, 1982).

In the meantime, the theory developed essentially along three lines: The MF theory of EA was expected to be exact for a model with infinite range interactions between all spins (the SK model: Sherrington and Kirkpatrick, 1975; Kirkpatrick and Sherrington, 1978). However, de Almeida and Thouless (1978) showed that the solution of the EA or SK model with a single spin glass order parameter q_{EA} is unstable below the Almeida–Thouless (AT) line in the field-temperature plane. After attempts of various authors, in particular Blandin (1978) and Sommers (1978, 1979), Parisi found a stable solution of the SK model characterized by the order parameter function $q(x)$ (Parisi, 1980). The physics of this function was finally clarified considerably by Mézard et al (1984a,b, 1985). Already earlier, Thouless et al derived equations which did not rely on the replica trick (TAP, 1977). These equations turned out to have a number of locally stable solutions or states below T_f which increases exponentially with the number of spins (Bray and Moore, 1980; de Dominicis et al, 1980; Tanaka and Edwards, 1980). Obviously, these solutions could not be described by a single order parameter. A dynamical theory which describes transitions between these states and which reduces in the static limit to Parisi's results was in turn developed by Sompolinsky (1981). Along and above the AT line, these dynamics become considerably simpler (Sompolinsky and Zippelius, 1981, 1982).

The SK model was devised for Ising spins but can easily be extended to vector spins. Gabay and Toulouse (1981) observed that in vector spin glasses, in addition to the AT line, there exists a second line in the field-temperature plane below which the transverse spin components freeze.

In a second line of theoretical development, one tried to extract the information about the EA model with short-range interactions by means of Monte Carlo simulations. Earlier MC simulations reproduced the cusp in the susceptibility, the observed broad maximum in the specific heat (Binder and Schröder, 1976, Binder, 1977) and the field dependence of the isothermal and thermoremanent magnetizations (Kinzel, 1979). They also yielded some evidence that the AT line derived in MF theory is of dynamic origin and vanishes in the limit of infinite times (Kinzel and Binder, 1983). Already these early MC simulations showed that it might be extremely difficult to reach thermal equilibrium at low temperatures. This problem stimulated the construction of a special purpose computer with which larger systems and longer times could be investigated (Ogielski, 1985). The new computer also allowed a fairly reliable determination of the critical exponents of three-dimensional Ising spin glasses.

This leads us to the third main activity in the theory of spin glasses: The question whether or not there is a sharp phase transition in a given system can be answered by means of renormalization group calculations. Since spin glasses are random systems, these calculations are highly non trivial and have to be done numerically. First attempts to determine the lower critical dimension of Ising spin glasses with short range interactions did not lead to conclusive results (Young and Stinchcombe, 1975, 1976; Southern and Young, 1977; Kinzel and Fischer, 1978). Considerable progress was achieved with the work of McMillan (1984) and by Bray and Moore (1986), and today the existence of a sharp transition in three-dimensional Ising spin glasses is fairly well established. Data for 3d Heisenberg spin glasses with short-range or RKKY interactions are so far less conclusive, but it seems probable that the lower critical dimensionality for short-range interactions is greater than 3.

On the experimental side, a possible phase transition could only be proved or disproved after one concentrated on the nonlinear susceptibility. Earlier data on AgMn and amorphous AlGd already indicated such a singularity (Monod and Bouchiat, 1982; Barbara et al, 1981; Malozemoff et al, 1982) but only more recent experiments gave reliable data for the critical exponents of AgMn, CuMn, $Eu_xSr_{1-x}S$, AlGd, and other spin glasses (see Chapter 8, Table 2). There remains the question why these systems should be Ising like, since otherwise the theory would not predict a finite freezing temperature. This is clearly related to the amount and origin of anisotropy in a specific spin glass, but this question is not settled in detail.

A new trend is now to consider systems which definitely are two- or three-dimensional Ising spin glasses such as $Fe_{0.3}Mg_{0.7}Cl_2$ or $Fe_{0.35}Mg_{0.65}Br_2$. Here, three dimensional sytems, in contrast to two dimensional ones, indeed show a finite temperature phase transition (Bertrand and Ferré, 1989). There is also a trend to study more complicated spin glasses. This brings us to the middle of current research, and perhaps in the future spin glasses will show new and unexpected properties, as they have so often in the past.

References

Abd-Elmeguid, M M, H Micklitz, R A Brand, and W Keune, 1986, Phys Rev B **33** 7833

Abragam, A, 1961, *The Principles of Nuclear Magnetism* (Clarendon, Oxford)

Abramowitz, M and I A Stegun, 1965, *Handbook of Mathematical Functions* (Dover, New York)

Abrikosov, A A, L P Gorkov, and I E Dzyaloshinskii, 1963, *Methods of Quantum Field Theory in Statistical Physics* (Prentice Hall, Englewood Cliffs)

Aharony, A and E Pytte, 1980, Phys Rev Lett **45** 1583

Alba, M, J Hammann, M Ocio, Ph Refrigier and H Bouchiat, 1987, J Appl Phys **61** 3683

Albrecht, H, E F Wasserman, F T Hedgcock, and P Monod, 1982, Phys Rev Lett **48** 819

Alexander, S, J Bernasconi, W R Schneider, and R Orbach, 1981, Rev Mod Phys **53** 175

Alloul, H, 1979, Phys Rev Lett **42** 603

Alloul, H, and F Hippert, 1983, J Magn Magn Mater **31–34** 1321

Alloul, H, and P Mendels, 1985, Phys Rev Lett **54** 1313

Alloul, H, P Mendels, P Beauvillain, and C Chappert, 1986, Europhys Lett **1** 595

Amit, D J, 1984, *Field Theory, the Renormalization Group and Critical Phenomena* (McGraw-Hill, New York)

Amit, D J, H Gutfreund, and H Sompolinsky, 1985a, Phys Rev A **32** 1007

Amit, D J, H Gutfreund, and H Sompolinsky, 1985b, Phys Rev Lett **55** 1530

Amit, D J, H Gutfreund, and H Sompolinsky, 1987a, Ann Phys (NY) **173** 30

Amit, D J, H Gutfreund, and H Sompolinsky, 1987b, Phys Rev A **35** 2293

Anderson, P W, B I Halperin, and C M Varma, 1972, Phil Mag **25** 1

Anderson, P W, and G Yuval, 1971, J Phys C **4** 607

Anderson, P W, G Yuval and D R Hamann, 1970, Phys Rev B **1** 4464

Arrott, A, 1965, J Appl Phys **36** 1093

Ashcroft, N W and D Mermin, 1976, *Solid State Physics* (Holt, Rinehart and Winston, New York)

Baberschke, K, P Pureur, A Fert, R Wendler, and S Senoussi, 1984, Phys Rev B **29** 4999

Baldi, P, and E B Baum, 1986, Phys Rev Lett **56** 1598

Barahona, F, R Maynard, R Rammal, and J P Uhry, 1982, J Phys A **15** 673

Barbara, B, A P Malozemoff, and Y Imry, 1981, Phys Rev Lett **47** 1852

Bausch, R, H K Janssen, and H Wagner, 1976, Z Phys B **24** 113

Beauvillain, P, C Chappert, and J P Renard, 1984, J Phys (Paris) Lett **45** L665

Beauvillain, P, C Chappert, J P Renard, and J Seiden, 1986, J Magn Magn Mater **54–57** 127

Beauvillain, P, C Dupas, J P Renard, and P Veillet, 1984, Phys Rev B **29** 4086

Beck, P A, 1972, J Less Common Metals **28** 193

Belanger, D P, A R King, V Jaccarino,and J L Cardy, 1983, Phys Rev B **28** 2522

Belanger, D P, A R King, and V Jaccarino, 1985, Phys Rev Lett **54** 577; Phys Rev B **31** 4538

Bennett, H S, and E Pytte, 1967, Phys Rev **155** 553

Berlin, T H, and M Kac, 1952, Phys Rev **86** 821

Berton, A, J Chaussy, J Odin, R Rammal, and R Tournier, 1982, J Phys (Paris) Lett **43** L153

Bertrand, D. and J Ferré, 1989, J Phys (Paris) Colloq (to be published)

Bertrand, D, A R Fert, M C Schmidt, F Bensamka, and S Legrand, 1982, J Phys C **15** L883

Beton, P H, and M A Moore, 1983, J Phys C **16** 1245

Beton, P H, and M A Moore, 1984, J Phys C **17** 2157

Bhatt, R N, and A P Young, 1985, Phys Rev Lett **54** 924

Bhatt, R N, and A P Young, 1986a, J Magn Magn Mater **54-57** 191

Bhatt, R N, and A P Young, 1986b, in *Heidelberg Colloquium on Glassy Dynamics*, Lecture Notes in Physics **275**, edited by J L van Hemmen and I Morgenstern (Springer, Berlin)

Bhatt, R N, and A P Young, 1988, Phys Rev B **37** 5606

Bhattacharya, S, S R Nagel, L Fleishman, and R Susman, 1982, Phys Rev Lett **48** 1267

Bieche, I, R Maynard, R Rammal, and J P Uhry, 1980, J Phys A **13** 2253

Binder, K, 1977, Z Phys B **26** 339

Binder, K, 1978, J Phys (Paris) **39** C6-1527

Binder, K, 1980a, in *Fundamental Problems in Statistical Physics V*, edited by E G D Cohen (North–Holland, Amsterdam)

Binder, K, 1980b, in *Ordering in Strongly–Fluctuating Condensed Matter Systems*, edited by T Riste (Plenum, New York), p 423

Binder, K, 1983, Z Phys B **50** 343

Binder, K, and K Schröder, 1976, Phys Rev B **14** 2142

Binder, K, and A P Young, 1986, Rev Mod Phys **58** 801

Birgeneau, R J, R A Cowley, G Shirane, and H Yoshizawa, 1984, J Stat Phys **34** 817

Birgeneau, R J, Y Shapira, G Shirane, R A Cowley, and H Yoshizawa, 1986, Physica B+C **37** 83

Blandin, A, 1961, Thesis, University of Paris

Blandin, A, 1978, J Phys (Paris) Colloq **C6-39** 1499

Bohn, H G, W Zinn, B Dorner, and A Kollmar, 1980, Phys Rev B **22** 5447

Bontemps, N, J Rajchenbach, R V Chamberlin, and R Orbach, 1984, Phys Rev B **30** 6514

Bontemps, N, J Rajchenbach, R V Chamberlin, and R Orbach, 1986, J Magn Magn Mater **54–57** 1

Börgermann, F-J, H Maletta, and W Zinn, 1986, Phys Rev B **35** 8454

Bouchiat, H, 1986, J Phys (Paris) **47** 71

Bray, A J, 1988, Comments Cond Mat Phys **14** 21

Bray, A J, and M A Moore, 1978, Phys Rev Lett **41** 1068

Bray, A J, and M A Moore, 1979, J Phys C **12** L441

Bray, A J, and M A Moore, 1980, J Phys C **13** L469

Bray, A J, and M A Moore, 1981, J Phys C **14** 2629

Bray, A J, and M A Moore, 1982, J Phys C **15** 3897

Bray, A J, and M A Moore, 1983, J Phys Soc Japan Suppl **52** 101

Bray, A J, and M A Moore, 1984, J Phys C **17** L463

Bray, A J, and M A Moore, 1985a, Phys Rev B **31** 631

Bray, A J, and M A Moore, 1985b, J Phys C **18** L139

Bray, A J, and M A Moore, 1986, in *Heidelberg Colloquium on Glassy Dynamics*, Lecture Notes in Physics **275**, edited by J L van Hemmen and I Morgenstern (Springer, Berlin)

Bray, A J, and M A Moore, 1987, Phys Rev Lett **58** 57

Bray, A J, M A Moore, and A P Young, 1986, Phys Rev Lett **56** 2641

Bray, A J, and S A Roberts, 1980, J Phys C **13** 5405

Bray, A J, H Sompolinsky, and C Yu, 1986, J Phys C **19** 6389

Bray, A J, and L Viana, 1983, J Phys C **16** 4679

Brodale, G E, R A Fisher, W E Fogle, N E Phillips, and J van Curen, 1983, J Magn Magn Mater **31–34** 1331

Brout, R H, 1965, *Phase Transitions* (Benjamin, New York)

Brout, R H, 1974, Phys Repts **10** 1

Caianiello, E R, 1961, J Theor Biol **1** 204

Callen, H B, and T Welton, 1951, Phys Rev **83** 34

Campbell, I A, 1981, Phys Rev Lett **47** 1473

Campbell, I A, P J Ford, and A Hamzić, 1982, Phys Rev B **26** 5195

Campbell, I A, S Senoussi, F Varret, J Teillet, and A Hamzić, 1983, Phys Rev Lett **50** 1615

Cannella, V, J A Mydosh, and J Budnick, 1971, J Appl Phys **42** 1689

Cannella, V, and J A Mydosh, 1972, Phys Rev B **6** 4220

Cannella, V, and J A Mydosh, 1974, Proceedings, International Conference on Magnetism, vol 2 (Nauka, Moscow), p 74

Carlson, J M, J T Chayes, L Chayes, J P Sethna, and D J Thouless, 1988, Europhys Lett **5** 355

Chakrabarti, A, and C Dasgupta, 1988, J Phys C **21** 1613

Chalupa, J, 1977a, Solid State Commun **22** 315

Chalupa, J, 1977b, Solid State Commun **24** 429

Chamberlin, R V, G Mazurkevich, and R Orbach, 1984, Phys Rev Lett **52** 867

Chayes, J T, L Chayes, J P Sethna, and D J Thouless, 1986, Commun Math Phys **106** 41

Chen, H H, and S K Ma, 1982, J Stat Phys **29** 717

Ching, W Y, and D L Huber, 1977a, Phys Lett A **59** 383

Ching, W Y, and D L Huber, 1977b, in *Magnetism and Magnetic Materials 1976*, AIP Conference Proceedings **34**, edited by J J Becker and G H Lander (AIP, New York), p 370

Chowdhury, D, 1986, *Spin Glasses and Other Frustrated Systems* (World Scientific, Singapore)

Chudnovsky, E M, W M Saslow, and R A Serota, 1986, Phys Rev B **33** 251

Coles, B R, B V B Sarkissian, and R H Taylor, 1978, Phil Mag **37** 489

Cooper, J R, L Nonveiller, P J Ford, J A Mydosh, 1980, J Magn Magn Mater **15–18** 181

Cortes, C, A Krogh and J A Hertz, 1987, J Phys A **20** 4449

Courtens, E, 1984, Phys Rev Lett **52** 69

Cragg, D M, and D Sherrington, 1982, Phys Rev Lett **49** 1190

Cragg, D M, D Sherrington, and M Gabay, 1982, Phys Rev Lett **49** 158

Dahlberg, E D, M Hardiman, R Orbach, and J Souletie, 1979, Phys Rev Lett **42** 401

de Almeida, J R L, R C Jones, J M Kosterlitz, and D J Thouless, 1978, J Phys C **11** L871

de Almeida, J R L, P Mottishaw, and C De Dominicis, 1988, J Phys A **21** L693

de Almeida, J R L, and D J Thouless, 1978, J Phys A **11** 983 de Courtenay, N, A Fert, and I A Campbell, 1984, Phys Rev B **30** 6791

de Courtenay, N, H Bouchiat, H Hurdequint, and A Fert, 1986, J Phys (Paris) **47** 1507

De Dominicis, C, 1978, Phys Rev B **18** 4913

De Dominicis, C, M Gabay, T Garel, and H Orland, 1980, J Phys (Paris) **41** 923

De Dominicis, C, and I Kondor, 1983, Phys Rev B **27** 606

De Dominicis, C, and I Kondor, 1984, J Phys (Paris) Lett **45** L205

De Dominicis, C, and I Kondor, 1985a, in *Applications of Field Theory to Statistical Mechanics*, Lectures Notes in Physics **16**, edited by L Garrido (Springer, Berlin), p 93

De Dominicis, C, and I Kondor, 1985b, J Phys (Paris) Lett **46** L1037

De Dominicis, C, and I Kondor, 1989, J Phys A **22** L743

De Dominicis, C and L Peliti, 1978, Phys Rev B **18** 353

De Dominicis, C, and A P Young, 1983, J Phys A **16** 2063

de Nobel, J, and J J du Chantenier, 1959, Physica **25** 969

Derrida, B, 1980, Phys Rev Lett **45** 79

Derrida, B, 1981 Phys Rev B **24** 2613

Derrida, B, and J Vannimenus, 1980, J Phys C **13** 3261

Dieny, B, 1985, Thesis, University of Grenoble

Dieny, B, and B Barbara, 1985, J Phys (Paris) **46** 293

Dieny, B, and B Barbara, 1986, Phys Rev Lett **57** 1169

Doussineau, P, A Levelut, and W Schön, 1988, Z Phys B **73** 80

Dubiel, S M, K H Fischer, Ch Sauer, and W Zinn, 1987, Phys Rev B **36** 360

Dzyaloshinskii, I E, 1958, J Phys Chem Solids **4** 241

Dzyaloshinskii, I E, 1980, in *Modern Trends in the Theory of Condensed Matter*, Lecture Notes in Physics **115**, edited by A Pekalski and J Przystawa (Springer, Berlin), p 204

Dzyaloshinskii, I E, 1983, Sov Phys JETP Letters **37** 190

Dzyaloshinskii, I E, and G E Volovik, 1978, J Phys (Paris) **39** 693

Edwards, S F, and P W Anderson, 1975, J Phys F **5** 965

Edwards, S F, and R C Jones, 1978, J Phys A **9** 1595

Efetov, K B, and Larkin, A I, 1977, Zh Eksp Teor Fiz **72** 2350 [Sov Phys JETP **45** 1236]

Eiselt, G, J Kötzler, H Maletta, D Stauffer, and K Binder, 1979, Phys Rev B **19** 2664

Elderfield, D, 1983, J Phys A **16** L439

Elderfield, D, 1984, J Phys A **17** L307, L517

Elderfield, D, and D Sherrington, 1982a, J Phys A **15** L437

Elderfield, D, and D Sherrington, 1982b, J Phys A **15** L513

Elderfield, D, and D Sherrington, 1983a, J Phys C **16** L497

Elderfield, D, and D Sherrington, 1983b, J Phys C **16** 4865

Elderfield, D, and D Sherrington, 1984, J Phys C **17** 5595

Emmerich, K, E Lippelt, R Neuhaus, H Pinkvos, Ch Schwink, F N Gugax, A Hintermann, A Schenck, W Studer, and A J van der Waal, 1985, Phys Rev B **31** 7226

Erwin, R W, J W Lynn, J J Rhyne, and H S Chen, 1985, J Appl Phys **57** 3473

Erzan, A, and E J S Lage, 1983, J Phys C **16** L555

Feigelman, M V, and A M Tsvelik, 1979, Zh Eksp Teor Fiz **77** 2524 [Sov Phys JETP **50** 1222]

Felten, G, and Ch Schwink, 1984, Solid State Commun **49** 233

Fernandez, J F and R Medina, 1979, Phys Rev B **19** 3561

Ferré, J, J Rajchenbach, and H Maletta, 1981, J Appl Phys **52** 1697

Fert, A, and P M Levy, 1980, Phys Rev Lett **44** 1538

Fert, A, P Pureur, F Hippwer, K Baberschke, and F Bruss, 1982, Phys Rev B **26** 5300

Fischer, K H, 1970, Springer Tracts in Modern Physics **54** (Springer, Berlin), p 2

Fischer, K H, 1976, Solid State Commun **18** 1515

Fischer, K H, 1978, Phys Rep **47** 225

Fischer, K H, 1979, Z Phys B **34** 45

Fischer, K H, 1980, Z Phys B **39** 37

Fischer, K H, 1981a, Z Phys B **42** 27

Fischer, K H, 1981b, Z Phys B **42** 245

Fischer, K H, 1983a, Z Phys B **50** 107

Fischer, K H, 1983b, Z Phys B **53** 215

Fischer, K H, 1983c, Phys Status Solidi B **116** 357

Fischer, K H, 1984, Z Phys B **55** 317

Fischer, K H, 1985, Phys Status Solidi B **130** 13

Fischer, K H, 1987, Phys Rev B **36** 6963

Fischer, K H, and J A Hertz, 1983, J Phys C **16** 5017

Fischer, K H, and W Kinzel, 1984, J Phys C **17** 4479

Fischer, K H, and A Zippelius, 1986, Prog Theor Phys Suppl **87** 165

Fisher, D S, 1986, Phys Rev Lett **56** 416

Fisher, D S, and D A Huse, 1986, Phys Rev Lett **56** 1601

Fisher, D S, and D A Huse, 1988a, Phys Rev B **38** 373

Fisher, D S, and D A Huse, 1988b, Phys Rev B **38** 386

Fisher, D S, and H Sompolinsky, 1985, Phys Rev Lett **54** 1063

Fishman, S, and A Aharony, 1979, J Phys C **12** L729

Fogle, W E, J D Boyer, R A Fisher, and N E Phillips, 1983, Phys Rev Lett **50** 1815

Foiles, C L, 1978, Phys Lett A **67** 214

Ford, P J, Cooper, J R, and Jungfleish, N, 1973, Solid State Commun **13** 857

Ford, P J, and J A Mydosh, 1976, Phys Rev B **14** 2057

Forster, D, 1975, *Hydrodynamic Fluctuations, Broken Symmetry, and Correlation Functions* (Benjamin, New York)

Fradkin, E, B A Huberman, and S H Shenker, 1978, Phys Rev B **18** 4789

Fu, Y, and P W Anderson, 1986, J Phys A **19** 1605

Fyodorov, Y V, I Y Korenblit, and E F Shender, 1987, J Phys C **20** 1835

Gabay, J, T Garel, and C De Dominicis, 1982, J Phys C **15** 7165

Gabay, M, and G Toulouse, 1981, Phys Rev Lett **47** 201

Gardner, E, 1987, Europhys Lett **4** 481

Geschwind, S, G Devlin, and S L McCall, 1987, Phys Rev Lett **58** 1895

Ginzburg, S L, 1978, Zh Eksp Teor Fiz **15** 1497 [Sov Phys JETP **48** 756)]

Glauber, R J, 1963, J Math Phys **4** 294

Goldberg, S M, P M Levy, and A Fert, 1985, Phys Rev B **31** 3106

Goldberg, S M, and P M Levy, 1986, Phys Rev B **33** 291

Goldberg, S M, P M Levy, and A Fert, 1986, Phys Rev B **33** 276

Goltsev, A V, 1984a, J Phys A **17** 237

Goltsev, A V, 1984b, J Phys A **17** L241

Granberg, P, L Sandlund, P Nordblad, P Svedlindh, and L Lundgren, 1988, Phys Rev B **38** 7097

Green, J E, M A Moore, and A J Bray, 1983, J Phys C **16** L815

Grinstein, G, 1976, Phys Rev Lett **37** 944

Grinstein, G, 1985, in *Fundamental Problems in Statistical Mechanics VI*, edited by E G D Cohen (Elsevier, Amsterdam)

Grinstein, G, A N Berker, J Chalupa, and M Wortis, 1976, Phys Rev Lett **36** 1508

Gross, D J, and M Mézard, 1984, Nucl Phys B **240** 431

Gross, D J, I Kanter, and H Sompolinsky, 1985, Phys Rev Lett **55** 304

Grossmann, S, F Wegner, and K H Hoffmann, 1985, J Phys (Paris) Lett **46** L575

Gullikson, E M, R Dalichaouch, and S Schultz, 1985, Phys Rev B **32** 507

Gullikson, E M, D R Fredkin, and S Schultz, 1983, Phys Rev Lett **50** 537

Guy, C N, 1982, J Phys F **12** 1453

Hagen, M, R A Cowley, S K Satija, H Yoshizawa, G Shirane, R J Birgeneau, and H J Guggenheim, 1983 , Phys Rev B **28** 2602

Halperin, B I, and P C Hohenberg, 1969, Phys Rev **177** 952

Halperin, B I, and W M Saslow, 1977, Phys Rev V **16** 2154

Halsey, T C, 1985, Phys Rev Lett **55** 1018

Hamida, J A, and S J Williamson, 1986, Phys Rev B **34** 8111

Harris, A B, T C Lubensky, and J H Chen, 1976, Phys Rev Lett **36** 415

Harris, R, M Plischke, and M J Zuckermann, 1973, Phys Rev Lett **31** 160

Hawkins, G F, and R L Thomas, 1978, J Appl Phys **49** 1627

Hawkins, G F, R L Thomas, and A M de Graaf, 1979, J Appl Phys **50** 1709

Hebb, D O, 1949, *The Organization of Behaviour* (Wiley, New York)

Henley, C L, H Sompolinsky, and B I Halperin, 1982, Phys Rev B **25** 5849

Hennion, B, M Hennion, F Hippert, and A P Murani, 1983, Phys Rev B **28** 5365

Hennion, B, M Hennion, F Hippert, and A P Murani, 1984, J Appl Phys **55** 1694

Hennion, M, I Mirebeau, B Hennion, B, S Lequien, and F Hippert, 1986, Europhys Lett **2** 393

Hertz, J A, 1978, Phys Rev B **18** 4875

Hertz, J A, 1983a, J Phys C **16** 1219

Hertz, J A, 1983b, J Phys C **16** 1233

Hertz, J A, G Grinstein and S A Solla, 1986a, in *Neural Networks for Computing*, AIP Conf Proc **151**, edited by J S Denker

Hertz, J A, G Grinstein and S A Solla, 1986b, in *Heidelberg Colloquium on Glassy Dynamics*, Lecture Notes in Physics **275**, edited by J L van Hemmen and I Morgenstern (Springer, Berlin)

Hertz, J A, A Khurana, and R A Klemm, 1981, Phys Rev Lett **46** 496

Hertz, J A, and R A Klemm, 1979, Phys Rev B **20** 316

Hertz, J A, and R A Klemm, 1983, Phys Rev B **28** 3849

Hinton, G E, and T Sejnowski, 1986, in *Parallel Distributed Processing*, edited by D E Rumelhart and J L McClelland (MIT Press, Cambridge and London), chapter 7

Hoekstra, F R, G J Nieuwenhuys, and J A Mydosh, 1985, Phys Rev B **31** 7349

Hoffmann, K H, S Grossmann, and F Wegner, 1985, Z Phys B **60** 401

Hohenberg, P C, and B I Halperin, 1977, Rev Mod Phys **49** 435

Holtzberg, F, J L Tholence, and R Tournier, 1977, in *Amorphous Magnetism II*, edited by R A Levy and R Hasegawa (Plenum, New York), p 155

Hopfield, J J, 1982, Proc Nat Acad Sci USA **79** 2554

Hopfield, J J, 1984, Proc Nat Acad Sci USA **81** 3088

Hopfield, J J, and D W Tank, 1985, Biol Cybernetics **52** 141

Hsu, T C, and J B Marston, 1987, J Appl Phys **61** 2074

Huberman, B A, and M Kerszberg, 1985, J Phys A **18** L331

Huse, D A, 1989, J Appl Phys (to be published)

Hüser, D, L E Wenger, A J van Duyneveldt, and J A Mydosh, 1983, Phys Rev B **27** 3100

Imbrie, J Z, 1984, Phys Rev Lett **53** 1747

Imbrie, J Z, 1985, Commun Math Phys **98** 145

Imry, Y, and S K Ma, 1975, Phys Rev Lett **35** 1399

Ioffe, L B, and M V Feigelman, 1987, Zh Eksp Teor Fiz Pis Red **44** 148 [Sov Phys JETP Lett **44** 189]

Kadanoff, L P, 1976, Ann Phys (NY) **100** 359

Kadanoff, L P, and P C Martin, 1962, Ann Phys **24** 419

Kandel, E R, and J H Schwartz, 1985, *Principles of Neural Science* (Elsevier, Amsterdam)

Kanter, I, and H Sompolinsky, 1987, Phys Rev Lett **58** 164

Kasuya, T, 1956, Prog Theor Phys **16** 45

Khurana, A, 1982, Phys Rev B **25** 452

Khurana, A, and J A Hertz, 1980, J Phys C **13** 2715

Kinzel, W, 1979, Phys Rev B **19** 4595

Kinzel, W, 1986a, Phys Rev B **33** 5086

Kinzel, W, 1986b, in *Heidelberg Colloquium on Glassy Dynamics*, Lecture Notes in Physics **275**, edited by J L van Hemmen and I Morgenstern (Springer, Berlin)

Kinzel, W, and K Binder, 1981, Phys Rev B **24** 2701

Kinzel, W, and K Binder, 1983, Phys Rev Lett **50** 1509

Kinzel, W, and K Binder, 1984, Phys Rev B **29** 1300

Kinzel, W, and K H Fischer, 1977, Solid State Commun **23** 687

Kinzel, W, and K H Fischer, 1978, J Phys C **11** 2775

Kirkpatrick, S, 1977, Phys Rev B **16** 4630

Kirkpatrick S, C D Gelatt Jr, and M P Vecchi, 1983, Science **220** 671

Kirkpatrick, S, and D Sherrington, 1978, Phys Rev B **17** 4384

Kirkpatrick, S, and G Toulouse, 1985, J Phys (Paris) **46** 1277

Kittel, C, 1964, *Quantum Theory of Solids* (Wiley, New York)

Klein, M W, and R. Brout, 1963, Phys Rev **132** 2412

Kogut, J B, 1979, Rev Mod Phys **51** 659

Kondo, J, 1969, *Solid State Physics*, vol 23 (Academic Press, New York)

Koper, G J M, and H J Hilhorst, 1988, J Phys (Paris) **49** 429

Korenblit, I Y and E F Shender, 1985, Sov Phys JETP **62** 1030

Kosterlitz, J M, and D J Thouless, 1973, J Phys C **6** 1181

Kosterlitz, J M, D J Thouless, and R C Jones, 1976, Phys Rev Lett **36** 1217

Kotliar, G, P W Anderson, and D L Stein, 1983, Phys Rev B **27** 602

Kotliar, G, and H Sompolinsky, 1984, Phys Rev Lett **53** 1751

Kotliar, G, H Sompolinsky, and A Zippelius, 1987, Phys Rev B **35** 311

Kouvel, J S, 1960, J Phys Chem Solids **16** 107; J Appl Phys **31** 1425

Kouvel, J S, 1961, J Phys Chem Solids **21** 1961

Krey, U, 1980, Z Phys B **38** 243

Krey, U, 1981, Z Phys B **42** 231

Krey, U, 1982, in *Proceedings of the Conference on Applied Magnetism, Bad Nauheim, April 1982,* edited by H Mende (Verlag Stahleisen, Düsseldorf)

Krogh, A, and J A Hertz, 1988, J Phys A **21** 2211

Kuffler, S W, J G Nichols, and A R Martin, 1984, *From Neuron to Brain,* 2nd edition (Sinauer Associates, Sunderland MA)

Kutasov, D, A Aharony, E Domany, and W Kinzel, 1986, Phys Rev Lett **56** 2229

Laborde, O, and P Radhakrishna, 1973, J Phys F **3** 1731

Landau, L D, and E M Lifshitz, 1969, *Statistical Physics,* 2nd ed (Pergamon Press, London)

Lauer, J, and W Keune, 1982, Phys Rev Lett **48** 1850

Lecomte, G V, H von Löhneysen, and E F Wassermann, 1983, Z Phys B **50** 239

Lee, P A, and T M Rice, 1979, Phys Rev B **19** 3970

Levitt, D A, and R E Walstedt, 1977, Phys Rev Lett **38** 178

Levy, P M, and A Fert, 1981, Phys Rev B **23** 4667

Levy, L P, and A T Ogielski, 1986, Phys Rev Lett **57** 3288

Li, T C, 1981, Phys Rev B **24** 6579

Lifshitz, I M, 1962, Zh Eksp Teor Fiz **42** 1354 [Sov Phys JETP **15** 939]

Little, W A, 1974, Math Biosci **19** 101

Lundgren, L, P Nordblad, and P Svedlindh, 1986, Phys Rev B **34** 8164

Lundgren, L, P Svedlindh, and O Beckman, 1981, J Magn Magn Mater **25** 33

Lundgren, L, P Svedlindh, and O Beckman, 1982a, J Phys F **12** 2663

Lundgren, L, P Svedlindh, and O Beckman, 1982b, Phys Rev B **26** 3990

Lundgren, L, P Svedlindh, P Nordblad, and O Beckman, 1983, Phys Rev Lett **51** 911

Ma, S K, 1976, *Modern Theory of Critical Phenomena* (Benjamin, New York)

Ma, S K, and G F Mazenko, 1975, Phys Rev B **11** 4077

Mackenzie, N D, and A P Young, 1981, J Phys C **14** 3927

Mackenzie, N D, and A P Young, 1982, Phys Rev Lett **49** 301

Mackenzie, N D, and A P Young, 1983, J Phys C **16** 5321

Mahan, G D, 1981, *Many-particle Physics* (Plenum, New York)

Maletta, H, and P Convert, 1979, Phys Rev Lett **42** 108

Maletta, H, and W Felsch, 1979, Phys Rev B **20** 1245

Maletta, H, and W Felsch, 1980, Z Phys B **37** 55

Maletta, H, W Zinn, H Scheuer, and S M Shapiro, 1981, J Appl Phys **52** 1735

Malozemoff, A P, and Y Imry, 1981, Phys Rev B **24** 489

Malozemoff, A P, Y Imry, and B Barbara, 1982, J Appl Phys **53** 7672

Manheimer, M A, S M Bhagat, and H S Chen, 1983, J Magn Magn Mater **38** 147

Maritan, A, and A L Stella, 1986, J Phys A **19** L259

Marshall, W, 1960, Phys Rev **118** 1519

Martin, D L, 1979, Phys Rev B **20** 368

Martin, D L, 1980a, Phys Rev B **21** 1902

Martin, D L, 1980b, Phys Rev B **21** 1906

Martin, P C, E D Siggia, and H A Rose, 1973, Phys Rev A **8** 423

Mattis, D C, 1976, Phys Lett A **56** 421

Mayanovic, R A, R J Sladek, and U Debska, 1988, Phys Rev B **38** 2787

McCullough, W S, and W Pitts, 1943, Bull Math Biophys **5** 115

McMillan, W L, 1984, J Phys C **17** 3179

McMillan, W L, 1985, Phys Rev B **31** 340

Mehta, M L, 1967, *Random Matrices and the Statistical Theory of Energy Levels* (Academic, New York/London)

Mendels, P, H Alloul, and M Ribault, 1987, Europhys Lett **3** 113

Meschede, O, F Steglich, W Felsch, H Maletta, and W Zinn, 1980, Phys Rev Lett **44** 102

Metropolis, N, A W Rosenbluth, M N Rosenbluth, A H Teller, and E Teller, 1953, J Chem Phys **21** 1087

Meyer, C, F Hartmann–Boutron, Y Gros, and I A Campbell, 1985, J Magn Magn Mater **46** 254

Meyer, C, F Hartmann–Boutron, and Y Gros, 1986, J Phys (Paris) **47** 1395

Mézard, M, and G Parisi, 1985, J Phys (Paris) Lett **46** L771

Mézard, M, and G Parisi, 1986, J Phys (Paris) **47** 1285

Mézard, M, G Parisi, N Sourlas, G Toulouse, and M Virasoro, 1984a, Phys Rev Lett **52** 1156

Mézard, M, G Parisi, N Sourlas, G Toulouse, and M Virasoro, 1984b, J Phys (Paris) **45** 843

Mézard, M, G Parisi, and M A Virasoro, 1985, J Phys (Paris) Lett **46** L217

Mézard, M, G Parisi, and M A Virasoro, 1987, *Spin Glass Theory and Beyond* (World Scientific, Singapore)

Mezei, F, 1980, in *Neutron Spin Echo,* Lecture Notes in Physics **128**, edited by F Mezei (Springer, Berlin), p 3

Mezei, F, 1981, in *Recent Developments in Condensed Matter Physics*, edited by J R Devreese (Plenum, New York), vol 1, p 679

Mezei, F, 1982, J Appl Phys **53** 7654

Migdal, A A, 1975, Zh Eksp Teor Fiz **69** 1457 [Sov Phys JETP **42**, 743 (1975)]

Mills, D L, and P Lederer, 1966, J Phys Chem Solids **27** 1805

Monod, P, and Y Berthier, 1980, J Magn Magn Mater **15-18** 149

Monod, P, and H Bouchiat, 1982, J Phys (Paris) Lett **43** 145

Monod, P, J J Prejean, and B Tissier, 1979, J Appl Phys **50** 7324

Moorjani, K, and J M D Coey, 1984, *Magnetic Glasses* (Elsevier, Amsterdam)

Morgenstern, I, and K Binder, 1979, Phys Rev Lett **43** 1615

Morgenstern, I, and K Binder, 1980a, Phys Rev B **22** 288

Morgenstern, I, and K Binder, 1980b, Z Phys B **39** 227

Morgenstern, I, and H Horner, 1982, Phys Rev B **504**

Morgownik, A F J, and J A Mydosh, 1981, Phys Rev B **24** 5277

Morgownik, A F J, and J A Mydosh, 1983, Solid State Commun **47** 321

Moriya, T, 1960, Phys Rev Lett **4** 51

Motoya, K, S M Shapiro, and Y Muraoka, 1983, Phys Rev B **28** 6183

Morris, B W, and A J Bray, 1984, J Phys C **17** 1717

Morris, B W, S G Colborne, M A Moore, A J Bray, and J Canisius, 1986, J Phys C **19** 1157

Mulder, C A M, A J van Duyneveldt, and J A Mydosh, 1981, Phys Rev B **23** 1384

Mulder, C A M, A J van Duyneveldt, and J A Mydosh, 1982, Phys Rev B **25** 515

Müller, K A, K W Blazey, J C Bednorz, and M Tagashiga, 1987, Physica B **148** 149

Murani, A P, and A Heidemann, 1978, Phys Rev Lett **41** 1402

Murani, A P, 1983, Phys Rev B **28** 432

Murnick, D E, A T Fiory, and W J Kossler, 1976, Phys Rev Lett **36** 100

Mydosh, J A, P J Ford, M P Kawatra, and T E Whall, 1974, Phys Rev B **10** 2845

Nagata, S, P H Keesom, and H R Harrison, 1979, Phys Rev B **19** 1633

Nakanishi, 1981, Phys Rev B **23** 3514

Nattermann, T and J Villain, 1988, Phase Transitions **11** 5

Néel, L, 1961, *Physique des Basses Températures* (Gordon and Breach, New York)

Nemoto, K, 1987, J Phys C **20** 1325

Nemoto, K, and H Takayama, 1985, J Phys C **18** L529

Nemoto, K, and H Takayama, 1986, J Magn Magn Mater **54–57** 135

Nigam, A K, and A K Majumdar, 1983, Phys Rev B **27** 495

Nordblad, P, P Svedlindh, P Granberg, and L Lundgren, 1987, Phys Rev B **35** 7150

Novak, M A, O G Symko, and D J Zheng, 1986, Phys Rev B **33** 343

Nozières, P, 1974, J Low Temp Phys **17** 31

Ogielski, A T, 1985, Phys Rev B **32** 7384

Ogielski, A T, 1986, Phys Rev Lett **57** 1251

Ogielski, A T, and I Morgenstern, 1985, Phys Rev Lett **54** 928

Ogielski, A T, and D L Stein, 1985, Phys Rev Lett **55** 1634

Olive, J A, A P Young, and D Sherrington, 1986, Phys Rev B **34** 6341

Omari, R, J J Préjean, and J Souletie, 1983, J Phys (Paris) **44** 1069

Onsager, L, 1936, J Am Chem Soc **58** 1486

Orland, 1985, J Phys (Paris) Lett **46** L763

Orwell, G, 1946, *Animal Farm* (Martin, Secker and Warburg, London)

Owen, J, M Brown, W D Knight, and C Kittel, 1956, Phys Rev **102** 160

Palmer, R G, 1982, Adv Phys **31** 669

Palmer, R G, D L Stein, E Abrahams, and P W Anderson, 1984, Phys Rev Lett **53** 958

Palmer, R G, and C M Pond, 1979, J Phys F **9** 1979

Palumbo, A C, R D Parks, and Y Yeshurn, 1982, J Phys C **15** L837

Parga, N, G Parisi, and M A Virasoro, 1984, J Phys (Paris) Lett **45** L1063

Parga, N and M A Virasoro, 1986, J Phys (Paris) **47** 1857

Parisi, G, 1979, Phys Rev Lett **43** 1754

Parisi, G, 1980, J Phys A **13** 1101, 1887

Parisi, G, 1983a, Phys Rev Lett **50** 1946

Parisi, G, 1983b, J Phys (Paris) Lett **44** L581

Parisi, G, and N Sourlas, 1979 , Phys Rev Lett **43** 744

Parisi, G, and G Toulouse, 1980, J Phys (Paris) Lett **41** L361

Paulsen, C, J Hamida, S J Williamson, and H Maletta, 1984, J Appl Phys **55** 1652

Paulsen, C C, S J Williamson, and H Maletta, 1987, Phys Rev Lett **59** 128

Prejean, J J, M Joliclerc, and P Monod, 1980, J Phys (Paris) **41** 427

Pytte, E, and J Rudnick, 1979, Phys Rev B **19** 3603

Rajchenbach, J, and N Bontemps, 1983, J Phys (Paris) Lett **44** L799

Rammal, R, G Toulouse, and M A Virasoro, 1986, Rev Mod Phys **58** 765

Reed, P, 1978, J Phys C **11** L979

Refrigier, P, E Vincent, J Hammann, and M Ocio, 1987, J Phys (Paris) **48** 1533

Reger, J D, and K Binder, 1985, Z Phys B **60** 137

Reger, J D, W Kinzel, and K Binder, 1984 Phys Rev B **30** 4028

Reger, J D, and A P Young, 1988, Phys Rev B **37** 5493

Rhyne, J J, 1986, Physica B **136** 30

Riera, R and Hertz, J A, 1988 (unpublished)

Roberts, S A, 1982, J Phys C **15** 4155

Roberts, S A, 1983, J Phys C **16** 5465

Roberts, S A, and A J Bray, 1982, J Phys C **15** L527

Ruderman, M A, and C Kittel, 1954, Phys Rev **96** 99

Salamon, M B, and R M Herman, 1978, Phys Rev Lett **41** 1506

Salamon, M B, and J L Tholence, 1983, J Magn Magn Mater **21-34** 1375

Saslow, W M, 1980, Phys Rev B **22** 1174

Saslow, W M, 1982a, Phys Rev Lett **48** 505

Saslow, W M, 1982b, Phys Rev B **26** 1483

Saslow, W M, 1983, Phys Rev B **27** 6873

Schreckenberg, M, 1985, Z Phys B **60** 483

Schröder, A, J Fischer, H von Löhneysen, W Bauhofer, and U Steigenberger, 1989, J Phys (Paris) Colloq (to be published)

Schuster, H G, 1981, Z Phys **45** 99

Shapiro, S M, G Aeppli, H Maletta, and K Motoya, 1986, Physica B **137** 96

Shapiro, S M, C R Fincher, A C Palumbo, and R D Parks, 1981a, Phys Rev B **24** 6661

Shapiro, S M, C R Fincher, A C Palumbo, and R D Parks, 1981b, J Appl Phys **52** 1729

Sherrington, D, and S Kirkpatrick, 1975, Phys Rev Lett **35** 1972

Sibani, P, 1986, Phys Rev B **34** 3555

Singh, G P, M von Schickfuss, and H Maletta, 1983, Phys Rev Lett **51** 1791

Singh, R R P, and S Chakravarty, 1986, Phys Rev Lett **57** 245

Singh, R R P, and S Chakravarty, 1987a, J Appl Phys **61** 4095

Singh, R R P, and S Chakravarty, 1987b, Phys Rev B **36** 546, 559

Sommers, H J, 1978, Z Phys B **31** 301

Sommers, H J, 1979, Z Phys B **33** 173

Sommers, H J, 1983a, J Phys A **16** 447

Sommers, H J, 1983b, Z Phys **50** 97

Sommers, H J, 1985, J Phys (Paris) Lett **46** L779

Sommers, H J ,1987, Phys Rev Lett **58** 1268

Sommers, H J, C De Dominicis, and M Gabay, 1983, J Phys A **16** L679

Sommers, H J, and W Dupont, 1984, J Phys C **17** 5785

Sommers, H J, and K H Fischer, 1985, Z Phys B **58** 125

Sompolinsky, H, 1981, Phys Rev Lett **47** 935

Sompolinsky, H, 1984, Phil Mag **50** 285

Sompolinsky, H, G Kotliar, and A Zippelius, 1984, Phys Rev Lett **52** 392

Sompolinsky, H, and A Zippelius, 1981, Phys Rev Lett **47** 359

Sompolinsky, H, and A Zippelius, 1982, Phys Rev B **25** 6860

Soukoulis, C M, K Levin, and G S Grest, 1982, Phys Rev Lett **48** 1756

Soukoulis, C M, K Levin, and G S Grest, 1983a, Phys Rev B **28** 1495

Soukoulis, C M, G S Grest, and K Levin, 1983b, Phys Rev B **28** 1510

Soukoulis, C M, and G S Grest, 1984, J Appl Phys **55** 1661

Souletie, J, and R Tournier, 1969, J Low Temp Phys **1** 95

Sourlas, N, 1988, Europhys Lett **6** 561

Southern, B W, 1976, J Phys C **9** 4011

Southern, B W, and A P Young, 1977, J Phys C **10** 2179

Spano, M L, and J J Rhyne, 1985, J Appl Phys **57** 3303

Stauffer, D, and K Binder, 1979, Z Phys B **34** 97

Stauffer, D, and K Binder, 1981, Z Phys B **41** 237

Suzuki, M, 1977, Prog Theor Phys **58** 1151

Suzuki, M, and R Kubo, 1968, J Phys Soc Japan **24** 51

Svedlindh, P, L Lundgren, P Nordblad, and H S Chen, 1987, Europhysics Lett **3** 243

Tanaka, F, and S F Edwards, 1980, J Phys F **10** 2471

Taniguchi, T, H Matsuyama, S Chikazawa, and Y Miyako, 1983, J Phys Soc Japan **52** 4323

Taniguchi, T, Y Miyako, and J L Tholence, 1985, J Phys Soc Japan **54** 220

Temesvári, T, I Kondor, and C De Dominicis, 1988, J Phys A **21** 1145

Teitel, S, and E Domany, 1985, Phys Rev Lett **55** 2176

Tholence, J L, 1980, Solid State Commun **35** 113

Tholence, J L, and R Tournier, 1974, J Phys (Paris) **35** C4-229

Thouless, D J, P W Anderson, and R G Palmer, 1977, Phil Mag **35** 593

Thouless, D J, J R L de Almeida, and J M Kosterlitz, 1980, J Phys C **13** 3271

Toulouse, G, 1977, Commun Phys **2** 115

Toulouse, G, 1979, Phys Repts **49** 267

Toulouse, G, 1980, J Phys (Paris) Lett **41** L447

Toulouse, G, and P Pfeuty, 1977, *Introduction to the Renormalization Group and its Applications* (Wiley, New York)

Toulouse, G, and J Vannimenus, 1980, Phys Repts **67** 47

Tustison, R W, 1976, Solid State Commun **19** 1075

Uemura, Y J 1981, Hyperfine Interact **8** 739

Uemura, Y J, K Nishiyama, T Yamazaki, and R Nakai 1981, Solid State Commun **39** 461

Uemura, Y J, T Yamazaki, R S Hayano, R Nakai, and C Y Huang, 1980, Phys Rev Lett **45** 583

van Duyneveldt, A J, and C A Mulder, 1982, Physica B **114** 82

van Hemmen, J L, 1986, in *Heidelberg Colloquium on Glassy Dynamics*, Lecture Notes in Physics **275**, edited by J L van Hemmen and I Morgenstern (Springer, Berlin)

van Hemmen, J L, and I Morgenstern (eds), 1983, *Heidelberg Colloquium on Spin Glasses*, Lecture Notes in Physics **192** (Springer, Berlin)

van Hemmen, J L, and I Morgenstern (eds), 1986, *Heidelberg Colloquium on Glassy Dynamics*, Lecture Notes in Physics **275** (Springer, Berlin)

Viana, L, and A J Bray, 1983, J Phys C **16** 6817

Viana, L, and A J Bray, 1985, J Phys C **18** 3037

Villain, J, 1977a, J Phys C **10** 1717

Villain, J, 1977b, J Phys C **10** 4793

Villain, J, 1978, J Phys C **11** 745

Villain, J, 1984, Phys Rev Lett **52** 1543

Vincent, E, and J Hammann, 1987, J Phys C **20** 2659

Vincent, E, J Hammann, and M Alba, 1986, Solid State Commun **58** 57

Violet, C E, and R J Borg, 1965, Phys Rev **149** 540

Violet, C E, and R J Borg, 1967, Phys Rev **162** 608B

Volovik, G E, and I E Dzyaloshinskii, 1978, Zh Eksp Teor Fiz **75** 1102 [Sov Phys JETP **48** 555]

von Löhneysen, H, 1981, Phys Repts **79** 163

von Molnar, S, B Barbara, T R McGuire, and R Gambino, 1982a, J Appl Phys **53** 2350

von Molnar, S, T R McGuire, R Gambino, and B Barbara, 1982b, J Appl Phys **53** 7666

Walker, L R, and R E Walstedt, 1977, Phys Rev Lett **38** 514

Walker, L R, and R E Walstedt, 1980, Phys Rev B **22** 3816

Walstedt, R E, and L R Walker, 1981, Phys Rev Lett **47** 1624

Wannier, G H, 1950, Phys Rev **79** 357

Wendler, R, P Pureur, A Fert, and K Baberschke, 1984, J Magn Magn Mater **45** 185

Wenger, L E, 1983, in *Heidelberg Colloquium on Spin Glasses*, Lecture Notes in Physics **192**, edited by J L van Hemmen and I Morgenstern (Springer, Berlin) p 60

Wenger, L E, and P H Keesom, 1975, Phys Rev B **11** 3497

Wenger, L E, and P H Keesom, 1976, Phys Rev B **13** 4053

White, R M, 1983, *Quantum Theory of Magnetism* (McGraw-Hill, New York)

Wilson, K G, 1975, Rev Mod Phys **47** 733

Wolff, W F, and J Zittartz, 1982, Z Phys B **49** 229

Wolff, W F, and J Zittartz, 1983a, Z Phys B **50** 131

Wolff, W F, and J Zittartz, 1983b, in *Heidelberg Colloquium on Spin Glasses*, Lecture Notes in Physics **192** edited by J L van Hemmen and I Morgenstern (Springer, Berlin) p 252

Wong, P-Z, P M Horn, R J Birgeneau, C R Safinya, and G Shirane, 1984, Phys Rev Lett **45** 1974

Yang, C N, and R L Mills, 1954, Phys Rev **96** 191

Yeshurun, Y, and H Sompolinsky, 1982, Phys Rev B **26** 1487

Yoshizawa, H, S Mitsuda, H Aruga, and A Ito, 1987, Phys Rev Lett **59** 2364

Yosida, K, 1957, Phys Rev **106** 893

Youm, D, and S Schultz, 1986, Phys Rev B **34** 7958

Young, A P, 1983a, Phys Rev Lett **50** 917

Young, A P, 1983b, Phys Rev Lett **51** 1206

Young, A P, 1984, J Phys C **17** L517

Young, A P, and A J Bray, and M A Moore, 1984, J Phys C **17** L149

Young, A P, and R B Stinchcombe, 1975, J Phys C **8** L535

Young, A P, and R B Stinchcombe, 1976, J Phys C **9** 4419

Ziman, J M, 1960, *Electrons and Phonons.* (Clarendon, Oxford)

Zimmermann, J E, and F E Hoare, 1960, J Phys Chem Solids **17** 52

Index

Printed in the United States
By Bookmasters